MACROSCALE MODELS OF FLOW THROUGH HIGHLY HETEROGENEOUS POROUS MEDIA

Theory and Applications of Transport in Porous Media

Series Editor:
Jacob Bear, *Technion – Israel Institute of Technology, Haifa, Israel*

Volume 16

The titles published in this series are listed at the end of this volume.

Macroscale Models of Flow Through Highly Heterogeneous Porous Media

by

Mikhail Panfilov

Russian Academy of Sciences,
Oil & Gas Research Institute,
Moscow, Russia

KLUWER ACADEMIC PUBLISHERS

DORDRECHT / BOSTON / LONDON

A C.I.P. Catalogue record for this book is available from the Library of Congress.

ISBN 0-7923-6176-8

Published by Kluwer Academic Publishers,
P.O. Box 17, 3300 AA Dordrecht, The Netherlands.

Sold and distributed in North, Central and South America
by Kluwer Academic Publishers,
101 Philip Drive, Norwell, MA 02061, U.S.A.

In all other countries, sold and distributed
by Kluwer Academic Publishers,
P.O. Box 322, 3300 AH Dordrecht, The Netherlands.

Printed on acid-free paper

Printed in the Netherlands.

PREFACE

The book was planned in such a manner that two basic goals would be reached. On the one hand, the goal was to show some new results in the field of modeling transport through highly heterogeneous media, based on the homogenization theory. Multiple new mathematical models of transport are presented herein, studying their properties, developing methods to compute effective parameters of the averaged media, simulation of cell problems, using new models to simulate some practical problems. High heterogeneity being subjected to the homogenization procedure, generates non-local phenomena and then gives a possibility to develop a new, non-local (or "dynamic"), theory of transport in porous media. For instance, a model of a two-phase flow with dynamic relative permeabilities and capillary pressures is one of the results of this book.

On the other hand, the goal was to illustrate the homogenization method as a powerful tool to deduce new models in continuum mechanics. Classical application of the homogenization method was devoted to justify the structure of the effective parameters of composite materials. Such a very narrow domain of the method application was determined by the fact that classical objects of the homogenization theory, such as diffusion or Laplace equations, had the property to be invariable with respect to scale transformations of the system. Due to this property, this type of equation remains invariable after homogenization, and the real problem is defining the structure of homogenized coefficients.

In systems that do not satisfy the property of scale invariance, the role of homogenization becomes totally different. Homogenization becomes a tool to deduce new mathematical models, usually considerably more complicated than the original equations. All the contents of the book illustrate this idea, where high heterogeneity should be considered as one of the brightest examples of systems that do not satisfy the scale invariance property. Nonlocality resulting from homogenization of such systems is an example of a new, rather general, phenomenon that can be expected to be generated by such tools.

From this point of view, the author hopes that this book may evoke some interest in specialists not only in mechanics of porous media, but also in general continuum and fluid mechanics, particularly in the field of the elasticity theory of composite media.

Since highly heterogeneous media constitute one of the basic components of underground water and oil reservoirs, the multiple applications of the developed theory are directly related to hydrogeology and reservoir engineering. Some very close phenomena are studied in the theory of heat transport through composite materials, and in physico-chemical chromatography.

This book is based on results obtained by the author during research performed in the Russian Academy of Sciences, in the Oil and Gas Research Institute of Moscow. I thank all my young colleagues, students and post-graduates, who have taken part in some numerical simulations.

Basic results concerning two-phase flow in double porosity media and in pseudo-cavity media have been obtained during the author's stay as invited professor at Saint-Etienne University, from 1995-1997, and in Laboratoire Energétique et Phénomènes de Transport de Bordeaux, in 1998. The author is grateful to Professor Alain Bourgeat, Dr. Michel Quintard and Dr. Henri Bertin for a fruitful collaboration in this area.

The author would like to thank Dr. Roland Lenormand from the Institut Français du Pétrole of Paris, for his significant research and personal support.

TABLE OF CONTENTS

INTRODUCTION

Scale Transformations, Scale Invariance and Averaging

One of the basic problems of theoretical hydrodynamics consists of studying what new flow property can arise if the characteristic scale of the system is changed. A mathematical point in the equations of continuum mechanics always corresponds to some finite elementary physical volume which contains many physical particles. This elementary volume forms the characteristic scale of the system. Only parameters averaged over the elementary volume enter into the flow equations. Change of scale means not only geometrical variation in the elementary volume size, but also the necessity to define new averaged parameters over new elementary volumes. Thus, the change of the equation scale is equal to averaging.

The size of the elementary volume is not strictly defined and varies within some limits. The limits within which types of equation remain invariable are called *limits of scale invariance.* The property to keep the mathematical type with respect to scale variations is called *scale invariance.* Note that the notion of scale invariance will be used in a wide sense. Only the type of equations is required to be retained, while the value of the equation coefficients can be changed, and non-basic new terms can arise after a change of scale. (In a stronger context, scale invariance means invariance of the overall equation, including the parameter values.)

It is clear that some new quality can result from the procedure of scale transformations, i.e., averaging, if the system is not invariable with respect to these transformations. The well-known transformation of kinetic equations of statistical mechanics into Navier-Stokes equations after a change of scale confirms that the arising of a new quality in the rescaled system is really possible. Another well-known effect consists of homogenization of the convection transport equation of the first order that leads to the arising of a new dispersion equation of the second order, with dispersion parameters dependent on the transport velocity.

On the other hand, classical objects of the averaging theory, such as Laplace's or the diffusion equation, or equations of elasticity theory with oscillating parameters, are typical examples of scale invariable systems. Any changes in characteristic scale retain the mathematical type of the mentioned systems. Due to this fact the classical problem of the homogenization theory consisted in deducing some mathematically proved relations for homogenized coefficients.

Non-Locality as a Result of Scale Transformations

What can we expect to obtain in case of systems which are not scale invariable? Do some general classes of equations exist when arising of new qualitative effects after up-scaling could be predicted *a priori*? Some physicists assert that averaging can generate nonlinearity. In the opinion of the author of this book, it is a mistake. Nonlinearity can be deduced by averaging if averaging is performed in some coarse approximated way, or the original equations already contain the nonlinearity.

However, wide research experience shows another very general phenomenon to arise from the averaging procedure. This phenomenon is *non-locality.* The phenomenon of non-locality in time, i.e. long-term memory, is known in mechanics as a result of the homogenization of processes with a short-term memory at the microscale. For example, homogenization of a viscous-elastic equation describing a system with short-term memory yields a long-term memory equation.[1]

In this book non-locality in time is always associated with the notion of memory. Due to this fact, the term *strongly nonlocal* in time will be applied to the phenomena of long-term memory described by integral or integro-differential equations. the term *weakly nonlocal* will be applied to the phenomena of short memory usually described either by integral equations with degenerating kernels, or by local equations but of a higher order.

Non-Locality and High Heterogeneity

An even stronger result can be obtained: non-locality can be generated by averaging equations which do not possess any memory on the microscale. In the simplest case, in a system without any memory at the microscale, some delay effects (the short-term memory) can arise after averaging. To produce such a phenomenon the structure of the medium must be complicated enough. In particular, the delay may be generated as the result of high difference in the physical properties of various components of the medium.

[1] Bakhalov, N.S. and Panasenko, G.P. Homogenization: Averaging Processes in Periodic Media. Kluwer Academic Publishers, Dordrecht, 1989.

In fact, it is known that the simplest linear diffusion equation being averaged may lead to an equation of a new type, which is integro-differential and describes the long-term memory effect. The necessary condition is the high heterogeneity of the medium. The medium is called highly heterogeneous if it consists of two or more components with contrasting transport properties.

Thus, working with highly heterogeneous media and using averaging methods, we can hope to obtain new mathematical models for flow processes. One of the basic new phenomena which is expected to be arisen is non-locality.

The Problem of Modeling the Flow through Porous Media

All the abovementioned ideas are directly related to the actual problems of modeling flow through porous media.

If the art of deducing new mathematical models in this field of knowledge was estimated as an abstract kind of activity ten years ago, its role is actually changing considerably. Classical models of flow have appeared powerless in some areas having industrial importance, for instance, in modeling the flow through media with heterogeneous wettability, three-phase flow, flow through fractured-porous media, double-porosity media, two-phase flow with variation in displacement direction and others.

One of the global attempts to improve classical models of flow is related to the idea to introduce time-dependent parameters into the flow equations. This sort of activity is known as development of *dynamic models* of flow. Time-dependent parameters should be introduced instead of parameters which are steady-state within the framework of the classical theory. New, dynamic theory will operate with dynamic relative permeabilities, dynamic capillary pressure, dynamic absolute permeability, etc. This formal operation is certainly justified at the physical level. Steady-state parameters conform to the equilibrium state of the system, which can only be really reached at long duration. The real behavior of the system is non-equilibrium, therefore time-dependent. This is absolutely true, especially for two or three-phase mixtures where the characteristic time of fluid redistribution within the pore space is huge compared with the time of a laboratory or an in-situ experiment. For a sample of natural porous medium with 20-25 cm of length the characteristic time of flow stabilization is in the order of several days.

Hence, introduction of non-equilibrium phenomena is a rational, physically justified way to construct improved models of transport through porous media. But the non-equilibrium phenomena are the phenomena with memory.

Improvement of the Flow Models Using Averaging

Remembering the conclusion in the previous paragraph, we obtain the following result. Averaging, or scale transformations (up-scaling), or homogenization, being applied to systems not satisfying the scale invariance property, like highly heterogeneous media for instance, may be considered as a powerful tool with which to construct new mathematical theories of transport in porous media, which are expected to be based on the non-local behavior of the system.

Nonlocality may be displayed in various forms and is always accompanied with the arising of new qualitative modifications in governing equations after averaging, which are shown in this book. They are: increasing of the order for ordinary differential equations, change of the type for PDE, transition from PDE to integral-differential equations, transition from a parabolic equation to an elliptic equation with the second time derivative, arising from delay differential equations, and so on. Dispersion, i.e., diffusion dependent on transport velocity, which is usually a weak effect, becomes of the first order in highly heterogeneous media. In case of nonlinear two-phase flow, interaction of nonlinearity and nonlocality generates new hydrodynamic models which are rich in physical effects.

Short-term in Homogenization Series and Weak nonlocality

Deduction of new models containing nonlocal effects may be regarded as a good result from a theoretical view, but the memory term is a very poor object for numerical simulations. The problem is how to replace such objects by some approximate local operators. In this book, the method of replacing strong nonlocality by weak nonlocality is illustrated.

The close problem arises when the nonlinear equation is homogenized with a generation of nonlocal phenomena on the macroscale. The linking of nonlinearity and nonlocality constitutes a very complicated phenomenon, which is usually very hard to explicitly describe. Science prefers to separate these two phenomena and to deal with only one of them. However, introduction of time-dependent functions in the theory of two-phase flow through porous media constitutes a typical example when such linking of nonlinearity and non-locality is formulated as the research objective.

To solve this problem the *method of weak non-locality* is suggested in this book, which consists in choosing such systems where only weak nonlocality can be generated and then extrapolating the results onto the strongly non-local systems in an approximate way. It will be shown that linking of nonlinearity and weak nonlocality is usually performed in constructive ways and gives origin to very fruitful theories.

However, weak nonlocality, or short-term memory, is defined as a phenomenon with a short time of storing in memory. Hence to keep the short-term memory effects in the macroscale equations, it is necessary to keep some short-terms in the homogenization series. This is the second point whereby the role of homogenization methods may be considerably enlarged.

The classical mathematical approach consists of defining the limit of the microscale equations, i.e., the zero term of asymptotic expansions, when the heterogeneity scale tends to zero, and to prove convergence. From the point of view of physics, the limit is often trivial, but some very important physical effects are hidden in the first approximations.

As stated, the requirement to keep the first terms in a homogenization series becomes necessary for instance when working with nonlinear systems, which generate nonlocal phenomena on the macroscale.

Objects of Research

For flow in a porous medium, which is described by many spatial scales associated with hierarchical heterogeneity of the internal structure, the problem of scale transformation for governing models of flow becomes dominant. Transition along sequential space scales is termed as *up-scaling*. According to some traditions, two characteristic spatial scales are distinguished in a porous medium. The first is a *pore-scale* that corresponds to a rather large assembly of pores (a statistically representative). The next spatial scale is formed by a composite structure of the porous medium, which may be constructed of several types of porous rocks with distinct hydrodynamic conductivity. The size of one block forms a *block-scale* of the porous medium.

Transition from a scale much smaller than the size of one pore to scale of a rather large assembly of pores leads to converting Stokes equations into Darcy's law. Such a transition is rather well studied in the literature, but any complication of the flow leads to problems that remain unresolved up to now. Flow at small but nonzero Reynolds numbers and two-phase flow are typical examples of such open problems.

This book deals with another class of objects. The next step in the hierarchy of scale transitions is studied, i.e., transition from the pore-scale into the block-scale. The problem is to then deduce the macroscale equations, i.e., equations averaged over the heterogeneity when the microscale flow is described by Darcy's law. When the hydraulic properties of the medium components are distinct but of the same order, the scale-up problem seems to be rather trivial: change of scale is formal and does not lead to any modifications in flow equations; the problem is to rescale the coefficients. In this book *highly heterogeneous media* are studied, i.e., media composed of two types of rock with highly distinct hydraulic properties. High heterogeneity disturbs the scale invariance of flow equations in any type of flow. Even in the simplest case of diffusion transport, the long-term memory phenomena arise on the macroscale.

When the scaling invariance is disturbed, we first need to determine the structure of the macroscale equations and to compute their effective parameters. Many of the classical methods of averaging, such as renormalization techniques, become useless in this situation. The nonlinear character of flow equations in the two-phase case does not allow other widely developed methods, such as Dyson's operator method, or Feynman diagrammatic techniques to be used. Due to these circumstances, another approach is applied in this book, which is based on asymptotic homogenization[1],[2],[3],[4]. The basic idea of this method is that the averaged medium can be obtained from a real medium if we tend the heterogeneity scale $\epsilon = l/L$ to zero. Hence, the zero term of asymptotic expansion corresponds to the averaged model. In the highly heterogeneous case, this method requires some modification.

Within the framework of this book mathematical aspects of the method are regarded as secondary. Moreover, the mathematical results are not always considered sufficient results from the point of view of mechanics. First of all, the result of homogenization is not accepted as a sufficient result, if the variables are not completely split, i.e., the macroscale equations do not contain the microscale variable, while the cell problem does not contain the macroscale variable. Such requirements for the result are found to be too strong. A unique way to satisfy them consists in thorough examination of the complex physical mechanisms of the process, especially in the case of two-phase flow, partitioning the system on elementary classes, numerical or experimental investigation of the features for each class, and, lastly, homogenization of each class separately. Generalized macroscale models covering several classes are obtained as some combination of particular models strictly deduced for each class.

Structure of the Book

The structure of the book is as follows. In the first chapter, one-phase flow described by a linear diffusion equation in a double porosity medium is homogenized. The second chapter is devoted to convection-diffusion transport in double porosity media with a non-periodic velocity field and with various orders of Peclet number. In the third chapter, an analytical method to calculate the dispersion tensor in media of lattice structure is developed.

[2]Bensoussan, A., Lions, J.-L. and Papanicolaou, G. *Asymptotic analysis for periodic structures*, Amsterdam: North-Holland, 1978.

[3]Sanchez-Palensia, E. *Non-homogeneous media and vibration theory.* Volume 129 of *Lecture Notes in Physics*, Springer Verlag, Berlin, 1980.

[4]Zhikov, V.V., Kozlov, S.M. and Oleinik, O.A. *Homogenization of Differential Operators.* Nauka, Moscow, 1991.

The fourth chapter is devoted to non-compressible two-phase flow with capillary forces in double porosity media. In the fifth chapter, the same type of flow is examined in pseudo-cavity media which are formed by a lowly permeable connected matrix and highly permeable inclusions and are close to purely cavity media where the inclusions are replaced by the concavities. In the last section of the first chapter some approaches are briefly examined to average the flow in disordered media.

ONE PHASE DARCY'S FLOW IN DOUBLE POROSITY MEDIA

1.1. ACTUAL STATE OF THE PROBLEM

The basic object of examination of this book is associated with highly heterogeneous media where the permeability is contrasting in various elements of the domain. Construction of the averaged models in this case requires to solve several specific basic problems, as determination of the averaging method and a priori classification of media allowing to obtain constructive results of homogenization.

1.1.1. METHODS OF AVERAGING

All averaging methods can be classed onto three groups, as coefficient, phenomenological and structural averaging.

Coefficient Averaging
In this case the type of averaged equations are known a priori, and the problem consists in construction of the effective coefficients. The choice of the macroscale equations is any, for example by analogy to available microscale model. The review of such methods is given in papers [71], [143], [141], [1], [94].

Among them the following methods should be noted.

Variational methods , [24], [25], [160], [66], [57], allow to obtain two-sided estimations for the effective tensors, being based on the variational principles of mechanics: the principle of minimum for potential energy and the Kastiliano principle of minimum for the additional work.

Method of diluted systems (or virial decomposition method), [56], [143], [142], is based on the expansion of the effective tensors into series with respect to the concentration of inclusions; it is least suitable for densely packed media.

Effective medium theory , [53], [49], [84], [11], [33], [74], [86], is based on selection of an elementary cell in the form of any inclusion surrounded with the homogeneous effective medium. One distinguishes the variants of this method for both the discrete and the continuous structures [98].

Method of correlation approximation , [88], [143], is applicable to the systems with weak heterogeneity (small difference between the conductivities of the medium components). It is based on presentation of the effective tensor as a sum of the average value and the correlation term which takes into account the pair interactions. The closure problem is easy solved due to the assumption about weak heterogeneity.

Method of singular approximation , [153], [58], is based on representing the fluctuation components using Green's function for the basic effective operator on the element of heterogeneity and on rejecting the nonsingular part of the derivative from the Green function. This is equivalent to neglecting the long-range interactions.

Strong isotropy method , [30], is applied when the field is assumed to be homogeneous within the given component of the medium.

Renormalization method , [2], [27], [96], [131], [134], [140], [72], [73], is based on the idea of scale invariance [41], that is common to a disordered self-similar field. Renormalization-group transformation (RG) translates a discrete field in the continuous one [28]. For a grid of conductors with characteristic scale l the RG-transformation reconstructs the grid consistently into a larger grid of the scale $2l, 4l, 8l, ...$, until the averaged conductivity A does not vary. The last grid consists of one conductor; its conductivity is equal to A. In papers [2], [131] the RG-transformation is reduced to a rather elementary procedure, namely, to calculation the total conductivity of an elementary circuit of five conductors that is a trivial problem in the theory of electric circuits.

Method of finite electrical circuits , [151], is used when the medium is replaced approximately by some equivalent system of conductors connected in parallel and in series.

Capillary network simulation , [54], [92], [1], [46], [90], is based on the direct numerical solution of the appropriate transport problem formulated at the level of Poiseuille equation in each pore channel.

Percolation method [139], [106], [145], replaces the real medium by a system of non-connected channels; their structure and volume fraction are defined from the universal correlations of the percolation theory obtained numerically basing on the network models of porous media. The numerical methods using general cellular automata algorithms [76] are now in development. They allow to use the grids of the size up to 22 millions of cells with the help of special processors [146].

All these methods are equally applied to both the periodic and the random media.

For regular 2D structures the methods of exact solution have been developed, which are based on the theory of complex variables [50], [93]. Ror random media the volume of results obtained is more small, [49].

It is necessary to note the papers where the effective conductivity or the diffusion tensor are determined on fractal media [133], [157].

Method of Coexisting Continuums

It is based on phenomenological approach to describe the heterogeneous system. Each component of the system is considered a priori as "smeared" across the overall continuum but having reduced density, so that its physical characteristics can be considered as continuous, smooth, e.t.c. Two variants of the method may be noted.

Classical method , [45], [77], [85], [99], [100], [130], [137], [4], assumes the conservation laws are written for averaged values introduced a priori. Additional relations between them are established from the principles of nonequilibrium thermodynamics. This approach has been applied to porous media in papers [102], [101], [147], [148], [12], [129], [94].

Variational method [138] allows to deduce the macroscale model from the variational equation of continuum mechanics. This method needs to postulate the form of dissipation, kinetic, and internal energy functions written through averaged variables chosen a priori.

Structural Averaging Methods

These methods suggest some algorithms allowing to derive the averaged equations proceeding from the available equations at the micro-level. The basic problem of averaging consists in construction of the constitutive relations, which relate the oscillating components of the heterogeneous field to its average value. Two groups of methods are developed where this problem is solved.

Methods of stochastic averaging. These methods, which have been developed in statistical physics, [3], [23], [163], [89], [97], where the macroscale equations describing behavior of continuum media are deduced by averaging over multi-particle ensembles.

The central point consists in introduction of multi-particle distributions describing the probability of occurrence for the particles under given dynamic condition, deduction the sets of Bogolubov's equations, and breaking them according to various physical reasons (kinetic and hydrodynamic approximations). Similar methods were applied in mechanics of disperse systems, for example, in papers [34], [135], [59].

Asymptotic methods of homogenization. The idea of homogenization consists in that the effective media can be obtained from the real if the scale of heterogeneity $\varepsilon=l/L$ tends to zero. Really, the lower ε, the structure of the medium becomes more fine-grained. In the limit, each cell transforms into the point, and we obtain a homogeneous medium with internal structure of each point. Therefore the zero term of asymptotic expansions determine the form of the averaged equation. Certainly, the weak convergence only may be ensured.

Formalism of the method was developed in papers [15], [14], [13], [26], basic convergence theorems were proved in papers [162], [136]. Along with this, the mathematical homogenization theory was developing other basic points such as prove of existence and uniqueness of the cell problems, reduction of the periodic cell problem to a boundary-value problem, construction of complete asymptotic expansion, estimations for residual terms of expansions, studying the tensor properties of effective coefficients, generalization of the theory in case of discontinuous fields, close to periodic, nonlinear equations, e.t.c. The most full description of the homogenization theory can be found in monographs and reviews [16], [17], [161], [92], [136], [1], [26], [39].

Basic results has been obtained for periodic fields. The theory for random media is developing in [78], [79], [80], [81], [82], [83], [110], [113], [158], [159].

1.1.2. AVERAGING OF PROCESSES IN HIGHLY HETEROGENEOUS MEDIA

High Heterogeneity

One of the more significant problem of the homogenization theory deals with high heterogeneity. The medium is high heterogeneous if it is composed by two or more alternating sub-domains having contrasting transport properties. The double porosity or the fractured-porous media are typical examples.

More accurate definition for highly heterogeneous medium is the following.

Let us examine a periodic heterogeneous field characterized by two dimensionless parameters, "the heterogeneity scale" ε and "the heterogeneity degree" ω. The last is defined as a ratio between the minimal and the maximal values of the oscillating function. For a porous medium the parameter ω is equal to the ratio of permeabilities of two various components of the medium.

A field is highly heterogeneous when:

- $\omega << 1$.

A field is fast-oscillating and highly heterogeneous when:

- $\omega << 1$,
- $\varepsilon << 1$.

In general case four situations can be observed:

i) weak heterogeneity: $\omega << 1$, ε is anyone;
ii) fast oscillating, moderate heterogeneity: $\omega \sim 1$, $\varepsilon << 1$;
iii) large-scale heterogeneity: $\omega \sim 1$, $\varepsilon \sim 1$;
iv) fast-oscillating, high heterogeneity: $\omega << 1$, $\varepsilon << 1$.

The first class is elementary, as the effective characteristics of the averaged fields coincide with the simple arithmetic mean value.

The second class is a classical object of the averaging theories [17], [136].

The third class concerns the non-homogenizable systems and will not be examined herein at all.

The fourth class is a matter of essential interest to this book. Methods of averaging for this class were developed in papers [108], [109], [111], [119], [120], [121], [122], [125] where it was noted that classical homogenization does not work for some types of highly heterogeneous media. The raison consists in that the equations of process are degenerating when ω becomes very large or very small. In these cases the averaged equations either does not exists, or the microscale equations do not converge to the macroscale model even in the weak sense.

High Heterogeneity, Disconnected Field of High Conductive Inclusions
The averaged behavior of the system depends considerably on the connectivity degree of the highly conductive component. In papers [109], [111] the case was analyzed when the high conductive subsystem is disconnected with mono-directed infinite cylinders-fibres. Behavior of such a system was

shown to be depending on the parameter

$$\pi \equiv \frac{\varepsilon^2}{\omega} \tag{1.1}$$

In [109] three classes of systems were stood out depending on the ratio:

$$\begin{aligned} &1)\quad \pi \ll 1 \qquad (\omega \gg \varepsilon^2),\\ &2)\quad \pi \sim 1 \qquad (\omega \sim \varepsilon^2),\\ &3)\quad \pi \gg 1 \qquad (\omega \ll \varepsilon^2), \end{aligned} \tag{1.2}$$

where the first case is homogenized in the custom way: first, the asymptotic expansion is developed at $\varepsilon \to 0$ with fixed parameter ω, the consequent expansion at $\omega \to 0$ is constructed after. (In [109] the parameter $\omega_p \equiv \frac{1}{\omega}$ was used with asymptotic at $\omega_p \to \infty$). In the second and the third cases "the multicomponent averaging" was proposed in [111] where averaging is made over each medium component, instead of averaging over the whole cell.

The results are the following.

i) If the conductivity of non-connected fibres is very great ($\pi \gg 1$), than they operate as completely independent of each other. The low permeable connected matrix is removed from the process. Averaged equations constitute a system of ordinary differential equations (ODE) for all fibres.

ii) When the fibre conductivity becomes lower ($\pi \sim 1$), the low permeable matrix is already involved in the process. Its role is displayed in arising the exchange process between the fibres. Averaged equations constitute a system of ODE for all fibres related by the exchange terms, which are proportional to the difference between the mean "pressures".

iii) Consequent decreasing of the fibre conductivity, such that ($\pi \ll 1$), leads to a uniform averaged equation for the overall medium. The medium behaves as a uniform system. Simple averaging may be applied by developing expansions over ε at first, and over ω after. High heterogeneity influences the structure of the averaged coefficients only.

It is necessary to note that in the first and in the second cases any transport can not happen through the medium when the fibres have a finite small length. The system does not possess any averaged model.

High Heterogeneity, High Conductive Subdomain is Connected

This case in not symmetrical to the previous one. The basic transport being happened in highly conductive subdomain, it remains lower sensitive to conductivity degree of the dense fibres. For any length of fibres the system has an averaged model.

Such systems were examined in [119], [121], [123], [125], [120], where it has been shown independently that the usual homogenization became invalid for high degree of heterogeneity, or for large magnitudes of the parameter π. However the main result is that the form of the averaged model is always invariable and does not depend on the system class (1.2). The heterogeneity degree influences the structure of the averaged conductivity only. The condition of high heterogeneity is not sufficient to the exchange process arises. This phenomenon will be examined furthermore.

Note the papers [70], [71], [8], [9], [10], [47], [62], [144] where high heterogeneous media were studied too.

Multiple Heterogeneity

The transport in porous media is usually characterized by several heterogeneous fields: of porosity and permeability, of porosity and diffusion parameter, e.t.c. In classical homogenization all the heterogeneous fields are assumed to be similar, so their scale of oscillation ε and their heterogeneity degree ω are equivalent. This assumption is usual as for homogenization of moderately heterogeneous media [17], as for high heterogeneity [111].

Removing this assumption, we get a more complicated class of media where homogenization leads to more reach results.

A medium is named as *"multiply heterogeneous"* when the heterogeneity parameters of various fields describing this medium are dissimilar. The media of such a type were considered in papers [115], [117], [116], [118], [119], [123], [125], [120]. In [119], [123] it was shown that behavior of the medium described by two heterogeneous fields (porosity and permeability) with an equivalent period of oscillations ε but distinct heterogeneity degree, ω_K and ω_m, is defined by a parameter similar to (1.1):

$$\varepsilon_p \equiv \frac{\varepsilon^2 \omega_m}{\omega_K} \tag{1.3}$$

The following classification may be suggested with respect to this parameter.

$$\begin{array}{llll} i) & \varepsilon_p \leq \varepsilon^2 & \text{or} & \omega_K/\omega_m \geq 1 \\ ii) & \varepsilon^2 \ll \varepsilon_p \ll 1 & \text{or} & \varepsilon^2 \ll \omega_K/\omega_m \ll 1 \\ iii) & \varepsilon_p \sim 1 & \text{or} & \omega_K/\omega_m \sim \varepsilon^2 \\ iv) & \varepsilon_p \gg 1 & \text{or} & \omega_K/\omega_m \ll \varepsilon^2 \end{array} \tag{1.4}$$

The classical variant of homogenization can be applied only in the first case when the uniform averaged equation may be deduced for the overall medium. In other cases, non-uniform averaging has been proposed, which leads to the macroscale model in the form of two sub-systems, each of them is averaged over the appropriate component of the medium. Both the subsystems are related by the exchange terms.

Macroscale Models of One-Phase Flow through Highly Heterogeneous Media

Not all researches mentioned above are completed up to the closed macroscopic models. The following types of models may be noted.

Model with quasi-equilibrium mass exchange. The classical model derived from phenomenological approach [18] [19] considers the dense blocks and the highly permeable matrix (fractures) as two coexisting continuums with mass exchange between them described by an exchange term. The form of the exchange term is postulated as the difference between mean pressures, that corresponds to the quasi-equilibrium exchange flow. The form of the exchange term is a key problem. Equations for blocks and fractures are postulated to have the symmetrical form. Some complications of averaged models are associated to including deformation of rocks and consequent variations of fracture parameters.

In some variants of the process the blocks are considered only as a capacity element. In paper [44] such model is named as "kinetic model of first order". The models of this types are appropriate to describe diffusion with sorption in granular media when the stagnant zones are formed. They play the role of low permeable blocks [38]. In the original paper [19], a variant is suggested when the blocks correspond to the capacity elements, while the capacity of fractures is neglected.

The similar models were suggested later in [29], [36], [48], [51], [59], [60], [154].

The assumption about the quasi-equilibrium exchange flow was checked out in [68] where it has been shown to be valid, because the transient phenomena happen rapidly in real situations. Other result has been obtained in [107] where thorough numerical study has shown the exchange flow be not governed by equation [19] as qualitatively, as well as quantitatively.

Later we shall demonstrate that after reaching stationary state the averaged pressures in blocks and in fractures become practically equivalent, and the exchange flow can not be detected. Therefore, strictly speaking, the assumption about steady-state exchange flow is equivalent to the assumption that the block-matrix exchange does not exist.

Quasi-homogeneous model. In paper [104], the exchange flow has been shown to be rapidly established, as well as to be extremely insignificant. Thus the pressure in a highly heterogeneous medium is uniform for the blocks and for the fractures and may be described by one parabolic diffusion equation without any exchange terms. This conclusion has been confirmed in [124], [125] where averaging of one-phase flow was made in a medium of block structure. The exchange term has been shown to be of order ε^2 when the porosity and the permeability fields are similar.

Model with non-stationary exchange. In [42] the non-stationarity of the exchange flow was taken into account by presenting the exchange term in the form of a convolution of the pressure gradient at the block boundary. The form of this term was postulated, and closing relations for integrand function and parameters of exchange intensity have been determined by examination of a problem of the liquid flow from a single block under constant pressure at the block border.

In papers [119], [122], [125] a more general model with the close form of the exchange term has been derived by homogenization for one class of highly heterogeneous media (source-type media). In the same time, the model has been shown to be of other type for other classes of media. Later de Swaan's model has been obtained by homogenization in papers [10], [61] for one fixed ratio among the parameters of the medium. The models with non-stationary exchange was suggested in papers [35], [20], [21] where the solution of the cell problem has been brought either to some integral relation for the Laplace image, which was not more simple than the original microscale equations.

In papers [70], [71] a special type of media was examined when heterogeneity consists in thin impermeable rings having small orifices. Walking particle being penetrated inside a ring can not abandon it for a long time. Hence the rings play a role of traps leading to long memory effects on the macroscale. The averaged model of walking is similar to the model of flow through one of the classes of double porosity media.

Numerical models with two-scale grids. Another approach consists in using two grids of different scale for numerical simulation. The coarse grid corresponds to the macroscale. Each cell of the coarse grid is covered by a fine grid. The equation, which is considered as averaged over the fractures with the exchange term, is computed basing on the coarse grid. The exchange function results from the numerical solution of each fine-grid cell problem, which requires conditions of matching with the macroscale solution. Two types of matching conditions are considered.

The more widely used condition is that of equality of the pressure at the block border to the pressure averaged over the fractures [144], [87]. The same condition is introduced in the MINC-method (multiple interacting continua) [128], [156]. Its feature is the use of the microscale grid in the form of narrow layers forming a concentric system covering the cubic block.

In the paper [144] the such a model is named as *"microstructural matched model"*.

Another condition was suggested in [144]. The flow rate on the block border is assumed to be equal to the pressure difference at the interface. The corresponding model has been named as *"regularized micro-structural model"*. Correctness of Cauchy's problem for these models was studied in [62].

Model with non-stationary exchange and translation type of flow through the blocks. The model is derived in [119], [117], [116], [127] for highly heterogeneous media of translation-source type. The blocks are operating in two ways, as a non-stationary source, as well as they give a contribution to the translation flow through the medium. The exchange term is described by the second time derivative of pressure.

The reviews of models for fractured-porous media are available in papers [55], [51].

The applications to the well investigation in highly heterogeneous reservoirs have been developed in papers [35], [55], [40], [52], [104], [105]. The analytical solutions of reservoir engineering problems was constructed in several of papers mentioned above, as well as in [37], [91], [155]. Numerical solutions of problems formulated within the framework of the macroscale models have been developed in [43], [69] etc.

In this chapter the complete hierarchy of macroscale models of flow through double porosity media is derived.

1.2. SETTING OF THE PROBLEM. CLASSIFICATION OF MEDIA

1.2.1. PHYSICAL SETTING

Let us consider the one-phase weakly compressible flow in a weakly deformable medium made of a "highly permeable matrix" (medium II) and of a system of periodically located "dense blocks" (medium I). Each of such sub-domains Ω^i, $i = I, II$, is characterized by the rock permeability tensor of the second rank, K^i, and the porosity m^i. The viscosity μ of fluid is supposed to be constant. Let P be the fluid pressure, X be the space vector coordinate and t be the time. Let L be the macroscopic scale and l be the period of heterogeneity.

We suppose that the medium is *"fast oscillating"* and *"highly heterogeneous"*.

Fast Oscillating Heterogeneity
The first property means that the characteristic scale of heterogeneity l is much smaller than the macroscale L:

$$\varepsilon \equiv l/L << 1 \tag{1.5}$$

Hence the medium consists of a periodic spatial alternation of two subdomains with distinct properties, as depicted in Fig. 1.1.

Due to this fact, the homogenized description seems to be natural for such a system, and it is intended to be equivalent to the detailed microscale description in some integral sense.

High Heterogeneity
The property of high heterogeneity means that two components of the medium, the blocks and the matrix, have contrasted permeabilities, such that ω_K, the ratio between the permeabilities in medium I and in medium II, is a small parameter.

This property will be assumed to be valid for all the components of the tensor K_{ij}, i.e.:

$$\frac{K^I_{ij}}{K^{II}_{ij}} \sim \omega_K \ll 1, \quad \forall i, j=1, ..., n$$

The medium of such a type is named as *"double porosity"*. Another example of high heterogeneity is associated to fractured porous media, when the matrix is made of thin empty fractures. However more fine microscale

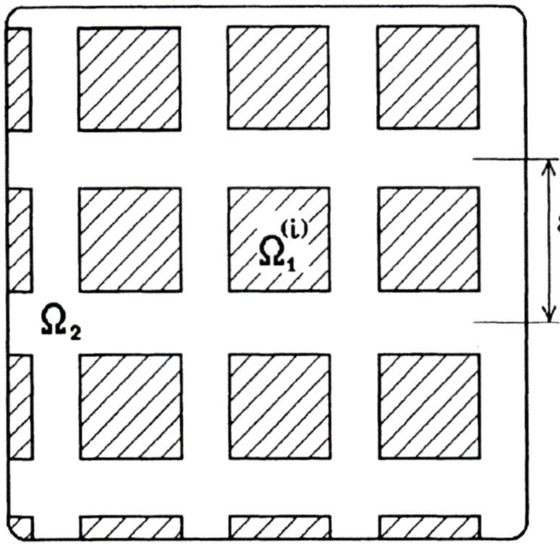

Figure 1.1. Highly heterogeneous medium

equations than Darcy's law must be applied to describe the flow inside the fractures. In this book we are studying the double porosity media only.

Let us assume the rock permeabilities K^I and K^{II}, as well as the porosities m^I and m^{II} to be constant.

Characteristic Parameters of the Process

Let us introduce the following characteristic magnitudes:

- Characteristic *macroscale* L which is equal to the linear size of the domain.

- Characteristic *microscale* l which is equal to the period of heterogeneity.

- Characteristic pressure P^* which can be equal to the maximal value of the initial function $P^0(X)$, for example.

- Mean permeabilities and porosities for each sub-domain Ω^i, $i=I, II$:

$$\overline{m}^i \equiv \frac{1}{|\Omega^i|}\int\limits_{\Omega^i} m^i(x)dx, \quad \overline{K}^i \equiv \frac{1}{|\Omega^i|}\max_{(j,k)}\int\limits_{\Omega^i} K^i_{jk}(x)dx$$

- Time of complete filling the matrix by the fluid

$$t_* \equiv \frac{\overline{m}^{II}\mu L^2}{\overline{K}^{II}P^*}$$

Parameters describing the *heterogeneity scale* and the *heterogeneity degree* are:

$$\varepsilon = \frac{l}{L}, \qquad \omega_K = \frac{\overline{K}^I}{\overline{K}^{II}}, \qquad \omega_m = \frac{\overline{m}^I}{\overline{m}^{II}}$$

1.2.2. MATHEMATICAL FORMULATION

Let examine a problem of perturbation of a fluid fulfilling a porous layer in the domain Ω. The initial characteristic pressure is $P^0(x)$.

Let us introduce the following notations : $p^i \equiv P^i/P^*$, $\tau \equiv t/t_*$, $x \equiv X/L$.

Flow Equations

Equations describing the one-phase flow in the reservoir $\Omega = \Omega^I \cup \Omega^{II}$, $\Omega \subset I\!R^s$, $s \geq 2$, at time $\tau \in (0,T)$ with respect to the pressure p may be written in the following form :

$$L_\varepsilon p \equiv \frac{\partial}{\partial x_i}\left(a_{ij}^{\omega_K}\left(\frac{x}{\varepsilon}\right)\frac{\partial p}{\partial x_j}\right) - b^{\omega_m}\left(\frac{x}{\varepsilon}\right)\frac{\partial p}{\partial \tau} = -f(x,\tau,\omega_K) \quad (1.6a)$$

$$[p]_\Gamma = 0 \;, \quad (1.6b)$$

$$\omega_K a_{ij}^I \frac{\partial p^I}{\partial x_j} n_i|_\Gamma = a_{ij}^{II}\frac{\partial p^{II}}{\partial x_j} n_i|_\Gamma \;, \quad (1.6c)$$

$$p(x,0) = p^0(x), \quad (1.6d)$$

We assume Ω to be a cube in $I\!R^s$ and the periodicity of boundary conditions. The subarea Ω^{II} is connected ("matrix"). The system of N not intersecting subareas Ω_i^I ("blocks") form a periodic structure. Γ is the interface between Ω^I and Ω^{II}; n is the external normal vector to the boundary; $[\cdot]$ is the symbol of jump.

Relations (1.6b) and (1.6c) require the pressure and flow rate to be continuous at Γ, the interface separating two media, Ω^I and Ω^{II}.

Tensor of the second rank $a^{\omega_K}(x,y)$ and the scalar $b(x,y)$ undergo the discontinuity of the first sort on Γ and are determined as :

$$a^{\omega_K}(x,y)_{ij} \equiv \frac{K_{ij}}{\overline{K}^{II}} \equiv \begin{cases} a_{ij}^{I,\omega_K} = \omega_K a_{ij}^I, & x \in \Omega^I \\ a_{ij}^{II,\omega_K} = a_{ij}^{II} & x \in \Omega^{II} \end{cases}$$

$$(1.7)$$

$$b^{\omega_m}(x,y) \equiv \frac{m}{\overline{m}^{II}} = \begin{cases} b^{I,\omega_m} = \omega_m b^I, & x \in \Omega^I \\ b^{II,\omega_m} = b^{II} & x \in \Omega^{II} \end{cases}$$

and they are periodic with respect to y. The matrix $\{a_{ij}(x)\}$ is positively defined for each x, the parameter $b(x)>0$, $\forall x$. All the parameters a_{ij} and b are periodic.

In the next equations, the following notation will be used:

$$a_{ij}=\begin{cases} a_{ij}^{I}, & x\in\Omega^{I} \\ a_{ij}^{II}, & x\in\Omega^{II} \end{cases} ; \qquad b=\begin{cases} b^{I}, & x\in\Omega^{I} \\ b^{II}, & x\in\Omega^{II} \end{cases} \tag{1.8}$$

When the permeability is piecewise constant, then all the functions a_{ij}^{I}, a_{ij}^{II}, are constant.

When the porosity is piecewise constant, then

$$b^{I}=b^{II}=1 \tag{1.9a}$$

When the permeability is scalar and piecewise constant, then

$$a^{I}=a^{II}=1 \tag{1.9b}$$

Unit Cell

Along with the area Ω we shall consider the map $Y=x/\varepsilon$ of its unit periodic element to the cell-period $Y=\{-1/2<y_i<1/2,\ i=1,\dots,s\}$, which is broken by a closed surface $\Gamma=\partial Y^{I}$ into two subareas Y^{I} (blocks) and Y^{II} ("matrix"), with $\Gamma\cap\partial Y=\emptyset$ (Fig. 1.2).

Limitations for the Parameters

According to the definition of a double porosity medium we impose the following assumptions on the range of parameters :

$$0<\omega_K\ll 1 \tag{1.10}$$

Contrarily to the last relation we assume the following for ω_m:

$$0<\omega_m\leq 1 \tag{1.11}$$

Restrictions on the variation of porosity b^{ω_m} and permeability a^{ω_K} are nonequivalent, that plays a considerable role on further analysis. The following relation is usually accepted in literature $\omega_m\sim\omega_K$ [16], [17] however this case is not typical. For underground reservoirs, small variation of porosity are accompanied by huge permeability oscillations. An example is demonstrated in Fig. 1.3 for the Tenguiz oil field located in Kazakhstan. The permeability is varying two-four orders, when the porosity varies two-three times.

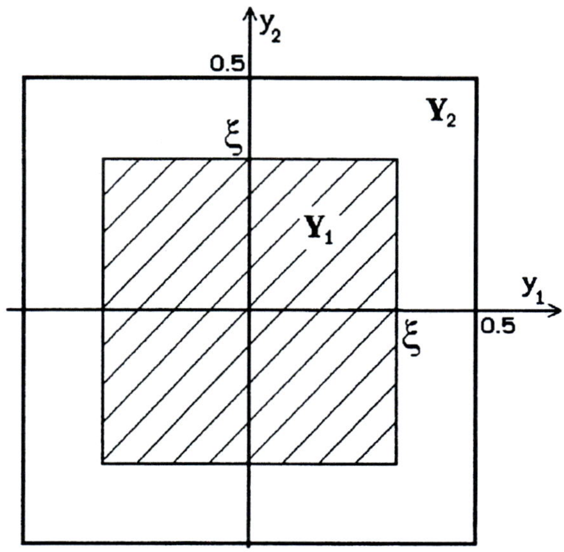

Figure 1.2. Structure of a heterogeneity cell Y

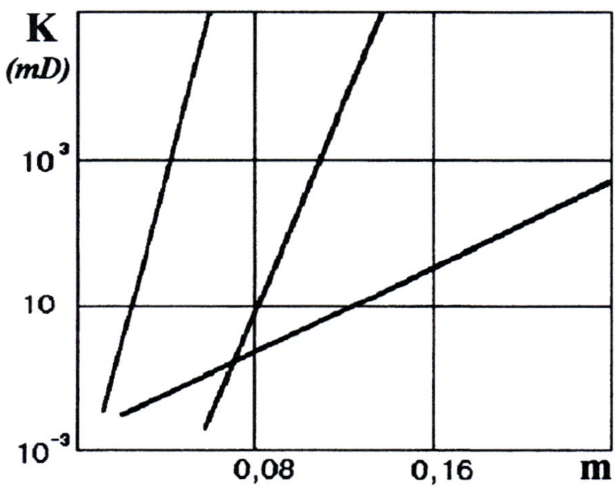

Figure 1.3. Correlation between permeability and porosity for the oil field Tenguiz

In general case, a medium being heterogeneous with respect to porosity and permeability may be homogeneous with respect to piezo-conductivity $æ=a/b$. The ratio of block and matrix piezo-conductivity is equal to $æ^I/æ^{II}=\omega_K/\omega_m$, therefore $æ^I \sim æ^{II}$ if $\omega_m \sim \omega_K$. The media of such a type will be named as piezo-homogeneous. When the behavior of the fields a^{ω_K} and b^{ω_m} is asymmetrical, the medium will be highly heterogeneous with respect to piezo-conductivity too.

It is required to construct the macroscale equation, which would describe the macroscale process with sufficient accuracy. Technically it is equivalent to construct the asymptotic expansion of the problem (1.6) when ε, $\omega_K \to 0$.

Note that the limit transition $\omega_K \to 0$ leads to a system with impenetrable blocks, similar to the perforated area in the elasticity theory. When $\omega_K = 1$ we get the custom system, which is not highly heterogeneous.

Multiple Heterogeneity

Various functions describing heterogeneous fields correspond to various terms of equation and are responsible for various physical components of flow. Comparative behavior of these functions defines the relative importance of some physical components. As to equation (1.6), the relative behavior of porosity $b^{\omega_m}(x)$ and permeability $a^{\omega_K}(x)$ determines the relative domination of one of two processes: non-stationarity of the pressure field caused by elastic properties of the system, and spatial variations of the pressure field associated to the directed current of flow. When the heterogeneous fields oscillate with various intensity, then one of the mentioned processes predominates on some components of the medium. Thus, the pressure field behaves in various way on various subdomains. However it remains continuous, that causes a third phenomenon as exchange flow, defining the submedia interaction.

In this case, the medium structure seams to be non-trivial, because it generates a new physical effect on the macroscale, which was named in [117] as "structural effect". We see that the structural effect is associated to the multiply heterogeneous structure of the medium.

Multiple heterogeneity of A-type is a system of several heterogeneous fields with equivalent heterogeneity scale ε, but with various heterogeneity degrees ω_K.

Multiple heterogeneity of B-type is a system of several heterogeneity fields with equivalent heterogeneity degree ω, but with various heterogeneity scales ε.

Multiple heterogeneity of C-type is a system of several heterogeneity fields with various heterogeneity degrees ω, and with various heterogeneity scales ε.

Trivial heterogeneity comprises all other types.

Hence, new physical effects are expected to be arising in multiply heterogeneous media. The intensity of new effects depends on the degree of multiple heterogeneity, whose measure is a difference in the scale properties of each of the fields.

Differences in the intensity of field's oscillations may be characterized quantitatively by the ratio of corresponding heterogeneity degrees:

$$\Lambda = \frac{\omega_K}{\omega_m}$$

Trivial heterogeneity corresponds to the case $\Lambda \sim 1$. All other cases fall into the category of multiple heterogeneity.

When several fields are distinct heterogeneity scales ε, averaging should be performed over the new scale, which is multiple of both the mentioned scales. Thus the type B of multiple heterogeneity may be reduced to the trivial heterogeneity, whereas the case C may be reduced to the type A. Thus, the type A is only studied in this book.

It is necessary to note, that the parameter Λ coincides with the ratio of block and matrix piezo-conductivities.

Note the multiply heterogeneous media are always highly heterogeneous respectively to one of parameters.

1.2.3. MEDIA CLASSIFICATION

Asymptotic Analysis of the System

Let us write equation (1.6) on a block and on the matrix separately, considering the process within one cell. Let us introduce new coordinates $\{y_i\}_{i=1}^{s}$ renormalizing them in such a way, that the new length of the cell Y be equal to 1: $y_i = x_i/\varepsilon$, $i=1,\ldots,s$. The equations (1.6) get the form:

$$\varepsilon_p b^I \frac{\partial p}{\partial \tau} - \frac{\partial}{\partial y_i}\left(a_{ij}^I \frac{\partial p}{\partial y_j}\right) = \varepsilon^2 f_*, \quad y \in Y^I$$

$$\varepsilon^2 b^{II} \frac{\partial p}{\partial \tau} - \frac{\partial}{\partial y_i}\left(a_{ij}^{II} \frac{\partial p}{\partial y_j}\right) = \varepsilon^2 f_*, \quad y \in Y^{II}$$

A new parameter arises:

$$\varepsilon_p = \frac{\varepsilon^2 \omega_m}{\omega_K} = \frac{\varepsilon^2}{\Lambda}$$

Its order determines the difference in the equations on the blocks and on the matrix. The parameter ε_p is the measure of the rate of perturbation propagation through a block. In fact, if $æ^I/æ^{II} = \omega_K/\omega_m$ is a ratio of the block and the matrix piezo-conductivity $æ^i$, than the coefficient $\varepsilon_p = (l^2/æ^I)/(L^2/æ^{II})$ is the ratio of perturbation propagation time for one block to that for the overall matrix.

Under conditions (1.10) the parameter ε_p can vary within a wide range from $\varepsilon_p \to 0$ to $\varepsilon_p \to \infty$.

Trivial heterogeneity corresponds to the case $\varepsilon_p \sim \varepsilon^2$.

Asymptotic analysis of equations (1.2.3), (1.2.3) offers the following media classification [119] including three main classes and two subclasses. Graphical illustration for three main classes is given in Fig. 1.4, where the left-hand part shows the structure of the fields $b^{\omega_m}(x)$ and $a^{\omega_K}(x)$, while the resulting pressure $p(x)$ field is illustrated on the right-hand part

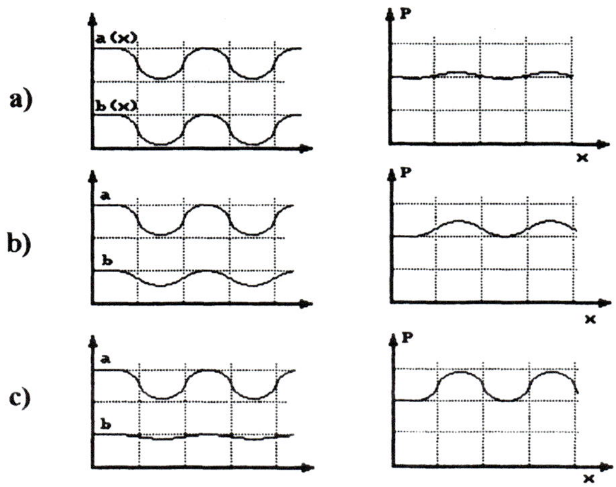

Figure 1.4. Three classes of highly heterogeneous media: a) translation-type, b) translation-source type, c) source-type

Elementary Flows in a Cell

We select two various geometrical types of flow which differ in geometrical structure of streamlines. The *"source flow"* is formed by a radial system of streamlines in the block, as shown in Fig. 1.5-a. The *"translation flow"* is formed by a system of streamlines on a block, which are oriented in the same direction as in the matrix, as shown in Fig. 1.5-b.

Translation-Type Media (æ-Homogeneous)

These media corresponds to the case of $\varepsilon_p \sim \varepsilon^2$. The heterogeneity degrees for porosity and permeability are equivalent: $\omega_K \sim \omega_m$, $\Lambda \sim 1$. Piezo-conductivities are identical, equations (1.2.3), (1.2.3) are equivalent and quasi-stationary, such that the non-stationarity is displayed only in a time boundary layer of the size $\sim \varepsilon^2$ at vicinity of the perturbation moment.

The medium is trivially heterogeneous. The problem refers to coefficiently averaged. Such medium behaves as homogeneous with respect to the pressure field, despite of high heterogeneity with respect to the porosity

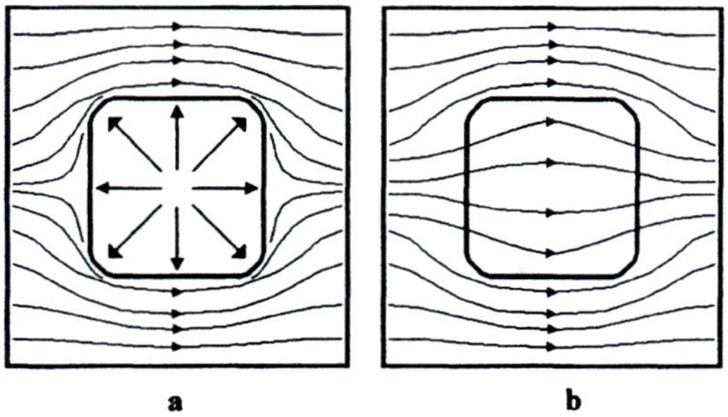

Figure 1.5. Source flow (a) and Translation flow (b) through a cell

and to the permeability. In frameworks of homogenization theory, these media have been examined in [124], where the block and matrix pressure difference ΔP has had been shown to be very small, of the order ε^2.

Averaged equations are derived by classical technique of Bakhvalov [16], [17]. In we takes into account the higher approximations and derive the averaged model of order $O(\varepsilon^2)$, than the exchange term can be obtained in the form similar to Barenblatt-Zheltov's one [19].

Only translation type of flow through each block is observed, that will be shown later.

Translation-Source Type Media (Weakly æ-Heterogeneous)

They belong to the case of $\varepsilon_p \sim \varepsilon$ or in more general sense: $\varepsilon^2 \ll \varepsilon_p \ll 1$. Oscillation intensity of porosity and permeability differ by one order, such that the ratio $\omega_K/\omega_m = \Lambda \sim \varepsilon_p$, hence the medium is multiply heterogeneous.

Due to a weak piezo-heterogeneity, the delaying of perturbation spreading arises inside the blocks. As result the pressure difference ΔP reaches the magnitude of order ε_p. Three sequential stages may be observed: i) perturbation spreading over the matrix, ii) perturbation spreading over the blocks, iii) quasi steady-state mode of flow.

The averaged model for such media has been derived first in [119] by nonuniform homogenization method. It differs from [19] in that, the exchange terms are weakly non-stationary, and their intensity depends on ΔP as well as on $\partial(\Delta P)\partial t$. Parabolic equation (1.6) goes over into elliptic with arising the second time derivative.

It will be shown later, that the flow in each block can be decomposed into the translation stream and the source flux.

Source-Type Media (æ-Heterogeneous)

They belong to the case of $\varepsilon_p \sim 1$. The intensity of porosity and permeability fields differs by two orders, such that $\Lambda \sim \varepsilon^2$. The ratio of the block and the matrix piezo-conductivities is the same. The medium is multiply heterogeneous.

The perturbation spreading over block is considerably delaying, therefore the difference ΔP becomes of order ~ 1. Process in the block is already non stationary, whereas it is steady-state in the matrix.

It will be shown later, that the flow in the block is similar to that propagating uniformly in all directions.

Averaged model has been derived in [119], [117] by nonuniform homogenization. It takes the form of a system of integro-differential equations corresponding to long-term memory effects.

This classification is not complete. Two subclasses should be described.

Source-Type Media with Boundary Layer (æ Highly Heterogeneous)
They correspond to the case of $\varepsilon_p \gg 1$ and belong to source-type class. However the blocks being almost impenetrable, a thin layer near the block border may be perturbed only, with the size $\xi \sim \sqrt{\varepsilon_p}$. Due to this fact a third spatial scale arises, which is equal to the boundary layer size and is less than the heterogeneity scale ε. The process inside each block is always non-stationary.

Reverse Translation-Type Media (or Reverse æ-Heterogeneity)
The system where $\varepsilon_p \ll \varepsilon^2$ or $\omega_m \ll \omega_K$ belongs to the case when either the block porosity is smaller than that of the matrix, while the permeabilities are of the same order, or the block permeability is higher than that of the matrix, while the porosities are of the same order.

In all cases the stabilization of process in blocks happens more rapidly than in the matrix, therefore the exchange flow directed from the blocks towards the matrix arises during some time. Behavior of such systems is very special, and will be examined in chapter 5.

All these situations are determined by different asymptotic expansions and various averaged models.

1.2.4. NUMERICAL PROOF OF EXISTENCE FOR ELEMENTARY FLOWS

Detection of two elementary flows within a cell is a rather important circumstance, that will be used in all further investigations. To be more sure in validity of our analysis, the numerical experiments were performed.

A 2D rectangular porous medium, 10×1, comprised a periodic sequence of ten square blocks located along the axis x_1. Thus, the heterogeneity scale ε was equal to 0.1. One-phase flow described by the microscale equations of this chapter was examined, with following boundary and initial conditions.

The initial pressure field was uniform in the whole domain and was equal to 1. At zero time moment, the pressure at the outside section (x_1=10) has been instantly reduced up to zero. This value of pressure was kept at the outlet section during all the process. At the inlet section (x_1=0), the pressure was always equal to 1. Two other sides of the rectangle was assumed to be impermeable, i.e. the normal derivative of pressure was equal to zero. To solve the problem the simple finite-difference method was applied.

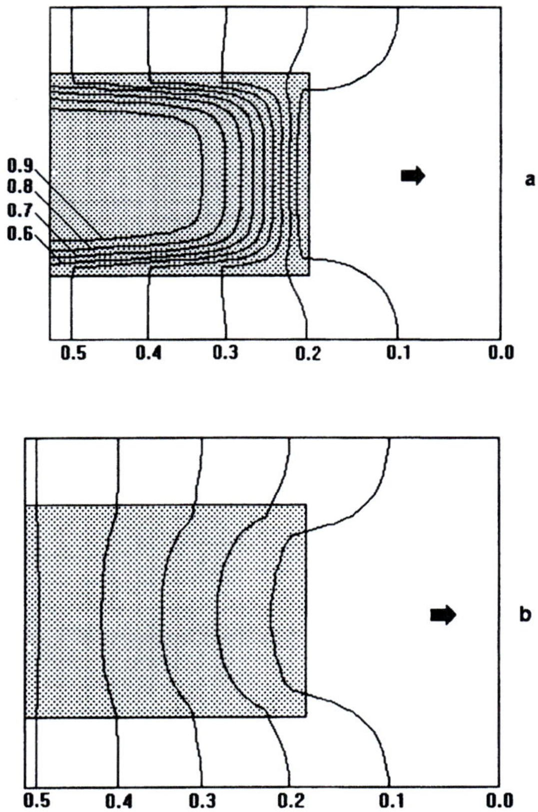

Figure 1.6. Numerical detection of two elementary flows: of source type (a) and of translation type (b)

The following parameters have been assumed: ω_K=0.01; ω_m=1; the volume fraction of blocks was equal to 0.5.

Fig. 1.6 illustrates some results of numerical simulations inside a central cell for two various instants of time. The central cell was chosen to avoid influence of boundary effects. The right-hand part of this cell is presented.

In this figure, the isobars are shown, which are orthogonal to the flow streamlines. The system of isobars in Fig. "a" explicitly demonstrates the source structure of flow.

The isobars in Fig. "b" correspond to translation type of elementary flow.

It is necessary to note that the evolution of the system is such that it passes through both the types of flow, starting by a source-type structure, which is transforming into the structure of translation type later.

1.3. UPSCALING FLOW IN HIGHLY HETEROGENEOUS MEDIA

1.3.1. PROBLEM OF UP-SCALING

Numerical Notion of Up-Scaling

The problem of up-scaling has been formulated according to the needs of petroleum engineers to obtain a tools allowing to use the coarse grids for numerical modelling of the underground reservoirs.

The classical problem of up-scaling is formulated in the following way.

Suppose we have the initial data on the porous field, forming a greed of the scale h_1. Hydrodynamic simulator performs the calculations with a scale h_2, when $h_2 \gg h_1$. We need to homogenize initial data within each cell of the hydrodynamic greed. Industrial application of this problem deals with elaboration of a numerical simulator performing up-scale procedure, to be included in the certified software like ECLIPS and other. Thus, the problem has an industrial importance.

Now the state of research is as follows. The essential problem consists in that the algorithm must be functioning for any structure of porous medium. The last is always irregularly, non-periodic, non-ergodic, non-stationary (in stochastic sense), non self-similar... Hence, all the situations which can be examined theoretically, are only particular cases.

Thus, the up-scale algorithm should satisfy to the following necessary conditions:

1) the medium structure is any, provided of the one limitation $h_1 \ll h_2$;

2) the algorithm has to make real simulations, so the theoretical derivations should be made in a manner ensuring the local problem be "solvable";

3) the algorithm must perform up-scaling for equations of one-phase flow, diffusion-convection transport of tracers, two-phase flow. In one-phase case only the permeability is to be homogenized. Convection-diffusion transport requests to homogenize the permeability and the dispersion parameter. Two-phase case is the must complicated: we need to homogenize the relative permeability functions and the capillary pressure function, which depend on the saturation.

A renormalization techniques applied to up-scale problem satisfies to all these conditions [72], [73]. The going from the lattice h_1 towards the lattice h_2 is performed step by step by iterations. Consequent steps correspond to going over into the lattice $2 * h_1$, $4 * h_1$, $8 * h_1$ etc, as shown in Fig. 1.7, where sequent steps are presented.

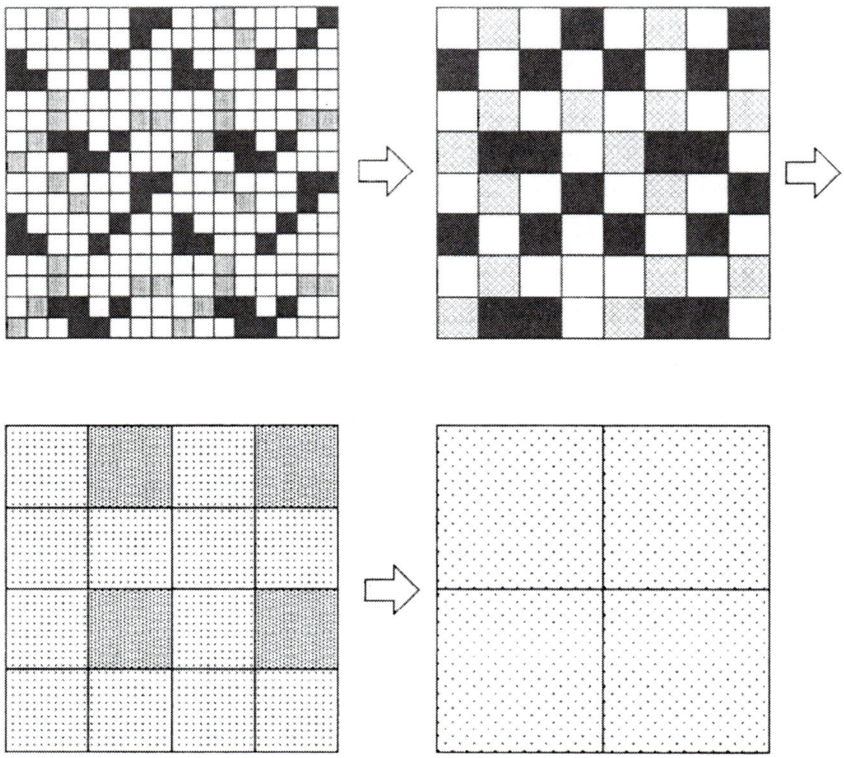

Figure 1.7. Procedure of up-scaling in renormalization technique

Averaging on each step ("elementary averaging") is a main item of the method. It is made in an approximate way. For one-phase case permeability for 4 cells if 2D or for 8 cells in 3D is obtained as a harmonic or geometric averaged value. As to homogenization of relative permeabilities, an approximate boundary problem on each cell of hydrodynamic greed is to be formulated, which is solved numerically.

The algorithm is functioning, however it has the following disadvantages.

i) The cell problem is artificial, it does not follow from some theoretical derivations; hence it is not matched with averaging procedure, with the original equations and with the macroscale equations. In several papers the elementary homogenization is shown to be invalid even in one-phase case. From this point of view, the renormalization algorithm seems to be less scientifically grounded, than the asymptotic homogenization technique.

ii) Basic hypothesis tells that the type of homogenized equations is the same as the microscale equations. Therefore all types of flow through highly heterogeneous media, where macroscale equations are always of different type respectively to the micro-scale, can not be studied by renormalization technique. In this case one needs first to deduce the form of macro-equations, and to homogenize the parameters after.

Thus, in case of highly heterogeneous media the problem of up-scaling is not solved.

Up-Scaling in Fluid Mechanics of Porous Media

The problem of up-scaling takes another aspect in continuum mechanics.

A mathematical point in transport equations corresponds to some finite physical volume called as elementary volume. Its dimension is not strongly defined and can be varying within some scales. Scaling invariance limits are equal to those, where the scale of elementary volume can be changed with keeping the form of modeling equations. We say "the model is written in scale l", when the length of order l is an upper limit of its scale invariance domain.

Considerable variations of the volume scale can change the equations. Upscale procedure deals with studying, how modeling equations are changed, when we pass from a lower scale to an upper one.

In a heterogeneous medium described by two spatial scales, the upscaling stands for passing from the micro-scale ε to the macro-scale 1. During such a transition, the scaling invariance is destroyed.

Mathematically, upscaling is performed in form of homogenization method. As the point functions in mathematical model are equal to averaged properties of elementary physical volume, then upscaling deals with homogenization. In the homogenization method, scale changes are made as homothetic compression of each heterogeneity cell into a point, as shown in Fig. 1.8, that is equal to tending ε to zero. This leads to a homogeneous, averaged medium.

From this point of view, in the simplest case, up-scaling is equivalent to homogenization.

Phenomena of modification of governing equations type after up-scaling are known in fluid mechanics. For example the system of equations of kinetics describing the motion of one fluid particle (molecule) with interactions between neighbors, goes over in Navier-Stokes equations of continuum mechanics. The last equations go over in Darcy's law for the flow through porous media.

This sequence may be continued in hierarchical porous media, where the heterogeneity is characterized by multiple scales.

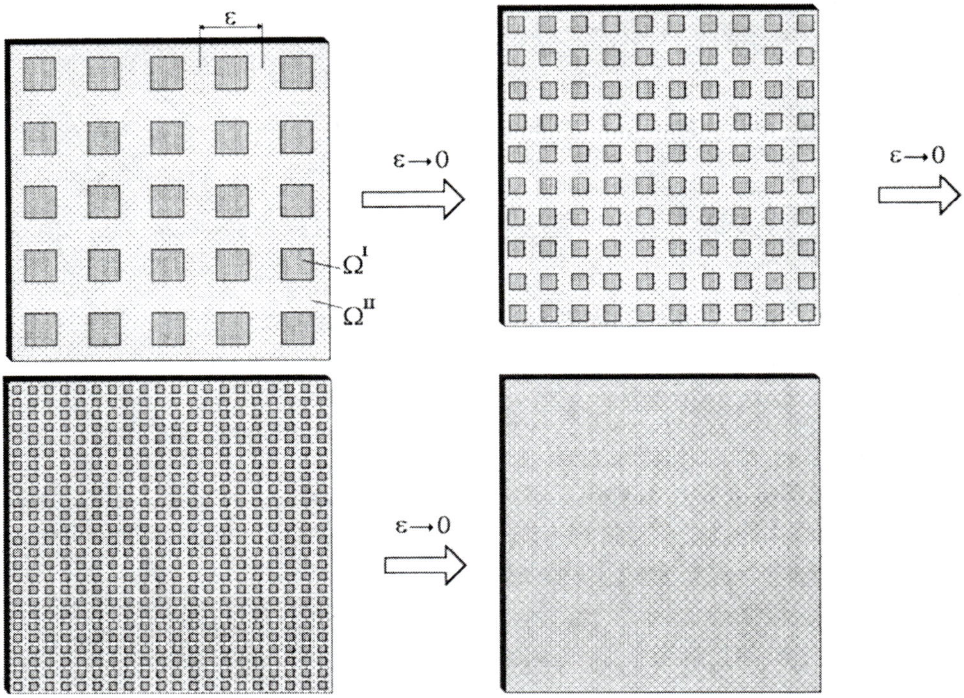

Figure 1.8. Procedure of up-scaling in homogenization technique

One of goal of this book is to derive the next mathematical model in the infinite hierarchical sequence describing the same flow on different space scales.

Up-Scaling Flow in Highly Heterogeneous Media

Upscaling and homogenization are equivalent only in case that homogenized model exists. The last depends on the degree of medium heterogeneity. Let the heterogeneous field be described by a function oscillating in space with a small period ε. Let the ratio between its maximum and minimum magnitudes reached in different parts of the heterogeneity cell be equal to ω. This parameter is a heterogeneity degree. The homogenized model exists, if the degree of heterogeneity is moderate, so $\omega \sim 1$. For highly heterogeneous media, where $\omega \ll 1$, or $\omega \gg 1$, the situation is different.

Let the oscillation amplitude be large, as $\omega \sim 1/\varepsilon$ for example. If $\varepsilon \to 0$, the amplitude is tending to infinity. So, the homogenized limit does not exist.

A simplest example is presented by a following highly oscillating function:

$$u(\frac{x}{\varepsilon},\omega) = \omega \sin \frac{x}{\varepsilon}, \qquad \omega \equiv \frac{1}{\varepsilon} \tag{1.12}$$

Such a catastrophe stands for the physical process can not be produced on some part of the cell after its compression. This corresponds to some type of degeneration for equations. The same singularity appears when $\omega \to 0$.

If we reduce the heterogeneity degree, we get an intermediate case, when the homogenized model exists, but appears to be not a best limit of upscale process. More strong convergence is ensured to other limit, which does not coincide with the simple homogenized value.

At least, in highly heterogeneous media, upscaling process is not equivalent to the usual homogenization. As consequence, usual methods of upscaling should be modified.

Hence, the problem of highly heterogeneous media consists in two items:
- 1) what is the limit of upscaling ?
- 2) what is a method to perform upscale process ?

A solution for the first problem was proposed in [119], [120], [127], where upscaling limit for highly heterogeneous media has been shown to be equal to *"nonuniform homogenized value"*. As to the second problem, a method of asymptotic expansion is used in this book. Combining the asymptotic expansion method with nonuniformly homogenized limit we obtain a *"nonuniform homogenization method"*, which will be illustrated below.

1.3.2. NONUNIFORM HOMOGENIZATION

Idea of Nonuniform Homogenization

The main point of the method concerns determination of the upscale limit for micro-scale equations.

The idea how to seek the upscale limit was proposed in [119], [120]. Let us examine the Laplace's equation for some function $u(x, x/\varepsilon; \omega)$ on Ω, where the conductivity coefficient is equal to 1 on Ω^{II} and to ω on Ω^{I}, $0<\omega\ll 1$. Let Y is an extended heterogeneity cell, whose volume is equal to 1. Two its parts Y^I and Y^{II} correspond to Ω^I and Ω^{II}. Example of the cell is illustrated in Fig. 1.9-a.

Examine two limiting magnitudes of ω.

When $\omega=1$, the medium is homogeneous or close to homogeneous, as shown in Fig. 1.9-b. Upscale limit of solution corresponds to simple averaging over the heterogeneity cell Y:

$$u(x, x/\varepsilon; \omega) \to \frac{1}{|Y|} \int_Y u(x, y) dy, \qquad \text{when } \varepsilon \to 0, \ \omega \to 1 \tag{1.13}$$

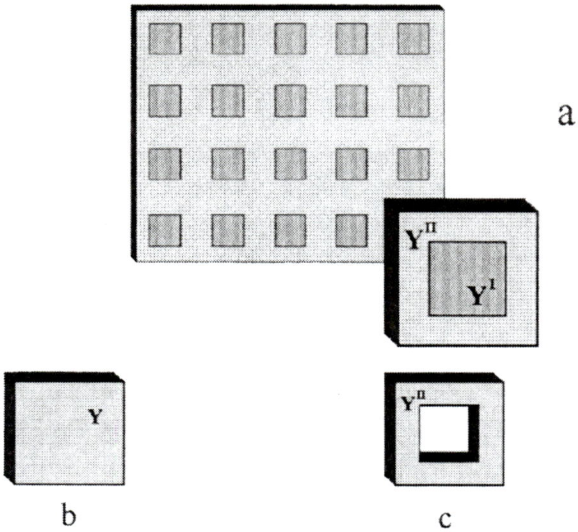

Figure 1.9. Unite cell of highly heterogeneous medium in case of $0<\omega<1$ (a), $\omega\sim 1$ (a), $\omega=0$ (a)

When $\omega=0$, the process is only determined on Ω^{II}, as seen from Fig. 1.9-c. It is natural to seek upscale limit of solution as averaging over Y^{II}:

$$u(x, x/\varepsilon) \to \frac{1}{|Y^{II}|}\int_{Y^{II}} u(x,y)dy, \qquad \text{when } \varepsilon\to 0,\ \omega\to 0 \tag{1.14}$$

It is easy to see from (1.13) and (1.14) that in general case, for any ω, upscale limit can be presented in the form of:

$$u(x, x/\varepsilon) \to \frac{1}{\omega|Y^{I}|+|Y^{II}|}\left(\omega\int_{Y^{I}} u(x,y)dy+\int_{Y^{II}} u(x,y)dy\right), \tag{1.15}$$
$$\text{when } \varepsilon\to 0,\ \forall\omega\in[0,1]$$

When ω tends to 1 or to 0, the right-hand part of (1.15) passes to (1.13) or (1.14) correspondingly.

Let us call the introduced value in (1.15) as *"nonuniform homogenized value"* and use the following notation for it:

$$\langle\chi_\omega u\rangle\equiv\omega\langle u\rangle_1+\langle u\rangle_2, \qquad \langle u\rangle_i\equiv\int\limits_{Y_i} udy,\ i=I, II \tag{1.16}$$

The use of such a generalized notion (1.16) of homogenization allows to develop generalizations for standard methods to produce upscaling for media with various degree of the heterogeneity ω.

Introduction to a Nonuniform Two-Scale Convergence Method
The two-scale convergence method developed by [103] and furthermore by [5], [6] is widely used actually, as an essential mathematical tool to solve the nonlinear problems of homogenization theory. It is based on the notion of two-scale convergence, which may be introduced in a following way.

Two-scale convergence. Let $C^{\infty}_{per}(Y)$ be a space of Y-periodic infinitely differentiable functions in R^n, $D(\Omega;\ C^{\infty}_{per}(Y))$ be a space of infinitely smooth and compactly supported functions in Ω with values in space $C^{\infty}_{per}(Y)$.

A sequence of functions $u^{\varepsilon}(x)$ in $L^2(\Omega)$ is said to two-scale converge to a limit $u_0(x,y)$ in $L^2(\Omega{\times}Y)$ if, for any function $\varphi(x,y)$ in $D(\Omega;\ C^{\infty}_{per}(Y))$, it satisfies the following equality:

$$\lim_{\varepsilon\to 0}\int_{\Omega} u^{\varepsilon}(x)\varphi(x,\frac{x}{\varepsilon})dx = \int_{\Omega}\int_{Y} u_0(x,y)\varphi(x,y)dxdy \tag{1.17}$$

Very weak restrictions are needed for a function to be two-scale converging. Essential theorem of Nguetsen says that from each bounded sequence $u^{\varepsilon}(x)$ in $L^2(\Omega)$, one can extract a subsequence, such that this subsequence two-scale converges to some limit $u_0(x,y){\in}L^2(\Omega{\times}Y)$.

The figure 1.10 illustrates the function $u^{\varepsilon}(x)$ and its two-scale limit $u(x,y)$.

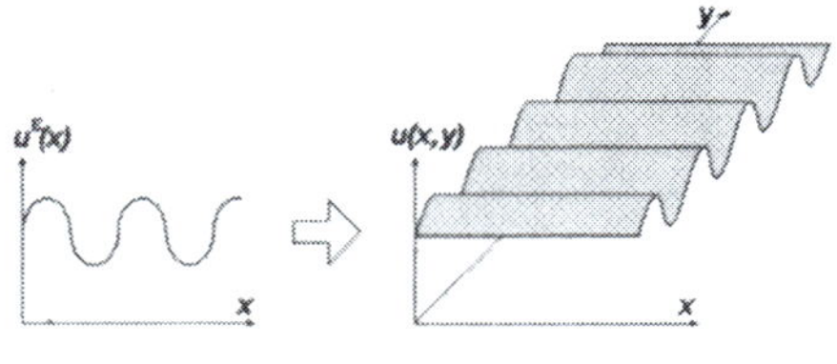

Figure 1.10. Illustration of two-scale convergence

Thus, oscillations by x are "transferred" on another variable y.

Further, one can show that this notion is intermediate between the weak and the strong convergence.

It is clear from (1.17), that notion of two-scale convergence is a convenient tool to operates with variational formulations.

Requestion for functions $u^\varepsilon(x)$ to be bounded cannot be satisfied in some cases of high heterogeneity like (1.12). In these situations some generalization of two-scale convergence should be developed. In this book we present an essential conceptual idea of nonuniform two-scale convergence, without any claims on mathematical contribution.

Two-scale convergence to a pair of functions. In case of high heterogeneity the main problem is to reduce high oscillations. From this point of view, the notion of two-scale convergence occurs useless, as one can see in Fig. 1.11.

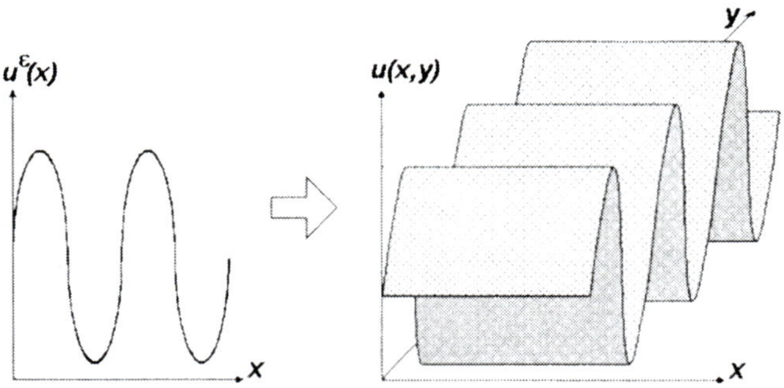

Figure 1.11. **Illustration of two-scale convergence for highly oscillating function**

Consider a sequence of functions $u^\varepsilon(x) \in L^2(\Omega)$ as well as the functions $U^I(x,y) \in L^2(\Omega) \times Y^I$ and $U^{II}(x,y) \in L^2(\Omega) \times Y^{II}$. Let us say the sequence $u^\varepsilon(x) \in L^2(\Omega)$ two-scale converges to a pair $U^I(x,y)$ and $U^{II}(x,y)$ on the partition $Y^I \cup Y^{II} = Y$ if, for any function $\varphi(x,y)$ in $D(\Omega;\, C^\infty_{per}(Y))$, it satisfies the following equality:

$$\lim_{\varepsilon \to 0} \int_\Omega u^\varepsilon(x) \varphi(x, \frac{x}{\varepsilon}) dx =$$

$$\int_\Omega \left[\int_{Y^I} U^I(x,y)\varphi(x,y) dx dy + \int_{Y^{II}} U^{II}(x,y)\varphi(x,y) dx dy \right] \quad (1.18)$$

This notion is closely related to the simple notion of two-scale convergence. Really if we introduce a new function $u(x,y) \in L^2(\Omega) \times Y$ such that:

$$u(x,y) = \begin{cases} U^I(x,y), & y \in Y^I \\ U^{II}(x,y), & y \in Y^{II} \end{cases}$$

then we obtain from (1.18) relation (1.17). The function $u(x, y)$ introduced in such a manner is a two-scale limit for $u^\varepsilon(x)$.

This formula gives a method to construct the pair U^I, U^{II}, if the simple two-scale limit $u(x, y)$ is known. They may be chosen as equivalent to the function $u(x, y)$ on corresponding subdomains of Y. The graphic interpretation of two-scale convergence to a pair of functions is displayed in Fig. 1.12.

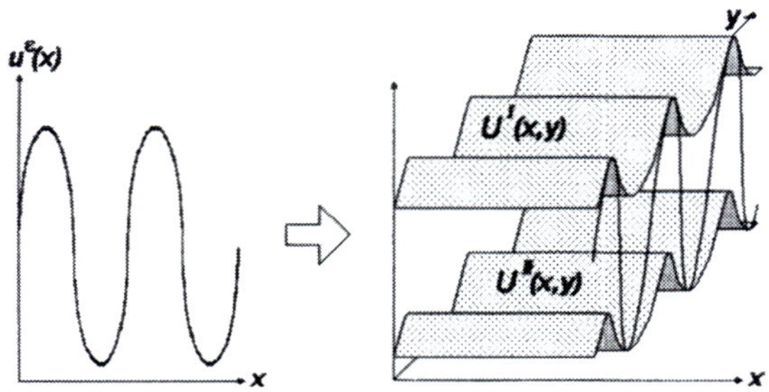

Figure 1.12. Illustration of two-scale convergence to a pair of functions

In contrast with the case (1.17), the function $U^I(x, y)$ does not need to be determined on the subdomain Y^{II} and may be any there, as well as the function $U^{II}(x, y)$ may be arbitrary on Y^I.

This property becomes more important for highly oscillating function $u^\varepsilon(x)$. The two-scale limit $u(x, y)$ is a highly oscillating function too, as illustrated in Fig. 1.11. However, we can chose the pair U^I and U^{II} in such a way that they would be weakly oscillating, as shown in Fig. 1.12.

Nonuniform two-scale convergence to a pair of functions. As last, the notion of nonuniform convergence to a pair of functions may be useful to apply in cases when some type of degeneration happens on a subdomain.

Consider functions $u^\varepsilon(x) \in L^2(\Omega)$. Let $\Omega^I \cup \Omega^{II} = \Omega$ be some partition of the domain Ω. Let $y = f(x)$ be an one-to-one image transforming the domain Ω in Y. Let Y^I and Y^{II} be the images of Ω^I and Ω^{II} on Y, then $Y^I \cup Y^{II} = Y$. Consider $U^I(x, y) \in L^2(\Omega) \times Y^I$, $U^{II}(x, y) \in L^2(\Omega) \times Y^{II}$. Let ω, $0 \le \omega \le 1$ be a real number.

The sequence $u^\varepsilon(x) \in L^2(\Omega)$ is said to nonuniformly two-scale converge to a pair $U^I(x, y)$ and $U^{II}(x, y)$ on the partition $Y^I \cup Y^{II} = Y$ if, for any function $\varphi(x, y)$ in $D(\Omega; C^\infty_{per}(Y))$, it satisfies the following equality:

$$\lim_{\varepsilon \to 0} \left(\omega \int_{\Omega^I} + \int_{\Omega^{II}} \right) u^\varepsilon(x) \varphi(x, \frac{x}{\varepsilon}) dx =$$

$$\left(\omega\int_{\Omega^I}+\int_{\Omega^{II}}\right)\frac{1}{\omega|Y^I|+|Y^{II}|}\left[\omega\int_{Y^I}U^I(x,y)\varphi(x,y)dy+\int_{Y^{II}}U^{II}(x,y)\varphi(x,y)dy\right]dx \qquad (1.19)$$

When ω=1, we get two-scale convergence to a pair of functions (1.18).

When ω=0, the subdomain Ω^{II} is completely removed from consideration.

Thus, formula (1.19) is a strict definition for an intuitive notion of nonuniform homogenization introduced in form of (1.15).

The method now is in elaboration. In this book we limit only by introducing the notion of nonuniform two-scale convergence to a pair of functions. Further developments will be published later.

It is necessary to note, that, more strictly speaking, the nonuniform convergence is closely related to a convergence with respect to a singular measure.

Nonuniform Two-Scale Asymptotic Expansion Method

Asymptotic expansion method is more simple to be developed. As usual in all forms of homogenization methods, the idea of two-scale expansions will be applied. It is known that any function allows an infinite set of asymptotic expansions, which differ in zero term. Zero term in asymptotic series of homogenization theory is equivalent to up-scale limit. As shown in previous paragraph, several different forms of limiting functions may be chosen as an up-scale limit, for example the function like a simple two-scale limit $u(x,y)$ in (1.17), or a pair of functions like $U^I(x,y), U^{II}(x,y)$ in (1.18).

Choosing various opportunities, we'll obtain various asymptotic series.

Basing on the idea, that nonuniform two-scale convergence to a pair of functions (1.19), which is more strict formulation of nonuniform homogenization (1.15), is a better variant for highly heterogeneous medium with degeneration of microscale equations on Ω^I, when the oscillation degree ω tends to 0, we chose the following form of asymptotic expansions, when $\varepsilon,\omega\to 0$:

$$u^\varepsilon(x,y;\omega)=u_0(x;\omega)+u'(x,y;\omega), \qquad y=x/\varepsilon \qquad (1.20)$$

where u_0 is the nonuniformly homogenized value, and u' is a function oscillating around u_0, such that:

$$u_0(x;\omega)=\frac{1}{|\chi_\omega Y|}\langle\chi_\omega u^\varepsilon(x,y)\rangle=\frac{1}{|\chi_\omega Y|}\left(\omega\langle u^\varepsilon\rangle_1+\langle u^\varepsilon\rangle_2\right), \qquad (1.21)$$

$$\langle\chi_\omega u'(x,y)\rangle=0 \qquad (1.22)$$

1.3.3. PHYSICAL INTERPRETATION OF NONUNIFORM AVERAGING

Relations (1.15) and (1.19) show that the averaging over the whole cell is reduced to averaging over each sub-domain of the cell with nonuniform weighting. This leads to the following interpretation for the nonuniform homogenization to a pair of functions given in Fig.1.12. Consequent steps of upscaling are shown there. Simultaneously to cell compression when ε tends to zero, we get a total medium decomposing onto two submedia. The smaller ε, the more continuous becomes each of them.

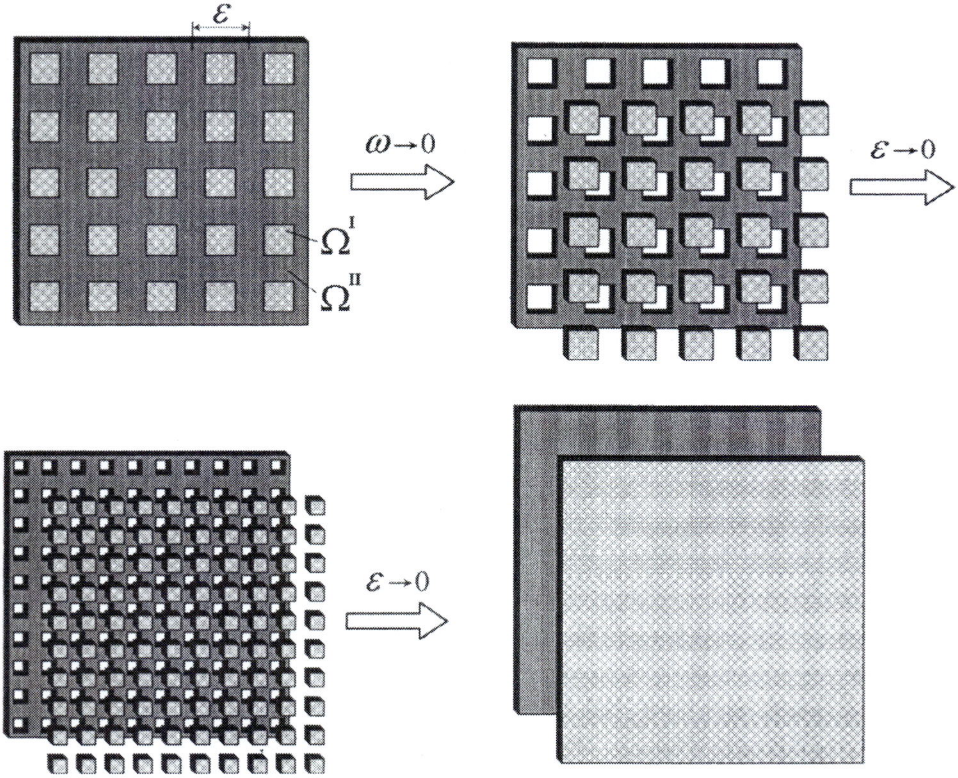

Figure 1.13. Sequential steps of nonuniform homogenization to a pair of functions

1.4. DERIVATION OF THE MACROSCALE EQUATIONS

Consider the problem (1.6). Its main feature is degeneration of the equation, when ω_K tends to zero. Physically, when ω_K=0, the blocks occur umpervious , the ellipticity property is lost on Y^I. When ω_K=1 the blocks and the matrix are indiscernible.

Therefore, it is natural to seek the up-scale limit when $\varepsilon \to 0$, in the form of nonuniform averaging (1.16).

The technique of the method will be illustrated on the example of Source-type media, where

$$\omega_m \sim 1, \qquad \omega_K \sim \varepsilon^2 \tag{1.23}$$

Thus, the problem becomes one-parametrical and dependent on ε only.

1.4.1. TWO-SCALE FORMULATION

Extension of the Problem and Regularization
Let us write (1.6) in the form:

$$L_\varepsilon^{\omega_K,\omega_m} p \equiv \varepsilon_p b^I \frac{\partial p}{\partial \tau} - \varepsilon^2 \frac{\partial}{\partial x_i}(a_{ij}^I \frac{\partial p}{\partial x_j}) = \varepsilon^2 f_*, \; x \in \Omega^I \tag{1.24}$$

$$L_\varepsilon^{\omega_K,\omega_m} p \equiv b^{II} \frac{\partial p}{\partial \tau} - \frac{\partial}{\partial x_i}(a_{ij}^{II} \frac{\partial p}{\partial x_j}) = f_*, \; x \in \Omega^{II} \tag{1.25}$$

where $\varepsilon_p \sim 1$.

Solution of equations (1.24) depends on ε in a singular way, because the limit of periodic functions $a(x/\varepsilon)$, $b(x/\varepsilon)$ is not defined. Regularization may be made, as usually, using two-scale formulation.

Let us introduce an extended function $\tilde{p}(x, y, \tau)$ such that its contraction at $y = x/\varepsilon$ would coincide with $p(x, \tau)$, so the singularity is postulated to be of the form x/ε and is presented as an independent variable.

Further the wave over $\tilde{p}$ will be omitted, and the symbol $p(x, y, \tau)$ will designate the extended two-scale function.

For the extended two-scale function, system (1.24) with condition (1.6d) get the form:

$$\tilde{L}p \equiv \begin{cases} (L_{yy} - \varepsilon_p b^I \partial_\tau)p + \varepsilon(L_{yx} + L_{xy})p + \varepsilon^2(L_{xx}p + f_*) = 0, \; y \in Y^I \\ L_{yy}p + \varepsilon(L_{yx} + L_{xy})p + \varepsilon^2\Big((L_{xx} - b^{II}\partial_\tau)p + f_*\Big) = 0, \; y \in Y^{II} \end{cases} \tag{1.26}$$

$$p(x,y,0)=p^{\circ}(x), \qquad [p]_\Gamma=0,$$

$$a_{ij}^{II}(y)\left(\frac{\partial p^{II}}{\partial y_j}+\varepsilon\frac{\partial p^{II}}{\partial x_j}\right)n_i\Big|_\Gamma = \\ =\frac{\omega_m\varepsilon^2}{\varepsilon_p}a_{ij}^{I}(y)\left(\frac{\partial p^{I}}{\partial y_j}+\varepsilon\frac{\partial p^{I}}{\partial x_j}\right)n_i\Big|_\Gamma, \tag{1.27}$$

Where

$$\begin{aligned}(L_{yx}+L_{xy})&\equiv\frac{\partial}{\partial y_i}(a_{ij}(y)\frac{\partial}{\partial x_j})+\frac{\partial}{\partial x_i}(a_{ij}(y)\frac{\partial}{\partial y_j}),\\ L_{yy}&\equiv\frac{\partial}{\partial y_i}(a_{ij}(y)\frac{\partial}{\partial y_j}),\quad L_{xx}\equiv\frac{\partial}{\partial x_i}(a_{ij}(y)\frac{\partial}{\partial x_j}),\\ \partial_\tau&\equiv\frac{\partial}{\partial\tau}\end{aligned} \tag{1.28}$$

n is a normal vector to the surface Γ in the cell Y, directed from Y^I to Y^{II}, $p^{(i)}\equiv p$ with $y\in Y^i$.

Y-periodicity of the function $p(x,y,\tau)$ is accepted as an additional condition needed to chose the unique solution after the problem extension.

The two-scale extended operator $\tilde{L}$ in (1.26) is introduced in such a manner, that its contraction $y=x/\varepsilon$ reduces it to the operator $\varepsilon^2 L_\varepsilon^{\omega_K,\omega_m}$.

Necessary Condition of Existence for Periodic Solution

Along with system (1.26) it is useful to introduce an integral relation which will allow to derive at once the form of averaged equations.

Let us designate:

$$L_0\equiv\begin{cases}L_{yy}-\varepsilon_p b^I\partial_\tau, & y\in Y^I\\ L_{yy}, & y\in Y^{II}\end{cases},\quad L_2\equiv\begin{cases}L_{xx}, & y\in Y^I\\ L_{xx}-b^{II}\partial_\tau, & y\in Y^{II}\end{cases} \tag{1.29}$$

Multiplying the first equation by ω_K, integrating it over Y^I, and the second over Y^{II}, summing the results and taking into account (1.27), we obtain:

$$\varepsilon_p\omega_K\frac{\partial}{\partial\tau}\langle b^I p\rangle_1=\varepsilon\langle\chi(\omega_K)(L_{yx}+L_{xy})p\rangle+\varepsilon^2\langle\chi(\omega_K)(L_2p+f_*)\rangle - \\ \varepsilon\int\limits_\Gamma(\omega_K a_{ij}^I\frac{\partial p^I}{\partial x_j}-a_{ij}^{II}\frac{\partial p^{II}}{\partial x_j})n_i dy$$

Here the Gauss theorem was used, as well the property of a periodic function p : $\int_{\partial Y} a_{ij}\frac{\partial p}{\partial n}dy=0$

The last equation may be simplified, if the following will be considered:

$$\langle (L_{yx}+L_{xy})p\rangle_k = \frac{\partial}{\partial x_i}\langle a^k_{ij}\frac{\partial p}{\partial y_j}\rangle_k + \int_{\partial Y^k} a^k_{ij}\frac{\partial p}{\partial x_j}n_i dy$$

Then we obtain

$$\omega_K \varepsilon_p \frac{\partial}{\partial \tau}\langle b^I p\rangle_1 = \varepsilon \frac{\partial}{\partial x_i}\langle \chi(\omega_K) a_{ij}\frac{\partial p}{\partial y_j}\rangle + \varepsilon^2 \langle \chi(\omega_K)(L_2 p + f_*)\rangle \tag{1.30}$$

This results from a y-periodicity of a_{ij} and p, then:

$$\int_{\partial Y} a_{ij}\frac{\partial p}{\partial x_j}n_i dy = \frac{\partial}{\partial x_i}\int_{\partial Y} a_{ij} p n_i dy = 0$$

Let us collect all terms of the same order in (1.30):

$$\frac{\partial}{\partial x_i}\langle a^{II}_{ij}\frac{\partial p}{\partial y_j}\rangle_2 + \varepsilon\left[\left\langle (L_{xx} - b^{II}\partial_\tau)p + f_*\right\rangle_2 - \omega_m \frac{\partial}{\partial \tau}\langle b^I p\rangle_1\right] +$$
$$\varepsilon^2 \frac{\omega_m}{\varepsilon_p}\frac{\partial}{\partial x_i}\langle a^I_{ij}\frac{\partial}{\partial y_j}\rangle_1 + \varepsilon^3 \frac{\omega_m}{\varepsilon_p}\langle L_{xx}p + f_*\rangle_1 = 0 \tag{1.31}$$

1.4.2. ASYMPTOTIC EXPANSIONS

Form of Expansions
Averaged equations result from asymptotic expansions when ε, $\omega_K \to 0$ for the problem (1.26), with relation between parameters (1.23).

According to previous considerations we assume the following form of the asymptotic expansions:

$$p(x, y, \tau; \varepsilon) = \sum_{k=0}^{\infty} \varepsilon^k p_k(x, y, \tau) \tag{1.32}$$

where

$$p_k(x, y, \tau) = p_{k0}(x, \tau) + p_{k1}(x, y, \tau),$$
$$\sum_{k=0}^{\infty} \varepsilon^k p_{k0}(x, \tau) = \langle \chi(\omega_K) p\rangle \tag{1.33}$$

Thus, in contrast to the usual schemes [17], [161], [79], [92], [136], [1] the oscillating component $p_{00}(x, \tau)$ is a nonuniformly averaged value (up to an order $O(\varepsilon)$).

An additional condition results from the last relation:

$$\sum_{k=0}^{\infty} \varepsilon^k \langle \chi(\omega_K) p_{k1} \rangle = 0 \tag{1.34}$$

Or, that is the same:

$$\langle p_{01}\rangle_2+\varepsilon\langle p_{11}\rangle_2+\sum_{k=2}^{\infty}\varepsilon^k[\langle p_{k1}\rangle_2+\frac{\omega_m}{\varepsilon_p}\langle p_{k-2,1}\rangle_1]=0 \tag{1.35}$$

Expansion of Equations and Boundary Conditions
Substitution of (1.32) in (1.26) leads to a following system:

$$L_0 p+\varepsilon(L_0 p_1+(L_{yx}+L_{xy})p_0)+\sum_{k=2}^{\infty}\varepsilon^k(L_0 p_k+$$

$$(L_{yx}+L_{xy})p_{k-1}+L_2 p_{k-2}+f_*\delta_{k2})=0, \quad y\in Y_i, \ i=1,2 \tag{1.36a}$$

$$\sum_{k=1}^{\infty}\varepsilon^k p_k(x,y,0)=0, \qquad \sum_{k=0}^{\infty}\varepsilon^k[p_k]_S=0 \tag{1.36b}$$

$$p_0(x,y,0)=p^\circ(x), \tag{1.36c}$$

$$a_{ij}^{II}(y)\frac{\partial p_0^{II}}{\partial y_j}n_i\Big|_\Gamma+\varepsilon a_{ij}^{II}\left(\frac{\partial p_1^{II}}{\partial y_j}+\frac{\partial p_0^{II}}{\partial x_j}\right)n_i\Big|_\Gamma+$$

$$\sum_{k=0}^{\infty}\varepsilon^k\left[a_{ij}^{II}\left(\frac{\partial p_k^{II}}{\partial y_j}+\frac{\partial p_{k-1}^{II}}{\partial x_j}\right)n_i\Big|_\Gamma-\frac{\omega_m}{\varepsilon_p}a_{ij}^{I}\left(\frac{\partial p_{k-2}^{I}}{\partial y_j}+\frac{\partial p_{k-3}^{I}}{\partial x_j}\right)n_i\Big|_\Gamma\right]=0 \tag{1.36d}$$

p_k are y-periodic, $\forall k$; here $p_k^{(i)}\equiv p_k$, when $y\in Y^i$.

In (1.36) all values with negative indexes are identical to zero.

Operators L_0 in the main term of expansions, (1.36), are of different type: it is parabolic on Y^I, and is elliptic on Y^{II}, according to (1.29).

Let derive also the expansion for integral correlation (1.31), by substituting relation (1.32) there:

$$\frac{\partial}{\partial x_i}\left\langle a_{ij}^{II}\frac{\partial p_0}{\partial y_j}\right\rangle_2+\varepsilon\left[\langle(L_{xx}-b^{II}\partial_\tau)p_0+f_*\rangle_2-\omega_m\frac{\partial}{\partial\tau}\langle b^I p_0\rangle_1+\frac{\partial}{\partial x_i}\langle a^{II}\frac{\partial p_1}{\partial y_i}\rangle_2\right]+$$

$$\sum_{k=-1}^{\infty}\varepsilon^{k+3}\left[\frac{\partial}{\partial x_i}\left\langle a_{ij}^{II}\frac{\partial p_{k+3}}{\partial y_j}\right\rangle_2+\langle(L_{xx}-b^{II}\partial_\tau)p_{k+2}\rangle_2-\omega_m\frac{\partial}{\partial\tau}\langle b^I p_{k+2}\rangle_1+\right.$$

$$\left.+\frac{\omega_m}{\varepsilon_p}\left(\frac{\partial}{\partial x_i}\left\langle a_{ij}^{I}\frac{\partial p_{k+1}}{\partial y_j}\right\rangle_1+\langle L_{xx}p_k\rangle_1\right)\right]=0 \tag{1.37}$$

$p_k\equiv 0$, when $k<0$.

Moreover conditions (1.34) take place.

1.4.3. BLOCK EXPANSION

Zero Approximation on the Block
It follows for zero approximation p_0 from (1.36), (1.37):

$$\begin{cases} (L_{yy}-\varepsilon_p b^I \partial_\tau)p_0 = 0, & y\in Y^I \\ p_0(x,y,0)=p^\circ, & p_0|_{\Gamma(Y^I)}=p_0|_{\Gamma(Y^{II})} \end{cases} \tag{1.38a}$$

$$\begin{cases} L_{yy}p_0 = 0, & y\in Y^{II} \\ a_{ij}^{II}(y)\frac{\partial p_0}{\partial y_j}n_i|_{\Gamma(Y^{II})} = 0, & \langle p_{01}\rangle_2=0 \end{cases} \tag{1.38b}$$

with an additional condition of y-periodicity, where

$$\frac{\partial}{\partial\tau}, \quad L_{yy}\equiv\frac{\partial}{\partial y_i}\left(a_{ij}(y)\frac{\partial}{\partial y_j}\right), \qquad \partial_\tau\equiv\frac{\partial}{\partial\tau}$$

Linear homogeneous problem (1.38a) has a trivial solution

$$p_0 = p_0(x,\tau) \equiv p_{00}(x,\tau), \quad y\in Y^{II}, \tag{1.39}$$

which is not determined yet.

Then the block problem (1.38) takes the form:

$$\begin{cases} (L_{yy}-\varepsilon_p b^I \partial_\tau)p_0 = 0, & y\in Y^I \\ p_0(x,y,0)=p^\circ \end{cases} \qquad p_0|_\Gamma=p_{00}(x,\tau) \tag{1.40}$$

the conditions of matching are satisfied too: $p_{00}(x,0)=p^\circ(x)$.

Solution of the Block Problem
That is a classical Dirichlet problem with heterogeneous conditions for a homogeneous equation of parabolic type. Its solution exists, is unique under rather wide restrictions [17], [161], [136], and parametrically depends on x.

The solution of (1.40) may be expressed through the Green function $G(y,\tau,\xi,\theta)$:

$$p_0(y,\tau;x)=p_{00}(x,\tau)-\int_0^\tau\int_{Y^I} G(y,\tau;\xi,\theta)\frac{\partial p_{00}(x,\theta)}{\partial\theta}d\xi d\theta, \quad y\in Y^I \tag{1.41}$$

where G is a solution of the problem:

$$\begin{cases} (L_{yy}-\lambda\partial_\tau)G= - \delta(y-\xi)\delta(\tau-\theta), \\ G\big|_{y\in\Gamma}=0 \end{cases} \tag{1.42}$$

Thus, the main term of the expansion, p_0, depends on the fast variable y. Its variations in time are described by a nonlocal operator (1.41).

The structure of Green's function (1.42) for some special forms of the domain Y^I will be studied later.

Spectral Expansion for Green's Function on the Block

A rather general presentation can be obtained for Green's function G (1.42) in the form of an expansion with respect to the eigenfunctions of the basic operator.

Due to the fact that only one characteristic scale of oscillation (ε) is significant in considered systems, the "micro-oscillations" of the parameters $a(y)$ and $b(y)$ within one block Y^I or within an elementary matrix Y^{II} have not any importance. Therefore the functions a and b may be considered as partially constant, then $a^k_{ij}(y), b^k(y)$=const, $k=I, II, i, j=1, ..., n$, moreover, $b^k\equiv 1$, according to (1.9a).

Then in problems (1.40), (1.42) the symbol L_{yy} designates the operator with constant coefficients:

$$L_{yy}\equiv\frac{\partial}{\partial y_i}\left(a_{ij}\frac{\partial}{\partial y_j}\right)$$

and the value:

$$\lambda\equiv\varepsilon_p b^I = \varepsilon_p \tag{1.43}$$

is also constant.

Let $\{R_k\}_{k=1}^{\infty}$ be the eigenfunctions, and $\{\mu_k\}$ be the eigenvalues of the operator $L^*\equiv - \Delta_{yy}$ in Y^I, i.e.

$$-L_{yy}R_k = \mu_k R_k, \quad y\in Y^I; \quad R_k\big|_\Gamma=0, \ \forall k\geq 1$$

Let the spectrum $\{\mu_k\}$ is simple. So we assume some limitations on the block form and on the constant values a_{ij}.

It can be deduced by usual technique:

$$G(y,\tau;\xi,\theta)=\sum_{k=1}^{\infty}\frac{R_k(y)R_k(\xi)}{\langle R_k^2\rangle_1}\exp\{\frac{-\mu_k(\tau-\theta)}{\lambda}\}, \tag{1.44}$$

such that:

$$p_0(x,y,\tau) = p_{00}(x,\tau) - \int_0^\tau \frac{\partial p_{00}(x,\theta)}{\partial\theta} \sum_{k=1}^{\infty} \frac{\langle R_k\rangle_1 R_k(y)}{\langle R_k^2\rangle_1} \exp\left\{\frac{-\mu_k(\tau-\theta)}{\lambda}\right\} d\theta \quad (1.45)$$

when $y \in Y^I$.

If permeability a is scalar, and Y^I is a sphere in R^3 of the radius ξ, then $a^I_{ij}=a^{II}_{ij}=1$, $\forall i,j$, according to (1.9b). Therefore the operator $L_{yy}=\Delta_{yy}$, where Δ_{yy} is the Laplacian. Then:

$$R_k = \sin(r\sqrt{\mu_k})/r, \qquad \mu_k=(\pi k/\xi)^2, \quad (1.46)$$

where r is a radial coordinate within one cell.

1.4.4. EXPANSION ON THE MATRIX

Zero Approximation on the Matrix

A closed system of two equations may be obtained on the subdomain Y^{II} for p_0, which relates it to p_1:

$$\langle (L_{xx}-b^{II}\partial_\tau)p_0+f_*\rangle_2 - \omega_m \frac{\partial}{\partial\tau}\langle b^I p_0\rangle_1 + \frac{\partial}{\partial x_i}\left\langle a^{II}_{ij}\frac{\partial p_1}{\partial y_j}\right\rangle_2 = 0, \quad (1.47a)$$

$$L_{yy}p_1 + (L_{yx}+L_{xy})p_0 = 0, \qquad y\in Y^{II} \quad (1.47b)$$

Here the first equation corresponds to the main term in relation (1.37), where the outer left-hand term is equal to zero due to (1.39), whereas the second term results from (1.36).

The problem for p_1 is supplemented by conditions resulting from (1.34) and (1.36):

$$\begin{cases} \left(\dfrac{\partial p_1^{II}}{\partial y_i} + \dfrac{\partial p_0^{II}}{\partial x_i}\right) n_i \Big|_\Gamma = 0, \quad \langle p_{11}\rangle_2 = 0, \\ p_1 \text{ is } y-\text{periodic} \end{cases} \quad (1.48)$$

The initial condition for p_1^{II} is lost.

The problem for p_1 on Y^{II} is close to a typical cell problem in methods [17], [136] excluding the condition $\langle p_{11}\rangle_2=0$, which results from the nonuniform averaged value accepted as a form for the main term of asymptotic expansions. This condition is in essence.

Cell Problem
The following representation takes place for the function p_1:

$$p_1(x,y,\tau)=p_{10}(x,\tau)+\psi_k^{II}(y)\frac{\partial p_{00}}{\partial x_k},\ y\in Y^{II}, \tag{1.49}$$

where the function $p_{10}(x,\tau)$ will be defined later, while the following problem results for ψ_k^{II}, $k=1,2,...$, from (1.48), (1.49):

$$\begin{aligned} &L_{yy}\psi_k^{II} = -\frac{\partial \alpha_*}{\partial y}, \qquad y\in Y^{II}, \\ &\left(\frac{\partial \psi_k^{II}}{\partial y_i}+\delta_{ik}\right)n_i\bigg|_{\Gamma}=0, \\ &\langle\psi_k^{II}\rangle_2=0, \qquad \psi_k^{II} \text{ is } y-\text{periodic} \end{aligned} \tag{1.50}$$

The last condition should be interpreted in a following way. Let us continue the function ψ_k^{II} in such a way, that it would be a constant on Y^I. Then periodicity of p_1 on Y leads to periodicity of continued function ψ_k^{II} on all Y, and on Y^{II} too.

Substituting (1.49) in (1.47a) we obtain for zero approximation on Y^{II}: $p_0=p_{00}$.

$$\frac{\partial}{\partial x_i}\left(A_{ij}^{II}\frac{\partial p_{00}}{\partial x_j}\right) - B^{II}\frac{\partial p_{00}}{\partial \tau} = \omega_m\frac{\partial}{\partial \tau}\langle b^I p_0\rangle_1-\langle f_*\rangle_2, \tag{1.51}$$

where

$$A_{ij}^{II}\equiv\left\langle a_{ik}^{II}\left(\frac{\partial \psi_j^{II}}{\partial y_k}+\delta_{jk}\right)\right\rangle_2, \quad B^{II}\equiv\langle b^{II}\rangle_2 \tag{1.52}$$

$\langle p_0 b^I\rangle_1$ is determined from (1.41).

Relation (1.51) is the equation averaged over the matrix.

Correctness of the Cell Problem
Consider the problem (1.50).

Let us seek for its solution in class of twice differentiable functions in Y^{II} and smooth in ∂Y^{II} ($\partial Y^{II}=\partial Y\cup S$).

Only two opportunities of behavior at the boundary ∂Y result from periodicity of the function $\psi_k^{II}(y)$ $(\forall k)$: $\psi_k^{II}=0$ on one part of the boundary, whereas $(\partial\psi_k^{II}/\partial y_i)n_i=0$ on another part. Then (1.50) is equivalent to a Neumann problem in Y^{II}, or to a mixed problem for Poisson's equation. Solution of both problems exists and is unique with additional integral condition $\langle\psi_k^{II}\rangle=0$ [152], [132].

1.4.5. FULL DETERMINATION OF THE FIRST APPROXIMATION

The main term of expansions (1.36) occurs sufficient to construct homogenized model of zero order $p_0(x,y,\tau)$ on the all cell Y and the oscillating component $p_{11}(x,y,\tau)$ of the first approximation in the matrix Y^{II}. Functions $p_{10}(x,\tau)$ from (1.36) and (1.48), as well as $p_{11}(x,y,\tau)$ in Y^I remain undefined. All consequent approximations remain undefined too.

Adding approximations of higher order we can precise the homogenized equations. Below we show an example of behavior for higher approximations.

First Approximation

First of all, let us derive the full definition for the first approximation p_1, i.e. the functions $p_{11}(x,y,\tau)$ in Y^I and $p_{10}(x,\tau)$. The final result is as follows.

The functions $p_{10}(x,\tau)$ and $p_{11}(x,y,\tau)$ in Y^I are determined from the system of equations:

$$A_{ij}^{II}\frac{\partial^2 p_{10}}{\partial x_i \partial x_j}-B^{II}\frac{\partial p_{10}}{\partial \tau}-\omega_m\frac{\partial}{\partial \tau}\langle b^I(p_{10}+p_{11})\rangle_1=$$
$$-\langle a_{ij}^{II}(\frac{\partial \psi_{km}^{II}}{\partial y_j}+\psi_k^{II}\delta_{jm})\rangle_2\frac{\partial^3 p_{00}}{\partial x_i\partial x_k\partial x_m}+$$
$$\langle a_{ij}\frac{\partial \varphi}{\partial y_j}+b\psi_i^{II}\rangle\frac{\partial^2 p_{00}}{\partial x_i\partial \tau}-\frac{\partial}{\partial x_i}\langle a_{ij}^{II}\frac{\partial p_{2f}}{\partial y_j}\rangle_2+$$
$$\frac{\omega_m}{\varepsilon_p}\int_0^\tau \frac{\partial^2 p_{00}(x,\theta)}{\partial x_i\partial \theta}\int_{Y^I}\int_{Y^I} a_{ij}^I(y)\frac{\partial G(y,\tau;\xi,\theta)}{\partial y_j}d\xi dy d\tau, \tag{1.53a}$$
$$p_{10}(x,0)=0 \tag{1.53b}$$
$$p_{11}(x,y,\tau)=\frac{\partial p_{00}}{\partial x_i}\tilde{\psi_i^{II}}-\int_0^\tau\int_{Y^I}G(y,\tau;\xi,\theta)\times$$
$$\Big\{\varepsilon_p b^I(\xi)\frac{\partial p_{10}(x,\theta)}{\partial \theta}-\frac{\partial}{\partial \xi_i}\left(a_{ij}^{II}(\xi)\frac{\partial p_0^I(x,\xi,\theta)}{\partial \xi_j}\right)-$$
$$\frac{\partial}{\partial x_i}\left(a_{ij}^I(\xi)\frac{\partial p_0^I}{\partial \xi_j}\right)-\frac{\partial p_{00}(x,\theta)}{\partial x_k}\frac{\partial}{\partial \xi_i}\left(a_{ij}^I(\xi)\frac{\partial \tilde{\psi_k^{II}}}{\partial \xi_j}\right)+$$
$$\varepsilon_p b^I(\xi)\frac{\partial p_{00}(x,\theta)}{\partial x_k\partial \theta}\tilde{\psi_k^{II}}(\xi)\Big\}d\xi d\theta,\quad y\in Y^I, \tag{1.53c}$$

where $\tilde{\psi_k^{II}}(y)\equiv\psi_k^{II}(y)|_{y\in\Gamma}$, p_0^I is defined from (1.41), G is Green's function (1.42), (1.44).

The y-periodic functions $\psi^{II}_{km}(y)$, $\varphi(y)$, $p_{2f}(x,y,\tau)$ and $p_{22}(x,y,\tau)$ are determined by the problems in Y^{II}:

$$\begin{cases} L_{yy}\psi^{II}_{km} = -\frac{\partial}{\partial y_m}(a^I_{kj}\psi^{II}_j) - a^I_{mj}\frac{\partial \psi^{II}_k}{\partial y_j} - a^I_{km} + \frac{1}{(1-\alpha)}A^{II}_{km}, \; y\in Y^{II} \\ \left(a^{II}_{ij}\frac{\partial \psi^{II}_{km}}{\partial y_j} + \psi^{II}_k\delta_{jm}\right) n_i\big|_\Gamma = 0, \; \langle \psi^{II}_{km}\rangle_2 = 0, \; k,m=1,...,n \end{cases} \tag{1.54a}$$

$$\begin{cases} L_{yy}\varphi = -b^{II}(y) + B^{II}/(1-\alpha), \; y\in Y^{II} \\ a^{II}_{ij}\frac{\partial\varphi}{\partial y_j} n_i\big|_\Gamma = 0, \quad \langle\varphi\rangle_2 = 0 \end{cases} \tag{1.54b}$$

$$\begin{cases} L_{yy}p_{2f} = -f_*(x,y,\tau) + \langle f_*\rangle_2/(1-\alpha), \; y\in Y^{II} \\ a^{II}_{ij}\frac{\partial p_{2f}}{\partial y_j} n_i\big|_\Gamma = 0, \; \langle p_{2f}\rangle_2 = 0 \end{cases} \tag{1.54c}$$

$$\begin{cases} L^0 p_{22} = \frac{\omega_m}{(1-\alpha)}\frac{\partial}{\partial\tau}\langle b^I p^I_0\rangle_1, \; y\in Y^{II} \\ a^{II}_{ij}\frac{\partial p_{22}}{\partial y_j} n_i\big|_\Gamma = -\frac{\omega_m}{\varepsilon_p} a^I_{ij}\frac{\partial p^I_0}{\partial y_j} n_i\big|_\Gamma, \; \langle p_{22}\rangle_2 = -\frac{\omega_m}{\varepsilon_p}\langle p^I_0 - p_{00}\rangle_1 \end{cases} \tag{1.54d}$$

where α is a volume fraction of blocs.

It is clear, these problems have unique solutions, that may be confirmed in the same way as in section 1.4.4.

Second Approximation

Procedure explained above gives, in the same time, the oscillating component for the second approximation in p_{21} in Y^{II}:

$$p_{21}(x,y,\tau) = \psi^{II}_{km}(y)\frac{\partial^2 p_{00}}{\partial x_i \partial x_j} + \psi^{II}_k(y)\frac{\partial p_{10}}{\partial x_k} - \varphi(y)\frac{\partial p_{00}}{\partial\tau} + p_{22}(x,y,\tau), \quad y\in Y^{II} \tag{1.55}$$

If a^I_{ij}, a^{II}_{ij}, b_*, f_* are constant, then $\varphi\equiv 0$, $p_{2f}\equiv 0$. However, even in this case the variables y and τ of the function p_{21} in Y^{II} can not be splitted, in contrast to the systems examined in [17], due to the component $p_{22}(x,y,\tau)$ describing the mass exchange with blocks, which is strictly non-stationary.

All higher approximations may be found in the analogous way.

1.5. MACROSCALE MODELS FOR SOURCE MEDIA

The classification of macroscale models is based on two types of elementary flows described in section 1.2.3. However, it is necessary to note that the limits of validity for each class are fuzzy, as these classes are superposed. To identify more exactly the limits of validity for each class, it is useful to select those cases, when each of the elementary flows is exhibited in the most bright form, as well as the case when both the elementary flows are entirely equilibrated. We obtain three key points:

$\omega_K=1$, which corresponds to the translation flow in the pure state;

$\omega_K=\varepsilon^2$, which corresponds to the source flow in pure state;

$\omega_K=\varepsilon$, which corresponds to a system, where the intensity of two elementary flows is equivalent.

Then we can confirm that three averaged models deduced in this section correspond to the following situations:

- Model of source flow: $\omega_K \ll \varepsilon$, or $\varepsilon_p \gg \varepsilon$;
- Model of translation flow: $\varepsilon \ll \omega_K \leq 1$, or $\varepsilon^2 \leq \varepsilon_p \ll \varepsilon$;
- Model of translation-source flow: $\varepsilon^2 \ll \omega_K \ll 1$, or $\varepsilon^2 \ll \varepsilon_p \ll 1$.

Along with these cases, a system with almost impenetrable blocks may be selected, that corresponds to $\omega_K \ll \varepsilon^2$.

In all cases the method of nonuniform averaging is used. The asymptotic expansions are developed with respect to two parameters: ε and ω_K, or ε and ε_p.

1.5.1. MODEL EQUATIONS

Let us introduce the pressures, averaged over subdomains Y^I and Y^{II} of a cell:

$$P^I = \frac{1}{\alpha}\langle p \rangle_1, \qquad P^{II} = \frac{1}{1-\alpha}\langle p \rangle_2, \tag{1.56}$$

where α is the volume fraction of the blocks.

It follows from the results of the previous section, that:

$$P^I = \frac{1}{\alpha}\langle p_0 \rangle_1 + O(\varepsilon), \qquad P^{II} = p_{00} + O(\varepsilon) \tag{1.57}$$

Then we obtain for p_0 in Y^I from relations (1.41) or (1.45), and for p_{00} from (1.51), with accuracy of order ε:

$$P^I(x,\tau) = P^{II}(x,\tau) - \int_0^\tau \frac{\partial P^{II}(x,\theta)}{\partial \theta} K(\tau-\theta)d\theta \tag{1.58a}$$

$$\frac{\partial}{\partial x_i}\left(A^{II}_{ij}\frac{\partial P^{II}}{\partial x_j}\right)-B^{II}\frac{\partial P^{II}}{\partial \tau}=B^{I}\frac{\partial P^{I}}{\partial \tau}-\langle f_*\rangle_2 \qquad (1.58b)$$

with a condition $P^i(x,0)=p^\circ(x)$.

The system (1.58) can be reduced to one integro-differential equation with respect to the pressure P^{II}:

$$\frac{\partial}{\partial x_i}\left(A^{II}_{ij}\frac{\partial P^{II}}{\partial x_j}\right)-B^{II}\frac{\partial P^{II}}{\partial \tau}=$$

$$B^{I}\frac{\partial}{\partial \tau}\left(\int_0^\tau \frac{\partial P^{II}}{\partial \theta}(1-K(\tau-\theta))d\theta\right)-\langle f_*\rangle_2 \quad (1.59)$$

The integro-differential operator in (1.59) describes mass exchange between the blocks and the matrix.

1.5.2. EFFECTIVE PARAMETERS

Effective Porosity and Permeability

The model is characterized by two scalar factors B^I, and B^{II}, one constant tensor $\{A^{II}_{ij}\}^n_{ij=1}$ and the function $K(\tau)$.

It is designated that

$$B^I\equiv\langle b^I\rangle_1=\omega_m\langle b^I\rangle_1$$

$$B^{II}\equiv\langle b^{II}\rangle_2=\langle b^{II}\rangle_2$$

The effective permeability tensor is:

$$A^{II}_{ij}\equiv\left\langle a^{II}_{ik}\left(\frac{\partial \psi^{II}_j}{\partial y_k}+\delta_{jk}\right)\right\rangle_2, \quad B^{II}\equiv\langle b^{II}\rangle_2 \qquad (1.60)$$

where the cell functions ψ^{II}_k, $k=1,...,n$ are defined as:

$$\begin{cases} \frac{\partial}{\partial y_i}\left(a^{II}_{ij}(y)\frac{\partial \psi^{II}_k}{\partial y_j}+\delta_{kj}\right)=0, & y\in Y^{II}, \\ a^{II}_{ij}\left(\frac{\partial \psi^{II}_k}{\partial y_j}+\delta_{kj}\right)n_i\Big|_\Gamma=0, & \\ \langle \psi^{II}_k\rangle_2=0, & \psi^{II}_k \text{ is } y-\text{periodic} \end{cases} \qquad (1.61)$$

In a specific case, when the function $a(y)$ is scalar and is even, i.e., the block area Y^I has central symmetry, it is possible to prove that the tensor A_{ij}^{II} is transformed into a scalar:

$$A^{II} \equiv \left\langle a(y)\frac{\partial \psi_1^{II}}{\partial y_1} + 1 \right\rangle_2. \tag{1.62}$$

Exchange Kernel

The kernel K of the integral operator has the following structure (when all the parameters a_{ij}^I, a_{ij}^{II}, b^I and b^{II} are constant):

$$K(\tau) = \frac{1}{\alpha}\sum_{k=1}^{\infty}\frac{\langle R_k\rangle_1^2}{\langle R_k^2\rangle_1}\exp\{\frac{-\mu_k\tau}{\lambda}\}d\theta, \tag{1.63}$$

here $K(0)=1$, as shown in Fig. 1.14.

The parameter λ is defined in (1.43).

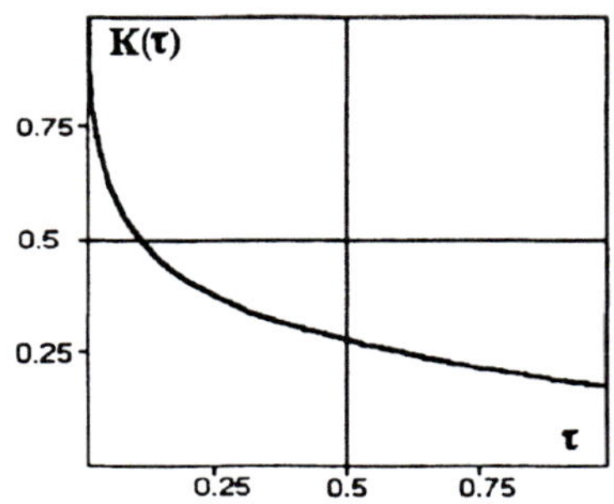

Figure 1.14. Exchange kernel $K(t)$ in the Source-type model

1.5.3. ABOUT THE MODELS OF HIGHER ORDER. PRINCIPLE OF SUFFICIENCY

It follows from (1.57), that system (1.58) is only the zero approximation for the averaged equations, and they may be infinitely updated. We call *"k-order averaged equation"*, if the residual terms are of the order $O(\varepsilon^{k+1})$. System (1.58) is a zero-order averaged model. Construction of higher order models may be produced in the same way.

The following principle of sufficiency gives solution for the problem, what order of model may be considered as sufficient.

The order of the averaged model and the order of the microscale equations should coincide. (1.64)

The physical sense of this postulate is rather obvious: the macroscale equations should be rather exact, but no more than the microscale.

It follows from this principle that in many cases, when the micro-equations contain small terms, we need to construct the macroequations of high order.

The typical example is related with the convection-diffusion equation, which will be studied in chapter 2, and which are very close to the models considered herein when the convective term is small. It follows from the principle of sufficiency, that the obtained averaged models of order zero for similar systems appear unsatisfactory and should contain small terms of the same order as the convective component in the microscale equations.

1.5.4. MODEL FOR SOURCE-TYPE MEDIA WITH QUASI-IMPENETRABLE BLOCKS

The model of such a type has been proposed in [114].

Consider a subclass of Source-type media, when the blocks are close to impenetrable:

$$\varepsilon_p \gg 1, \qquad \text{or} \qquad \omega_m \sim 1, \ \omega_K \ll \varepsilon^2 \tag{1.65}$$

Three-Scale Asymptotic Expansions

In this case, all flow inside a block is concentrated in a narrow space layer near the block border. The type of flow is source like. The size of this boundary layer is of order $\varepsilon_p^{1/2}$. A remaining part of the block is undisturbed during all the time of process.

Thus, the third scale arises, which is equal to $\varepsilon_p^{-1/2}$ and is smaller than the scale of medium heterogeneity ε.

Derivation of the macroscale model is based on developing three-scale asymptotic expansions when $\varepsilon \to 0$ and $\varepsilon_p \to \infty$.

Asymptotic expansions have the three-scale form in the boundary layer:

$$p^I = p_0^I(x, y, \xi, t) + \varepsilon p^{I,I}(x, y, \xi, t) + \ldots; \quad \xi \equiv (r - r^*)\sqrt{\varepsilon_p}$$

where r^* is the radial coordinates of block boundary (a block is assumed to be equivalent to a sphere).

Asymptotic expansion appears to be trivial far from the block boundary:

$$p(x, y, \tau) = p^\circ + O(\varepsilon + \exp(-\varepsilon_p^{-1/2})), \ y \in Y^I$$

Taking into account the first approximation, we can obtain more exact relation for the block pressure field inside a block:

$$p_0^I(x, r, t) = p_0^{II}(x, t) + (p^0 - p_0^{II})\Phi\left(\frac{1}{2}(r^* - r)\sqrt{\frac{\lambda' \sqrt{\varepsilon_p}}{t}}\right) \tag{1.66}$$

where $\lambda'=b^I/a^I$, $\Phi(u)=\frac{2}{\sqrt{\pi}}\int_0^u e^{-t^2}dt.$

Averaged Model

Averaging of these expansions over the whole block volume leads to a trivial result: $\langle p\rangle_1=p^\circ+O(\varepsilon+\varepsilon_p^{-1/2})$, because the boundary layer expansion, being concentrated in the small volume, does not give any contribution in the averaged value $\langle p\rangle_1$.

Thus, the averaged model coincides with the equations for media with entirely impenetrable blocks:

$$\begin{cases} P^I(x,\tau)=p^\circ+O(\varepsilon+\varepsilon_p^{-1/2}) \\ A_{ij}^{II}\frac{\partial^2 P^{II}}{\partial x_i \partial x_j}-B^{II}\frac{\partial P^{II}}{\partial \tau}=-\langle f_*\rangle_2+O(\varepsilon+\varepsilon_p^{-1/2}) \end{cases} \tag{1.67}$$

CONCLUSIONS

1. The average equations change the type, an integro-differential operator arises with a non-degenerating kernel describing the exchange process between blocks and matrix.

2. The effective conductivity of the medium coincides with the effective conductivity of the high conductive matrix. The blocks do not give any contribution in the translation flow through the cell. Therefore, only the source flow directed from the block center to its boundary is observed in the blocks.

1.6. MACROSCALE MODEL FOR MEDIA OF TRANSLATION TYPE

Media of such a type satisfy the condition

$$\varepsilon_p \sim \varepsilon^2, \quad \text{or} \quad \omega_K \sim \omega_m$$

They were investigated in papers [112], [124], with symmetrical condition $\omega_K \to \infty$, therefore the averaged equations may be obtained directly from the results of [17]. For these media it is possible to use the classical homogenization method. However it will be shown below that the same results may be obtained by nonuniform averaging.

1.6.1. ABOUT DEDUCTION OF THE MODEL

Structure of the Expansions

We shall consider the case of a macro and micro-isotropic medium. In problem (1.6), let us assume that $\omega_m \sim \omega_K$.

Asymptotic expansion when $\varepsilon, \omega_K \to 0$ will be constructed in the form (1.32), (1.33), where:

$$p_0 = p_0(x,\tau) = \sum_{n=0}^{\infty} \omega_K^n v_{on}(x,\tau) \tag{1.68a}$$

$$p_k = p_k(x,y,\tau) = \sum_{n=0}^{\infty} \omega_K^n v_{kn}(x,y,\tau), \quad k \geq 1 \tag{1.68b}$$

Relations (1.36), (1.37) are satisfied for p_k, with $\omega_m/\varepsilon_p = \omega_K/\varepsilon^2$.

Technique of Derivations

To construct the averaged model, it appears to be sufficient to determine the functions $v_{00}(x,\tau)$ and $v_{01}(x,\tau)$, and oscillating components (dependent on y) of the functions $v_{10}(x,y,\tau)$ on Y and $v_{11}(x,y,\tau)$ on Y^{II}.

The following results from (1.36), (1.37) for v_{00} and v_{01}:

$$\langle (L_{xx} - b^{II}\partial_\tau) v_{00} + f_* \rangle_2 + \frac{\partial}{\partial x_i} \left\langle a_{ij}^{II} \frac{\partial v_{10}}{\partial y_j} \right\rangle_2 = 0,$$

$$\langle (L_{xx} - b^{II}\partial_\tau) v_{01} \rangle_2 - \tau^S \frac{\partial}{\partial \tau} \langle b^I v_{00} \rangle_1 + \frac{\partial}{\partial x_i} \left\langle a_{ij}^{II} \frac{\partial v_{11}}{\partial y_j} \right\rangle_2 + \tag{1.69}$$

$$+ \frac{\partial}{\partial x_i} \left\langle a_{ij}^{I} \frac{\partial v_{10}}{\partial y_j} \right\rangle_1 + \left\langle L_{xx} v_{00} \right\rangle_1 = 0,$$

It is easy to derive for v_{10} and v_{11} from (1.36) and (1.37):

$$L_{yy}v^i_{1m}=-(L_{yx}+L_{xy})v^i_{0m},\quad y\in Y^i;\quad i=I,II,\quad m=0,1$$

$$\left(\frac{\partial v^{II}_{10}}{\partial y_i}+\frac{\partial v^{II}_{00}}{\partial x_i}\right)n_i\bigg|_\Gamma=0,\quad \langle v^{(*)}_{10}\rangle=0,$$

$$\left(\frac{\partial v^{II}_{11}}{\partial y_i}+\varepsilon\frac{\partial v^{II}_{01}}{\partial x_i}\right)n_i\bigg|_\Gamma=\left(\frac{\partial v^{I}_{10}}{\partial y_i}+\frac{\partial v_{00}}{\partial x_i}\right)n_i\bigg|_\Gamma, \tag{1.70}$$

$$\langle v^*_{11}\rangle_2=-\langle v^*_{10}\rangle_1$$

$$[v_{10}]_S=0$$

where $v^i_{km}=v_{km}$ when $y\in Y^i$, and the upper star designates the function component dependent on y.

All functions v_{ik} are y-periodic.

A representation follows for v_{10} and v_{11} from (1.70):

$$v^i_{10}(x,y,\tau)=\frac{\partial v^i_{00}}{\partial x_k}\psi^i_k(y)+w_{10}(x,\tau),\quad i=I,II$$
$$v^{II}_{11}(x,y,\tau)=\frac{\partial v_{01}}{\partial x_k}\psi^{II}_k(y)+\frac{\partial v_{00}}{\partial x_m}\psi^{II,I}_m(y)+w_{11}(x,\tau)$$

where the following takes place for y-periodic functions ψ^i_k and $\psi^{II,I}_k$, ($k=1,2,3$), ($i=I,II$):

$$\begin{cases} L_{yy}\psi^{II}_k=-\partial a^{II}_{kj}/\partial y_j,\ y\in Y^{II},\quad \langle\psi^{II}_k\rangle_2=0 \\ (\frac{\partial\psi^{II}_k}{\partial y_i}+\delta_{ik})n_i|_\Gamma=0 \end{cases} \tag{1.71a}$$

$$\begin{cases} L_{yy}\psi^{I}_k=-\partial a^{I}_{kj}/\partial y_j,\ y\in Y^{I}, \\ \psi^I_k|_\Gamma=\psi^{II}_k|_\Gamma \end{cases} \tag{1.71b}$$

$$\begin{cases} L_{yy}\psi^{II,I}_k=0,\ y\in Y^{II};\quad \langle\psi^{II,I}_k\rangle_2=0 \\ \frac{\partial\psi^{II,I}_k}{\partial y_i}n_i|_\Gamma=(\frac{\partial\psi^{I}_k}{\partial y_i}+\delta_{ik})n_i|_\Gamma \end{cases} \tag{1.71c}$$

1.6.2. MACROSCALE MODEL

We shall introduce the averaged pressure in the blocks, P^I, and in the matrix, P^{II}, according to (1.56). We get from (1.34), (1.68):

$$P^I = v_{00}(x,\tau) + \omega_K v_{01}(x,\tau) + O(\varepsilon + \omega_K^2)$$

Then we obtain from (1.68), (1.71) with accuracy $\sim(\varepsilon+\omega_K^2)$:

$$\begin{cases} P^I(x,\tau) = P^{II}(x,\tau), \\ \dfrac{\partial}{\partial x_i}\left(\left[A_{ij}^I + A_{ij}^{II} + A_{ij}^{II,I}\right]\dfrac{\partial P^{II}}{\partial x_j}\right) - (B^I + B^{II})\dfrac{\partial P^{II}}{\partial \tau} = -\langle f_* \rangle, \\ P^i(x,0) = p^0(x) \end{cases} \qquad (1.72)$$

where

$$\begin{cases} A_{ik}^I \equiv \omega_K \left\langle a_{ij}^I(y)\left(\frac{\partial \psi_k^I}{\partial y_j} + \delta_{jk}\right)\right\rangle_1, \\ A_{ik}^{II} \equiv \left\langle a_{ij}^{II}(y)\left(\frac{\partial \psi_k^{II}}{\partial y_j} + \delta_{jk}\right)\right\rangle_2, \\ A_{ij}^{II,I} \equiv \omega_K \left\langle a_{ij}^{II}(y)\frac{\partial \psi_k^{II,I}}{\partial y_j}\right\rangle_2, \\ B^I \equiv \omega_m \langle b^I \rangle_1, \qquad B^{II} \equiv \langle b^{II} \rangle_2 \end{cases} \qquad (1.73)$$

The cell functions $\psi_k^I(y)$, $\psi_k^{II}(y)$ and $\psi_k^{II,I}(y)$ are defined according to (1.71).

It is possible to show as in[17] for micro and macro-isotropic medium, that in the model (1.72) the residual term is of order $O(\varepsilon^2+\omega_K^2)$.

Parameters

$$A_{ij} \equiv A_{ij}^{II} + A_{ij}^I + A_{ij}^{II,I}, \quad B \equiv B^I + B^{II}$$

are the effective permeability and porosity averaged over the overall cell Y.

1.6.3. DECOMPOSITION OF TRANSLATION FLOW WITHIN A CELL

The total translation flow in a cell may be splitted onto three components, hierarchically dependent on each other. They are:

i) the "primary matrix flow", which corresponds to the component $\psi_i^{II}(y)$ describing the flow through the highly permeable matrix around the impenetrable block;

ii) the "translation block flow" $(\psi_i^I(y))$, which is caused by the primary flow in the matrix;

iii) the "secondary matrix flow" $(\psi_i^{II,I}(y))$, which is a correction to the primary matrix flow, caused by the translation flow inside the block.

Such a decomposition of flow determines the structure of the effective permeability tensor in (1.72) and (1.73)

$$A \equiv A^{II} + A^{I} + A^{II,I}$$

where:

i) A^{II} is the effective permeability of the matrix only, formed by the primary matrix flow;

ii) A^{I} is an additional term formed by the translation block flow;

iii) $A^{II,I}$ is the cross term describing the influence of the translation flow through the block on the primary flow in the matrix.

Decomposition of the translation flow allows to develop an effective method to calculate the averaged parameters of highly heterogeneous media, which is described in section 1.9.

CONCLUSIONS

1. Although the permeability and the porosity are highly heterogeneous, the medium behaves as homogeneous with practically indiscernible block and matrix pressures.

2. The macroscale and the microscale equations are of the same type.

3. High heterogeneity allows to split the structure of the effective permeability (1.73) into three components averaged either over the block or over the matrix only, that simplifies its calculation. That is a basic difference of the considered system from the case of moderate heterogeneity. Such a splitting has been proved first in [119].

1.7. MACROSCALE MODEL FOR TRANSLATION-SOURCE MEDIA

The media of such a type correspond to the following condition: $\varepsilon^2 \ll \varepsilon_p \ll 1$. In particular, it is possible to accept:

$$\omega_m \sim 1, \qquad \varepsilon^2 \ll \omega_K \ll 1 \tag{1.74}$$

The problem (1.6) with such conditions contains a time boundary layer $0 \leq \tau \leq \varepsilon_p$, during which the perturbation spans the blocks. It is the first stage of evolution. At the second stage ($\tau > \varepsilon_p$) the solution regularly depends on ε_p. Technically, it is possible to put that $\varepsilon_p \sim \varepsilon$ for the sake of definiteness.

1.7.1. ABOUT DEDUCTION OF THE MODEL

Structure of Expansions

At the first stage, i.e., $0 \leq \tau \leq \varepsilon_p$, let $\theta = \tau/\varepsilon_p$ be the new time of order one. The solution has a form (1.32), (1.33), with τ replaced by θ, where

$$p_0 = \begin{cases} v_{00}^I(x, y, \theta) + \sum_{n=1}^{\infty} \varepsilon_p^n v_{0n}^I(x, y, \theta), & y \in Y^I \\ v_{00}^{II}(x, \theta) + \sum_{n=1}^{\infty} \varepsilon_p^n v_{0n}^{II}(x, \theta), & y \in Y^{II} \end{cases} \tag{1.75}$$

$$p_k = \sum_{n=0}^{\infty} \varepsilon_p^n v_{kn}(x, y, \theta), \qquad k \geq 1$$

At the second stage, when $\tau \gg \varepsilon_p$, the asymptotic expansion is constructed in the form (1.32), (1.33) when $\varepsilon, \varepsilon_p \to 0$, where

$$p_0 = \begin{cases} v_{00}(x, \tau) + \sum_{n=1}^{\infty} \varepsilon_p^n v_{on}^I(x, y, \tau), & y \in Y^I \\ v_{00}(x, \tau) + \sum_{n=1}^{\infty} \varepsilon_p^n v_{on}^{II}(x, \tau), & y \in Y^{II} \end{cases} \tag{1.76}$$

$$p_k = \sum_{n=0}^{\infty} \varepsilon_p^n v_{kn}(x, y, \tau), \quad k \geq 1$$

Technique of Derivations for the First Time Stage

The solution is constructed in the form (1.32), (1.33), with τ replaced by $\theta = \tau/\varepsilon$, where:

$$p_0 = \begin{cases} v_{00}^I(x, y, \theta) + \sum_{n=1}^{\infty} \varepsilon_p^n v_{on}^I(x, y, \theta), & y \in Y^I \\ v_{00}^{II}(x, \theta) + \sum_{n=1}^{\infty} \varepsilon_p^n v_{on}^{II}(x, \theta), & y \in Y^{II} \end{cases}$$

$$p_k = \sum_{n=0}^{\infty} \varepsilon_p^n v_{kn}(x,y,\theta), \quad k \geq 1$$

To obtain the averaged system, it is sufficient to determine the functions v_{00}^I, v_{00}^{II}, v_{01}^{II}, v_{01}^I and a component of the function v_{10}^{II} which depends on y. We have:

$$v_{00}^{II}: \quad \langle b^{II}\rangle_2 \frac{\partial v_{00}^{II}}{\partial\theta} = -\omega_m \frac{\partial}{\partial\theta}\langle b^I v_{00}^I\rangle_1,$$

$$v_{01}^{II}: \quad \langle b^{II}\rangle_2 \frac{\partial v_{01}^{II}}{\partial\theta} = \langle (L_{xx} - b^{II}\partial_\tau)\rangle_2 v_{00}^{II} + \langle f_*\rangle_2 - \omega_m \frac{\partial}{\partial\theta}\langle b^I v_{01}^I\rangle_1 + \frac{\partial}{\partial x_i}\langle a_{ij}^{II} \frac{\partial v_{10}^{II}}{\partial y_j}\rangle_2,$$

$$v_{00}^I,\ v_{01}^I: \begin{cases} L_{yy} v_{0m}^I - b^I \frac{\partial v_{0m}}{\partial\theta} = 0, \\ v_{0m}^I\Big|_\Gamma = v_{00}^{II}, \quad v_{0m}^I\Big|_{\theta=0} = p^0\delta_{0m}, \; m=0,1 \end{cases} \tag{1.77}$$

The function v_{10}^{II} has the following structure:

$$\begin{aligned} v_{10}^{(i)}(x,y,\tau) &= \frac{\partial v_{00}}{\partial x_k}\psi_k^{(i)}(y) + w_{10}(x,\tau), \quad i=1,2 \\ v_{11}^{II}(x,y,\tau) &= \frac{\partial v_{01}}{\partial x_k}\psi_k^{II}(y) + \frac{\partial v_{00}}{\partial x_m}\psi_m^{II,I}(y) + w_{11}(x,\tau) \end{aligned} \tag{1.78}$$

where the following holds for y-periodic functions ψ_k^i and $\psi_k^{II,I}(k=1,..,s)$:

$$\begin{cases} L_{yy}\psi_k^{II} = -\partial a_{kj}^{II}/\partial y_j, \quad y \in Y^{II}; \quad \langle \psi_k^{II}\rangle_2 = 0 \\ \left(\frac{\partial \psi_k^{II}}{\partial y_i} + \delta_{ik}\right) n_i\Big|_\Gamma = 0 \end{cases} \tag{1.79}$$

$$\begin{cases} L_{yy}\psi_k^I = -\partial a_{kj}^{II}/\partial y_j, \quad y \in Y^I; \\ \psi_k^I\Big|_\Gamma = \psi_k^{II}\Big|_\Gamma \end{cases} \tag{1.80}$$

$$\begin{cases} L_{yy}\psi_k^{II,I} = 0, \quad y \in Y^{II}; \quad \langle \psi_k^{II,I}\rangle_2 = 0 \\ \frac{\partial \psi_k^{II,I}}{\partial y_i} n_i\Big|_\Gamma = \left(\frac{\partial \psi_k^I}{\partial y_i} + \delta_{ik}\right) n_i\Big|_\Gamma \end{cases} \tag{1.81}$$

Only slow components $w_{10}(x,\tau)$ and $w_{11}(x,\tau)$ remain undefined, however they are not needed to derive homogenized equations.

Further, it is easy to obtain for the averaged equations:

$$P^I(x,\theta)=\frac{1}{\alpha}\left[\langle v_{00}^I\rangle_1+\varepsilon_p\langle v_{01}^I\rangle_1\right]+O(\varepsilon+\varepsilon_p^2),$$
$$P^{II}(x,\theta)=v_{00}^{II}+\varepsilon_p v_{01}^{II}+O(\varepsilon+\varepsilon_p^2)$$

which are described by the model equivalent to (1.58), but with the order of residual terms $(\varepsilon+\varepsilon_p^2)$ and with the same expression for the kernel.

Technique of Derivations for the Second Time Stage

Let examine the second stage, when $\tau>\varepsilon_p$.

To construct the macroscale model it is sufficient to determine the functions v_{00}, v_{01}^I, v_{01}^{II} and the components of the functions v_{10}^{II} and v_{11}^{II} which depend on y.

The asymptotic expansion for the second stage gets the form: (1.32), (1.33) when $\varepsilon,\varepsilon_p\to 0$, where

$$p_0=\begin{cases} v_{00}(x,\tau)+\sum_{n=1}^{\infty}\varepsilon_p^n v_{0n}^I(x,y,\tau), & y\in Y^I \\ v_{00}(x,\tau)+\sum_{n=1}^{\infty}\varepsilon_p^n v_{0n}^{II}(x,\tau), & y\in Y^{II}\end{cases}$$

$$p_k=\sum_{n=0}^{\infty}\varepsilon_p^n v_{kn}(x,y,\tau),\quad k\geq 1$$

It takes place for them:

$$v_{00}:\quad \langle(L_{xx}-b^{II}\partial_\tau)\rangle_2 v_{00}^{II}+\langle f_*\rangle_2-\omega_m\langle b^I\rangle_1\frac{\partial v_{00}}{\partial\theta}=-\frac{\partial}{\partial x_i}\left\langle a_{ij}^{II}\frac{\partial v_{10}^{II}}{\partial y_j}\right\rangle_2,$$

$$v_{01}^{II}:\quad \langle(L_{xx}-b^{II}\partial_\tau)\rangle_2 v_{01}^{II}=\omega_m\frac{\partial}{\partial\tau}\langle b^I v_{01}^I\rangle_1-\frac{\partial}{\partial x_i}\left\langle a_{ij}^{II}\frac{\partial v_{11}^{II}}{\partial y_j}\right\rangle_2,$$

$$v_{01}^I:\quad L_{yy}v_{01}^I=b^I\frac{\partial v_{00}}{\partial\tau},\quad y\in Y_1,\quad v_{01}^I\Big|_\Gamma=v_{01}^{II},$$

$$v_{11}^{II}:\quad v_{11}^{II}=\frac{\partial v_{01}^{II}}{\partial x_i}\psi_k^{II}(y)+w_{11}(x,\tau)$$

the functions v_{10}^{II} and $\psi_k^{II}(y)$ have the same structure as earlier.

The following relationships result for the averaged pressures with exactness $O(\varepsilon_p^2\varepsilon)$:

$$P^I(x,\tau)=P^{II}(x,\tau)-\tau_p^S\frac{\partial P^{II}}{\partial\tau},$$

$$A_{ij}^{II}\frac{\partial P^{II}}{\partial x_i \partial x_j}-B\frac{\partial P^{II}}{\partial \tau}=-\langle f_*\rangle_2-\tau^S\frac{\partial^2 P^{II}}{\partial \tau^2},$$

$$\tau_p^S \equiv \frac{\varepsilon_p}{\alpha}\int_{Y^I}\int_{Y^I} G_1(y,\xi)b^I(\xi)d\xi dy,$$

$$\tau^S \equiv \omega_m\varepsilon_p\int_{Y^I}\int_{Y^I} G_1(y,\xi)b^I(\xi)b^I(y)d\xi dy,$$

$$B \equiv B^I+B^{II}$$

$G_1(y,\xi)$ is the Green function of the first boundary-value problem for the operator L_{yy} in Y^I:

$$L_{yy}G_1 \equiv \frac{\partial}{\partial y_i}\left(a_{ij}^I(y)\frac{\partial G_1}{\partial y_j}\right)=-\delta(y-\xi), \quad G_1\Big|_{y\in S}=0$$

1.7.2. MACROSCALE MODEL FOR THE FIRST TIME-STAGE

Let us introduce, as always, the averaged pressures (1.56). The following relation results for them from (1.76):

$$P^I(x,\theta) = \frac{1}{\alpha}\left[\langle v_{00}^I\rangle_1 + \varepsilon_p\langle v_{01}^I\rangle_1\right] + O(\varepsilon+\varepsilon_p^2)\ ,$$

$$P^{II}(x,\theta) = v_{00}^{II} + \varepsilon_p v_{01}^{II} + O(\varepsilon+\varepsilon_p^2)\ ,$$

As a result, the model for P^I and P^{II} coincides with (1.58), when τ is replaced by θ, and when the order of residual terms is $(\varepsilon+\varepsilon_p^2)$. For the exchange kernel we obtain the relation similar to (1.63):

$$K(\theta)=\frac{1}{\alpha}\sum_{k=1}^{\infty}\frac{\langle R_k\rangle_1^2}{\langle R_k^2\rangle_1}\exp\left(-\mu_k\theta/\tau_0^S\right),\ \theta\equiv\tau/\varepsilon_p\sim 1 \tag{1.82}$$

1.7.3. MACROSCALE MODEL FOR THE SECOND TIME-STAGE

It follows from (1.76) with accuracy $O(\varepsilon_p^2+\varepsilon^2+\varepsilon_p\varepsilon)$:

$$P^I(x,\tau)=P^{II}(x,\tau)-\tau_p^S\frac{\partial P^{II}}{\partial \tau}, \tag{1.83a}$$

$$\frac{\partial}{\partial x_i}\left(\left[A_{ij}^I+A_{ij}^{II}+A_{ij}^{II,I}\right]\frac{\partial P^{II}}{\partial x_j}\right)-B\frac{\partial P^{II}}{\partial \tau}=-\langle f_*\rangle_2-\tau^S\frac{\partial^2 P^{II}}{\partial \tau^2} \tag{1.83b}$$

where

$$B^I \equiv \omega_m\langle b^I\rangle_1, \quad B^{II} \equiv \langle b^{II}\rangle_2, \quad B \equiv B^I+B^{II} \tag{1.84}$$

The effective permeability $A_{ik}=A_{ij}^I+A_{ij}^{II}+A_{ij}^{II,I}$ is defined in the same way as (1.73), with cell problems (1.71).

Relaxation Time

The relaxation times are defined as:

$$\tau_p^S \equiv \frac{\varepsilon_p}{\alpha} \int_{Y^I} \int_{Y^I} G_1(y,\xi) b^I(\xi) d\xi dy \tag{1.85a}$$

$$\tau^S \equiv \omega_m \varepsilon_p \int_{Y^I} \int_{Y^I} G_1(y,\xi) b^I(\xi) b^I(y) d\xi dy \tag{1.85b}$$

where $G_1(y,\xi)$ is the Green function of the first boundary-value problem for the operator L_{yy} in Y^I:

$$L_{yy} G_1 \equiv \frac{\partial}{\partial y_i}(a_{ij}^{II}(y) \frac{\partial G_1}{\partial y_j}) = -\delta(y-\xi), \quad G_1|_{y\in\Gamma} = 0 \tag{1.86}$$

The value τ_p^S is the characteristic times of stabilization for the pressure field inside the block, while τ^S is the stabilization time for the exchange flow between the block and the matrix. Both these values are produced by the source flow and are related with the same process of the source flow stabilization (dampening). Although they are of the same order, it useful to note they are not fully equivalent.

The relaxation times can be expressed in other way, in terms of cell functions:

$$\tau^S = \varepsilon_p \left\langle b^I \varphi^I \right\rangle_1 \equiv \varepsilon_p \left\langle b^I \varphi^I \right\rangle_1 \tag{1.87a}$$

$$\tau_p^S = \frac{\varepsilon_p}{\omega_m \alpha} \langle \varphi^I \rangle_1 \tag{1.87b}$$

where the new cell function $\varphi^I(y)$ is the solution the following cell problem:

$$\begin{cases} \partial_i (a_{ij}^I \partial_j \varphi^I) = -\omega_m b^I, & y \in Y^I \\ \varphi^I|_\Gamma = 0 \end{cases} \tag{1.88}$$

If the porosity b^I is constant, then τ_p^S and τ^S are related by a linear dependence:

$$\tau^S = \omega_m \alpha b^I \tau_p^S \tag{1.89}$$

Explanation of the Results

The pressures P^I and P^{II} are linked with a simple kinetic relation with characteristic delaying time of exchange flow $\tau_p^S \sim \varepsilon_p$.

The exchange term has the following form:

$$q = -B^I \frac{\partial P^{II}}{\partial \tau} + \tau^S \frac{\partial^2 P^{II}}{\partial \tau^2} \tag{1.90}$$

q is the intensity of exchange between the blocks and the matrix.

It is easy to show by usual techniques of singular perturbations, [150], [126], that asymptotic expansions for two stages (1.83) are already defined and are automatically matched within the transient area, due to the successful choice of the expansion form (1.75), (1.76).

The character of the exact solution for all τ, as well as the interior and the exterior asymptotics are shown in Fig. 1.15, 1.16.

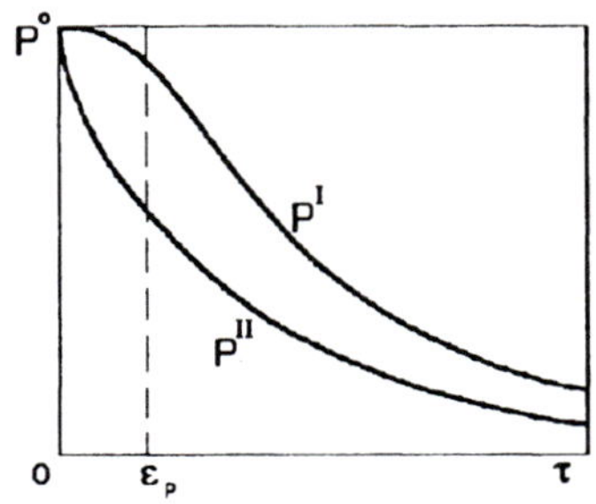

Figure 1.15. Pressure variation in the blocks, P^I, and in the matrix, P^{II}, for depletion of the source-type reservoir

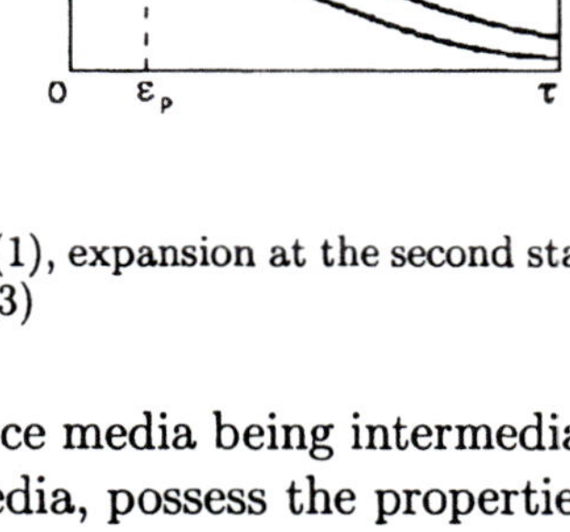

Figure 1.16. Exact solution (1), expansion at the second stage (2) for the block pressure, and for the matrix pressure (3)

The Translation-Source media being intermediate between Translation-type and Source-type media, possess the properties of both these essential classes, that leads to the following three nontrivial features of this type of systems.

First, the model is intermediate between the system of parabolic equations for Translation-type media and integro-differential system for Source-type media. The equation of transport through the matrix represents an elliptic equation with second time derivative, that is a unusual object in physics. Due to the change of type the problem of initial conditions arises, because the initial problem for an elliptic equation is unstable.

Secondly, a decomposition of the elementary flows is true, allowing to separate both the basic types of flow in the model and to formulate a rather general principle.

Third, a decomposition of the effective permeability onto three parts, or, that is the same, a decomposition of the translation-type flow onto three components remains valid, as well as in case of translation-type flow.

1.7.4. PROBLEM OF STABILITY FOR THE TRANSLATION-SOURCE MODEL

Elliptic Equation and Cauchy's Problem

The last equation in (1.83) contains a second temporal derivative and is elliptic, as $\tau^S>0$, and matrix A^{II}_{ij} and A^{I}_{ij} are positively defined. Therefore, it requires the second condition, while the classical initial problem is incorrect. Statement of only one condition

$$P^{II}(x,0) = p^{\circ} \tag{1.91}$$

is insufficient for uniqueness of the solution.

Statement of the similar condition for the pressure P^I is inadmissible, because expansion (1.6) is valid far from the initial point. In fact:

$$P^I(x,0) \equiv P^{II}(x,0) - \tau_p^S \left.\frac{\partial P^{II}}{\partial \tau}\right|_{\tau=0} \neq p^{\circ}$$

because of

$$\left.\frac{\partial P^{II}}{\partial \tau}\right|_{\tau=0} \neq 0$$

This fact becomes clear from Fig. 1.16, where behavior of the pressures P^I and P^{II} is shown for the process of reservoir exhaustion under constant rate.

It is known, that the initial problem for an elliptic equation is instable, because behaves as exponentially growing function of the time, which arises as the spectrum of the elliptic operator in (1.83b) contains the positive real eigen-values numbers $k>0$.

Stability Condition

The second necessary condition for the model obtained is to eliminate the exponentially increasing solutions:

$$|P^{II}| < M \exp(k\tau), \ \forall \tau \geq 0, \ \forall k > 0 \tag{1.92}$$

where $M>0$ is a constant.

That is a necessary stability condition.

For the right-hand part differing from zero on a finite support, this condition coincides with the requirement to the solution be bounded:

$$|P^{II}|<M<\infty, \forall\tau\geq 0 \tag{1.93}$$

where $M>0$ is some constant.

The strict results on this subject were obtained in paper [63], where the following was proved.

The problem

$$Lp \equiv \frac{\partial}{\partial x_j}(a_{ij}(x)\frac{\partial p}{\partial x_j}) = 0, \ x\in R^n,$$
$$p(x_1,\ldots,x_{n-1},0) = \varphi(x_1,\ldots,x_{n-1})$$

with a positively defined matrix a_{ij}, has a unique solution, if $|P|\leq C<+\infty$ and $\varphi(x\to\infty)=\varphi_0$, $|\varphi_0|<+\infty$, with $P(x\to\infty)\to\varphi_0$.

Thus, the problem (1.91), (1.93), (1.83b) has a unique solution in R^s, if the function $p^(x)$ has a final limit at $x\to\infty$. Behavior of the function $\langle f_*(x,\tau)\rangle$ in right-hand part of (1.83) is also limited: it should be either finite with respect to τ, or be tending to 0 faster, than any degree of τ, when $\tau\to\infty$.

In more general case, when the source function $f(x,\tau)$ is any, the condition of boundedness for the solution becomes insufficient or incorrect. For example, when a fluid is continuously injected in the porous medium under constant flow rate for any time moment, the pressure in the reservoir should grow as a linear function of the time up to infinity. However this growing can not be more rapid, that some power-value of the time; exponential growing is not physical and is not permitted.

Below in section 1.11.1, the analytical solutions of various boundary value problems are obtained for (1.83b) with conditions (1.91), (1.92) in the explicit form.

1.7.5. DECOMPOSITION OF THE TOTAL FLOW IN A CELL

The following relationships for the oscillating components $z_i\equiv p(x,y,\tau)-P^i(x,\tau)$ result from the asymptotic expansions for block and matrix pressures, with accuracy $O(\varepsilon^2+\varepsilon_p^2+\varepsilon\varepsilon_p)$.

$$z_1(x,y,\tau) \equiv p(x,y,\tau) - P^I(x,\tau) =$$
$$\left[\varepsilon\frac{\partial p_0}{\partial x_j}\psi_i^I(y) + \varepsilon\frac{\partial p_0}{\partial t}\phi^I(y)\right], \quad y\in Y^I \tag{1.94a}$$

$$z_2(x,y,\tau) \equiv p(x,y,\tau) - P^{II}(x,\tau) = \varepsilon\frac{\partial p_0}{\partial x_i}\psi_i^{II}(y), \quad y\in Y^{II} \tag{1.94b}$$

Where $\phi^I(y)=G_1(y,y)$, $G_1(y,\xi)$ is a Green function (1.86).

The following property results for considered media (*"principle of flow decomposition"*).

The total flow in a block is a superposition of two independent flows: Source-type flow described by a function $\phi^I(y)$, and Translation-type flow described by a function $\psi_i^I(y)$ from (1.71b). These two components additively enter in the relation for block pressure.

In the average model (1.83) the Source flow generates the exchange term; its intensity is proportional to $\partial^2 P^{II}/\partial\tau^2$, and is of order of τ^S. The source flow does not give any contribution to the effective conductivity of the matrix. The Translation flow generates a component A^I in the effective conductivity of the system and does not influence the exchange term.

1.8. GENERALIZED MACROSCALE MODEL

Instead of the set of various models deduced above, it is possible to deduce a generalized model, which is asymptotically true for all types of flow described above.

The generalization concerns the following aspects:

- the heterogeneity degree ω_K is any (but $\omega_K \leq 1$);

- the geometrical structure of heterogeneity is any, except one necessary conditions: the highly permeable subdomain Ω^{II} is connected.

Although the periodicity condition is kept, it is not necessary as it will be shown in chapter 6. The In the previous sections the blocks have been assumed to be disconnected. This assumption was only made to simplify some deductions, but it was not necessary. The system of low permeable blocks can be connected, but in such a way that the matrix remains always connected too.

1.8.1. ABOUT CONNECTIVITY OF THE MATRIX

Condition of connectivity for the matrix is absolutely necessary, because the basic cell problem (1.50) has solution only when this condition is true.

In fact, let us examine the problem (1.50):

$$\frac{\partial}{\partial y_i}\left(a_{ij}^{II}(y)\frac{\partial \psi_k}{\partial y_j}+\delta_{kj}\right)=0, \qquad y\in Y^{II}, \tag{1.95}$$

$$a_{ij}^{II}\left(\frac{\partial \psi_k}{\partial y_j}+\delta_{kj}\right)n_i\bigg|_{\Gamma}=0, \tag{1.96}$$

$$\langle\psi_k\rangle_2=0, \qquad \psi_k \text{ is } y-\text{periodic} \tag{1.97}$$

Let the domain Y^{II} is not connected, for example, as illustrated in Fig. ficonn for 2D case.

It is easy to show that the solution of the cell problem does not exist. In fact, let us integrate equation (1.95) over the segment Y_1^{II} which is bounded by the surface Y_1 that is a part of the external surface of the cell Y, and by the surface Γ_1 that is a part of the interface Γ. The following is true:

$$0=\int\limits_{Y_1^{II}}\frac{\partial}{\partial y_i}\left(a_{ij}^{II}(y)\frac{\partial \psi_k}{\partial y_j}+\delta_{kj}\right) = \{\text{due to the Gauss formula:}\} =$$

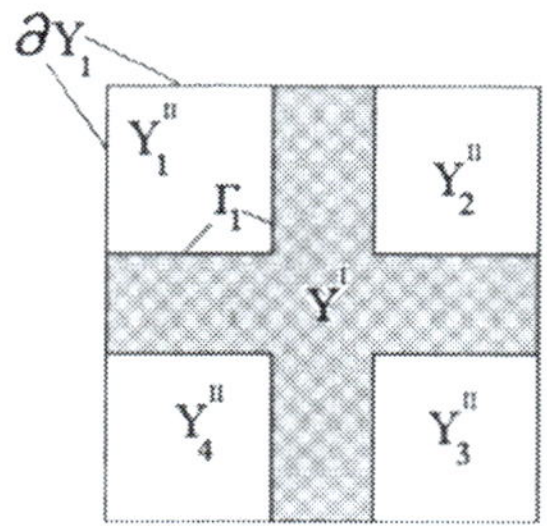

Figure 1.17. The disconnected form of the matrix Y^{II} within a cell

$$= \int_{\partial Y_1} a_{ij}^{II} \left(\frac{\partial \psi_k}{\partial y_j} + \delta_{kj} \right) n_i + \int_{\Gamma_1} a_{ij}^{II} \left(\frac{\partial \psi_k}{\partial y_j} + \delta_{kj} \right) n_i =$$

$$= \{\text{due to Eq. (1.96):}\} = \int_{\partial Y_1} a_{ij}^{II} \left(\frac{\partial \psi_k}{\partial y_j} + \delta_{kj} \right) n_i \neq 0$$

The last integral is not equal to zero, as for a periodic function such an integral is zero only over the whole surface ∂Y.

Then, connectivity of the matrix Y^{II} is necessary.

1.8.2. GENERALIZED MODEL

The generalized macroscale model covering all classes of flow may be obtained by keeping all the important terms arising in all particular models. Then we get:

$$\frac{\partial}{\partial x_i} \left(A_{ij} \frac{\partial P^{II}}{\partial x_j} \right) - B^{II} \frac{\partial P^{II}}{\partial \tau} =$$

$$B^I \frac{\partial}{\partial \tau} \left(\int_0^\tau \frac{\partial P^{II}}{\partial \theta} (1 - K(\tau - \theta)) d\theta \right) - \langle f_* \rangle_2 \quad (1.98a)$$

$$P^I(x,\tau) = P^{II}(x,\tau) - \int_0^\tau \frac{\partial P^{II}(x,s)}{\partial s} K(\tau - s) ds \quad (1.98b)$$

with the initial condition :

$$P^i(x,0) = p^\circ, \quad i = I, II$$

The kernel K of integral operator has the form (1.63):

$$K(\tau) = \frac{1}{\alpha} \sum_{k=1}^{\infty} \frac{\langle R_k \rangle_1^2}{\langle R_k^2 \rangle_1} \exp\{\frac{-\mu_k \tau}{\lambda}\} d\theta, \quad (1.99)$$

The effective permeability tensor A_{ij} is defined as

$$A_{ij}=A^{I}_{ij}+A^{II}_{ij}+A^{II,I}_{ij} \tag{1.100}$$

with A^I, A^{II} and $A^{II,I}$ defined according to (1.73), i.e.:

$$\begin{cases} A^{I}_{ik}\equiv\omega_K\left\langle a^{I}_{ij}(y)\left(\frac{\partial\psi^{I}_{k}}{\partial y_j}+\delta_{jk}\right)\right\rangle_1, \\ A^{II}_{ik}\equiv\left\langle a^{II}_{ij}(y)\left(\frac{\partial\psi^{II}_{k}}{\partial y_j}+\delta_{jk}\right)\right\rangle_2, \\ A^{II,I}_{ik}\equiv\omega_K\left\langle a^{II}_{ij}(y)\frac{\partial\psi^{II,I}_{k}}{\partial y_j}\right\rangle_2, \\ B^{I}\equiv\omega_m\langle b^{I}\rangle_1, \qquad B^{II}\equiv\langle b^{II}\rangle_2 \end{cases} \tag{1.101}$$

This model is true for any times $0\leq\tau\leq T$.

If ω_K tends to 1, then the parameter ε_p is small, and the kernel is transformed into the delta-function. Then the integral operator is transformed into the second order derivative with respect to the time, with a small parameter before it. Thus, Eqs. (1.83) yield the model for Translation-Source, or even for Translation media.

If omK tends to 0, then the parameters A^I and $A^{II,I}$ which are of order ω_K can be neglected. We obtain the model for Source media.

Thus, we can say, that the deviation of the generalized model from the exact results if of order $O(\varepsilon_*)$, where $\varepsilon_1=\min\{\omega_K,\varepsilon_p\}$.

1.8.3. GENERALIZED CELL PROBLEMS

In general case when the block boundary intersects the cell boundary, the boundary ∂Y^I does not coincide with the interface Γ separating two domains, Y^I and Y^{II}, as

$$\partial Y^I=\Gamma+\partial Y^I\cap\partial Y \tag{1.102}$$

Three sets of cell functions in (1.101) are defined as:

$$\begin{cases} \frac{\partial}{\partial y_i}\left(a^{II}_{ij}(y)\frac{\partial\psi^{II}_{k}}{\partial y_j}+\delta_{kj}\right)=0, \qquad y\in Y^{II}, \\ a^{II}_{ij}\left(\frac{\partial\psi^{II}_{k}}{\partial y_j}+\delta_{kj}\right)n_i\Big|_{\Gamma}=0, \\ \langle\psi^{II}_{k}\rangle_2=0 \end{cases} \tag{1.103a}$$

$$\begin{cases} \frac{\partial}{\partial y_i}\left(a^I_{ij}(y)\frac{\partial \psi^I_k}{\partial y_j}+\delta_{kj}\right)=0, & y\in Y^I, \\ \psi^I_k|_\Gamma=\psi^{II}_k|_\Gamma \end{cases} \tag{1.103b}$$

with periodicity condition referring to both the problems:

$$\psi_k \text{ is } y-\text{periodic}, \quad \psi_k\equiv\begin{cases} \psi^I_k, & y\in Y^I, \\ \psi^{II}_k, & y\in Y^{II} \end{cases} \tag{1.103c}$$

The third cell function is as follows:

$$\begin{cases} \frac{\partial}{\partial y_i}\left(a^{II}_{ij}(y)\frac{\partial \psi^{II,I}_k}{\partial y_j}\right)=0, & y\in Y^{II}, \\ a^{II}_{ij}\frac{\partial \psi^{II,I}_k}{\partial y_j}n_i\Big|_\Gamma=a^I_{ij}\left(\frac{\partial \psi^I_k}{\partial y_j}+\delta_{kj}\right)n_i\Big|_\Gamma, \\ \langle\psi^{II,I}_k\rangle_2=0, \\ \xi_k \text{ is } y-\text{periodic}, \quad \xi_k\equiv\begin{cases} 0, & y\in Y^I, \\ \psi^{II,I}_k, & y\in Y^{II} \end{cases} \end{cases} \tag{1.103d}$$

About the Generalized Periodicity Condition

In general case, the periodicity condition (1.103c) can not be splitted onto two conditions for each of the functions ψ^I or ψ^{II}. It means that the following holds in two opposite points y_+ and y_- of the cell boundary ∂Y (for any k):

$$\begin{aligned} &\psi^{II}_k|_{y_-}=\psi^{II}_k|_{y_+}, \text{ if } y_-,y_+\in\partial Y^{II}\cap\partial Y; \\ &\psi^I_k|_{y_-}=\psi^I|_{y_+}, \text{ if } y_-,y_+\in\partial Y^I\cap\partial Y; \\ &\psi^{II}_k|_{y_-}=\psi^I_k|_{y_+}, \text{ if } y_-\in\partial Y^{II}\cap\partial Y; \quad y_+\in\partial Y^I\cap\partial Y \end{aligned} \tag{1.104}$$

Only when the structure of heterogeneity is such that the boundary ∂Y does not contain the opposite points which belong to various subdomains, Y^I or Y^{II}, the periodicity condition (1.104) can be written separately for ψ^{II}_k and ψ^I_k.

In particular, when $\Gamma\cap\partial Y=\emptyset$, the periodicity condition in written only to the function Ψ^{II}_k.

1.8.4. COMPARISON WITH BARENBLATT-ZHELTOV-KOCHINA MODEL

The First Method of Reduction to the BZK-Model

Let us consider a medium of translation-source type at the second stage (1.83). The following relation can be deduced for the difference of pressures $\Delta P \equiv P^I - P^{II}$ from the first equation:

$$\Delta P = \tau_p^S \frac{\partial}{\partial \tau}(\Delta P) - \tau_p^S \frac{\partial P^I}{\partial \tau} + O(\varepsilon_p + \varepsilon)$$

Let us select a subclass of translation-source media, for which the source flow is slightly predominating ($\varepsilon^2 < \omega_K < \varepsilon$). Then the parameter ε_p is small. In section 1.6 it was shown that in this case the effect of translation flow through the block can be neglected. As the time $\tau_p^S \sim \varepsilon_p$ is small in (1.83), we can write the following expansion for (1.83) :

$$P^{II}(x,\tau) - P^I(x,\tau) = \tau_p^S \frac{\partial P^{II}}{\partial \tau} = \tau_p^S \frac{\partial}{\partial \tau}\left(P^I + \frac{\partial P^{II}}{\partial \tau}\right) = \tau_p^S \frac{\partial P^I}{\partial \tau} + O\left((\tau_p^S)^2\right) \tag{1.105}$$

Then the model (1.83) can be presented in the following form:

$$B^I \frac{\partial P^I}{\partial \tau} = -\frac{\Delta P}{T_*} + 0(\varepsilon_p + \varepsilon/\varepsilon_p)$$
$$B^{II} \frac{\partial P^{II}}{\partial \tau} - A_{ij}^{II} \frac{\partial^2 P^{II}}{\partial x_i \partial x_j} = \frac{\Delta P}{T_*} + 0(\varepsilon + \varepsilon_p^{-1/2}) \tag{1.106}$$

here $T_* \equiv \tau_p^S / B^I$.

Equations (1.106) coincide with the Barenblatt-Zheltov-Kochina model [19] (BZK-model). Three conclusions follow from this.

1. The BZK-model adequately describes the process only in media, which are transitional between translation-type and source-type at times $\tau > \varepsilon_p$.

2. The model (1.83) is of higher order of accuracy than the BZK-model, that follows from comparison the residual terms in (1.83) and (1.106).

3. System (1.83) can be reduced to a second order equation for P^{II}, whereas the BZK-model to an equation of third order.

The Second Method of Reduction to BZK-Model

Exponential approximation (1.118) for the exchange kernel allows to reduce integro-differential equation (1.59) or (1.98) to the differential one. As usual for the degenerating kernels, one simple differentiation of equation (1.59) gives the following result :

$$B^I \frac{\partial P^I}{\partial \tau} = \frac{P^{II} - P^I}{T_{**}} \tag{1.107}$$

where a new relaxation time is determined as :

$$T_{**}=\frac{\pi^2}{B^I\lambda_0} \tag{1.108}$$

the parameter λ_0 is defined in (1.110).

It is the similar equation to the first one in Two equations (1.98a) and (1.107) give the BZK-model. This result is similar to the system (1.106), but the delay times T_* and T_{**} are distinct.

Comparison of Two Methods

In the first case, the BZK-model is the exact asymptotic of the real process for $\tau \gg 0$.

In the second case, the BZK-model is the approximation for any time $\tau \in [0, \infty)$, as the exponential kernel has been deduced by least square fitting for the real kernel within the mentioned interval.

The difference of T_* and T_{**} can be easily explained. The real process described by nonlocal equations, can be considered as a relaxation process with the relaxation which is not constant. The value T_* is the limit relaxation time for $\tau to \infty$. Other value T_{**} is the mean relaxation time on the interval $\tau \in [0, T]$.

1.8.5. NONLOCAL PHENOMENA: ORIGIN AND DESCRIPTION

Arising of long-term memory in case of a source-type system, i.e., when $\omega_K=\varepsilon^2$, should be explained.

All the non-local in time phenomena are related to the delay of block operating with respect to the matrix. Due to condition of pressure continuity at the block-matrix interface, the averaged block pressure tends to be equal to that of the matrix. From this standpoint the matrix pressure may be considered as the equilibrium function, while the block pressure tends to reach the equilibrium state. Such a tendency to the equilibrium sate can be described in terms of the relaxation models, when the degree of non-equilibrium is weak, or by the models nonlocal in time for the high non-equilibrium.

Equilibrium Process

In the model of translation type, the time of perturbation propagation through a block and through the matrix within a cell is of the same order. Then the overall pressure field is equilibrated. As a result, the macroscale block and matrix pressures are equivalent:

$$P^I = P^{II}$$

Weakly Non-Equilibrium Process

For the medium of translation-source type, some small difference arises in times of perturbation propagation. Therefore the block pressure is delaying, but this delay is rather small. The block pressure is in weakly non-equilibrium state and can be described using the approach of the non-equilibrium thermodynamics. Then the rate of variations of a scalar non-equilibrium value is assumed to be proportional to the difference between this value and the equilibrium limit:

$$\frac{\partial P^{I}}{\partial \tau} = -\frac{1}{\tau^{S}}(P^{I} - P^{II})$$

where

Weak non-equilibrium means that the relaxation time τ^{S} is small. Then we get:

$$P^{I}{=}P^{II}{-}\tau^{S}\frac{\partial P^{I}}{\partial \tau}{\sim}P^{II}{-}\tau^{S}\frac{\partial P^{II}}{\partial \tau}+O(\tau^{S})$$

This relation coincide with that of translation-source model (1.83a).

Note that the weakly non-equilibrium system has only one eigen time; i.e., the relaxation process happens with a constant rate.

The last equation can be written if the following form:

$$P^{I}{=}P^{II}{-}\tau^{S}\int\limits_{0}^{\tau}\delta(\tau{-}\theta)\frac{\partial P^{II}}{\partial \theta}d\theta$$

where δ is the delta-function.

Highly Non-Equilibrium Process

In case of source media, the delay in block operating becomes considerable. The block pressure is in highly non-equilibrium state. The process of tending to the equilibrium state has variable rate and is characterized by a spectrum of the relaxation times. To describe such a process the following idea can be applied.

In the last equation for weakly non-equilibrium systems, the delta function can be considered as a limit of the exponential function:

$$\tau^{S}\delta(\tau) \sim \lim_{\tau^{S}\to 0} \exp\left\{-\frac{\tau}{\tau^{S}}\right\}$$

Then the integral relation can be written as:

$$P^{I}{=}\lim_{\tau^{S}\to 0}\left(P^{II}{-}\int\limits_{0}^{\tau}\exp\left\{-\frac{(\tau{-}\theta)}{\tau^{S}}\right\}\frac{\partial P^{II}}{\partial \theta}d\theta\right)$$

In highly non-equilibrium case, a number of relaxation times $\{\tau_k^S\}_{k=1}^{\infty}$ exist and all they are not small. Then the following relation can be suggested by the analogy:

$$P^I \sim P^{II} - \int_0^{\tau} \sum_{k=1}^{\infty} C_k \exp\left\{-\frac{(\tau-\theta)}{\tau_k^S}\right\} \frac{\partial P^{II}}{\partial \theta} d\theta$$

with some coefficient C_k.

This relation is similar to that deduced for the Source system (1.98).

1.9. CALCULATION OF THE MACROSCALE MASS-EXCHANGE PARAMETERS

The models (1.58) or (1.59), (1.83), (1.72), or the generalized model (1.98) are characterized by three scalar factors: B^I, B^{II} and τ_p^S, two constant tensors A_{ij}^{II}, A_{ij}^I, and the exchange kernel $K(\tau)$ (the kernel of the exchange operator). Definition (1.83) of the factors B^i is trivial. The remaining values are defined in terms of the boundary-value problems in a cell. In this chapter some analytical methods to calculate the exchange kernel and the relaxation times are examined.

1.9.1. PERFORMANCES OF THE EXCHANGE PROCESS

Exchange Kernel for Various Block Geometry

The main problem to practical use of the model (1.98) is to simplify the long-memory operator.

Simplifications result from some properties of the exchange kernel. General expression for $K(\tau)$ at any form of the block has a form (1.63).

If the parameters a_{ij}^{II}, a_{ij}^I are scalar and constant, the parameters b^{II} and b^I are constant, then all they are equal to 1, according to (1.9). If moreover, the blocks are of some regular shape, the following analytical results can be obtained for the exchange kernel.

For a sphere of dimensionless radius ξ (referred to a site of the cell):

$$K(\tau)=\frac{6}{\pi^2}\sum_{k=1}^{\infty}\frac{1}{k^2}\exp\{-\pi^2k^2\tau/\lambda_0\} \tag{1.109}$$

where

$$\lambda_0 \equiv \varepsilon_p\xi^2, \quad \xi=\left(\frac{3\alpha}{4\pi}\right)^{1/3} \tag{1.110}$$

α is the volume fraction of blocks.

For a cube with a dimensionless site h :

$$K(\tau)=\frac{8^3}{\pi^6}\sum_{k,m,n=0}^{\infty}\frac{\exp\{-(\tilde{k}+\tilde{m}+\tilde{n})\frac{\pi^2\tau}{\lambda_0'}\}}{\tilde{k}\tilde{m}\tilde{n}} \tag{1.111}$$

where

$$\lambda_0' \equiv \varepsilon_p h^2 \tag{1.112}$$

$h=\alpha^{1/3}$, $\tilde{k}=(2k+1)^2$, $\tilde{m}=(2m+1)^2$, $\tilde{n}=(2n+1)^2$.

For a parallelepiped $h_1 \times h_2 \times h_3$:

$$K(\tau) = \frac{8^3}{\pi^6} \sum_{k,m,n=0}^{\infty} \frac{\exp\{-h^2 (\frac{\tilde{k}}{h_1{}^2} + \frac{\tilde{m}}{h_2{}^2} + \frac{\tilde{n}}{h_3{}^2}) \frac{\pi^2 \tau}{\lambda_0'}\}}{\tilde{k}\tilde{m}\tilde{n}} \tag{1.113}$$

where $h^3 = \alpha = h_1 h_2 h_3 \leq 1, \quad h_i \leq 1, \forall i$.

Functions (1.109) – (1.113) calculated at the same magnitudes of α, ε_p, and $b^{II} = a^{II} = 1$, are displayed in Fig. 1.18 and 1.19.

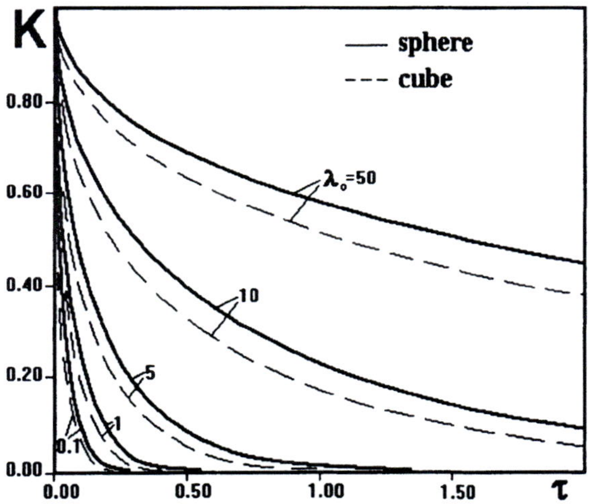

Figure 1.18. Exchange kernel $K(\tau)$ for the block in form of a sphere and a cube (a), for various magnitudes of the delaying parameter λ_0

Divergence of the curves $K(\tau)$ for a sphere and a parallelepiped with the equivalent volume, is less than 5%, if the ratio of the parallelepiped sites is not greater than 5:1. Thus, the approximation by a sphere may be sufficient for applications.

In case of strongly oblong blocks, the spanning by perturbation happens essentially in basic direction parallel to the short sites, therefore the approximation by a sphere or a cube with equivalent volume is unsuitable. It is clear from Fig. 1.19, where the ratio of the parallelogram sites is 1:1:10.

1.9.2. PROPERTIES OF THE EXCHANGE KERNEL

The higher a curve $K(\tau)$ is located on the plot, the slower exchange process happens.

The following properties of the exchange kernel do not depend on the block shape.

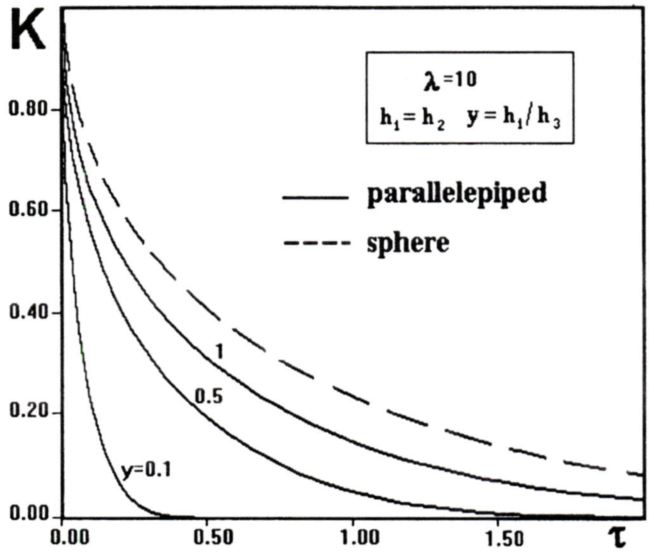

Figure 1.19. Exchange kernel $K(\tau)$ for the block in form of a parallelepiped for various magnitudes of the site ratio y

Behavior at $\tau \to 0$
1°.

$$K(0)=1 \tag{1.114}$$

Because

$$\sum_{k=1}^{\infty} \frac{1}{k^2} = \frac{\pi^2}{6}, \qquad \sum_{k,m,n=0}^{\infty} \frac{1}{\tilde{k}\tilde{m}\tilde{n}} = \frac{\pi^6}{8^3}$$

2°. Serieses (1.109) – (1.113) are fast converging at large τ. At small τ it is possible to use the equivalent form instead of (1.109):

$$K(\tau) = 1+\frac{3\tau}{\lambda_0} - \frac{6}{\sqrt{\pi\lambda_0}} \times$$
$$\times \left[\sqrt{\tau}+\sum_{k=1}^{\infty}\int_0^{\tau} \frac{\exp\{-k^2\lambda_0/\tau'\}}{\sqrt{\tau'}}d\tau'\right] \tag{1.115}$$

This relation is deduced with a known identity (for example [152], page 170):

$$\sum_{k=-\infty}^{+\infty} e^{-k^2\tau}=\sqrt{\pi/\tau}\sum_{k=-\infty}^{+\infty} e^{-k^2\pi^2/\tau},$$

by integrating from 0 up to τ and by using an even-value property of the kernel K.

3°. Asymptotics when $\tau \to 0$, or when $\varepsilon_p \to \infty$:

$$K(\tau)=1-\frac{6}{\sqrt{\pi\lambda_0}}\sqrt{\tau}+\ldots, \tag{1.116}$$

The derivative $dK/d\tau$ has a singularity in zero:

$$dK/d\tau \simeq O(\tau^{-1/2})$$

Behavior at $\tau \to \infty$

4°. Asymptotics when $\tau \to \infty$, or $\varepsilon_p \to 0$:

$$K(\tau) \sim \tau_p^S \delta(\tau)+O(\varepsilon_p^2) \tag{1.117}$$

where $\delta(\tau)$ is the delta-function, $\tau_p^S=\lambda_0/15$, the parameter λ_0 is defined in (1.110).

The relaxation time τ_p^S is exactly the same value, as in the model for the Source-Translation flow (1.83).

This relation follows from (1.115), in which the integral term is exponentially small.

Relation (1.117) can be deduced by integration by parts of the expressions $\int_0^T f(\theta)K(\tau-\theta)d\theta$, where $f(\tau)$ is any function integrable along the overall semiaxis.

As a result, the kernel $K(\tau)$ is weakly singular in zero, and is degenerating at large τ. In case that $a^I, b^I=1$ (see (1.9)), the kernel is determined only by one parameter

$$\lambda_0 \equiv \varepsilon_p \xi^2$$

that is exactly equal to the ratio of the perturbation propagation time across a block to that across the overall matrix.

1.9.3. APPROXIMATIONS FOR THE EXCHANGE KERNEL

Along with the above asymptotic relations, it is useful to deduce some approximations for the exchange kernel, which are valid for all the time values within the interval $\tau \in [0, T]$.

Then the following approximations can be suggested.

$$K(\tau) \simeq \frac{6}{\pi^2}\exp(-\pi^2\tau/\lambda_0) \tag{1.118}$$

This function is simply the first mode of series (1.109) with rescaled coefficient, such that it satisfies the initial condition (1.114).

The next relation has a higher degree of approximation, because it is deduced from a condition of mean-square equivalence with the exact function (1.109) within the interval $\tau\in[0,\infty]$.

$$K(\tau)\approx\frac{6}{\pi^2}\exp(-4\pi^2\tau/\lambda_0)+0.258\exp(-18.547\pi^2\tau/\lambda_0) \tag{1.119}$$

1.9.4. RELAXATION TIMES τ^S AND τ_P^S

In the model of translation-source type (1.83) the exchange process is characterized by two parameters: τ_p^S and τ^S. The same parameter, τ_p^S, characterizes the exchange kernel $K(\tau)$ at large τ, (1.117). For a concrete block geometry, explicit expressions may be obtained for them.

General Definition

According to (1.83), the parameters τ_p^S and τ^S are written through Green's function (1.86) of the first boundary-value problem for an operator

$$L_{yy}\equiv\frac{\partial}{\partial y_i}(a_{ij}(y)\frac{\partial}{\partial y_j})$$

on the block, which is equal to the Laplace operator at scalar and piece-constant coefficients.

The equivalent definition for these times is given by (1.87) in terms of the boundary-value cell problems (1.88).

The form of the cell function $\varphi^I(y)$ is exhibited in Fig. 1.20.

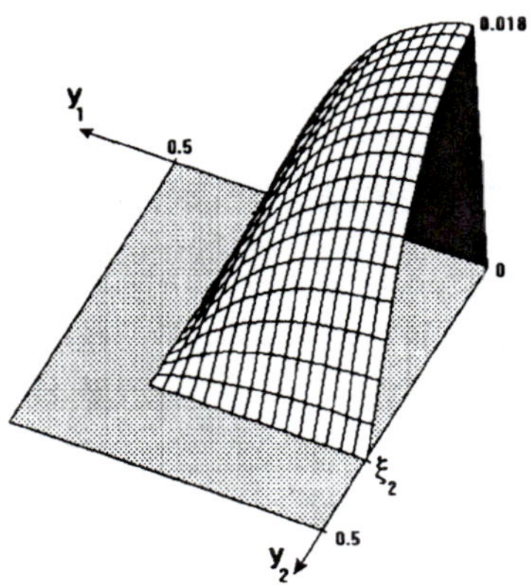

Figure 1.20. Cell function φ^I

The function is shown on a quarter of the cell $\{0 \le y_i \le 1/2,\ i=1,2$. If the block is symmetrical with respect to all the axis, the function $\varphi^I(y)$ is even with respect to each variable y_i.

Analytical Formulas
In case that $a^I, b^I=1$ (see (1.9)), the cell problems (1.87) or the eigenfunctions R can be analytically defined.

Eigenfunctions R and eigenvalues μ of an operator $L \equiv -\frac{\partial^2}{\partial y_i \partial y_i}$ are written down explicitly for some simple areas:

For a sphere $\{0 \le r \le \xi,\ r=\sqrt{y_1{}^2+y_2{}^2+y_3{}^2}\}$:

$$R_k = \sin(r\sqrt{\mu_k})/r, \quad \mu_k = (\pi k/\xi)^2$$

For a parallelepiped $\{0 \le y_i \le h_i,\ i=1,2,3\}$:

$$R_{kmn}(y) = \sin(\frac{\pi k y_1}{h_1})\sin(\frac{\pi k y_2}{h_2})\sin(\frac{\pi k y_3}{h_3}),$$

$$\mu_{kmn} = \pi^2(\frac{k^2}{h_1{}^2}+\frac{m^2}{h_2{}^2}+\frac{n^2}{h_3{}^2}),\ k,m,n \ge 1$$

Then τ_p^S is determined as:

For the block in the form of a sphere:

$$\tau_p^S = \frac{6\lambda_0}{\pi^4}\sum_{k=1}^{\infty}\frac{1}{k^4} = \frac{\lambda_0}{15} \tag{1.120}$$

the parameter λ_0 is defined in (1.110), or

$$\lambda_0 \equiv \varepsilon_p \xi^2, \quad \xi = \left(\frac{3\alpha}{4\pi}\right)^{1/3} \tag{1.121}$$

α is the volume fraction of blocks.

For a parallelepiped:

$$\tau_p^S = \frac{8^3\lambda_0'}{\pi^8}\sum_{k,m,n=0}^{\infty}\frac{\left(\frac{\tilde{k}}{h_1{}^2}+\frac{\tilde{m}}{h_2{}^2}+\frac{\tilde{n}}{h_3{}^2}\right)^{-1}}{h^2\tilde{k}\tilde{m}\tilde{n}}, \tag{1.122}$$

where h_i/h is a ratio of the $i-th$ parallelepiped site to the site of an equivalent cube;

$$\lambda_0' \equiv \varepsilon_p h^2 \tag{1.123}$$

$h=\alpha^{1/3}$, $\tilde{k}=(2k+1)^2$, $\tilde{m}=(2m+1)^2$, $\tilde{n}=(2n+1)^2$.

The delay time of mass-exchange τ^S in equation (1.83b) is determined in terms of τ_p^S, according to (1.89):

$$\tau^S = \frac{\lambda_0 \omega_m}{15} \tag{1.124}$$

Dependence on the Bloc Size

Dependence of the relaxations times on the volume fraction of blocks in case of an isotropic medium is shown in Fig. 1.21.

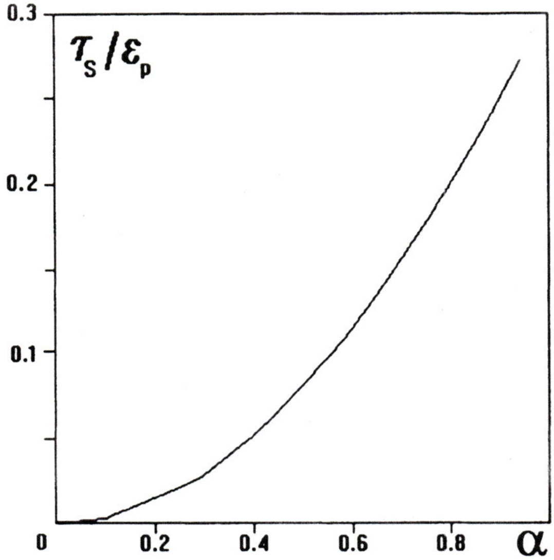

Figure 1.21. Relaxation time for the source flow, τ_p^S, versus volume fraction of blocks

1.10. EFFECTIVE PERMEABILITY OF HIGHLY HETEROGENEOUS MEDIA

1.10.1. NUMERICAL SIMULATION OF EFFECTIVE PERMEABILITY. DECOMPOSITION METHOD

Let us examine the general definition of the macroscale permeability defined as (1.100), (1.101) in terms of the cell functions $\psi_k^{II}(y)$, $\psi_k^{I}(y)$ and $\psi_k^{II,I}(y)$, k=1, ..., n being the solutions of the cell problems (1.103).

The decomposition of the tensor A_{ij} into three parts, (1.100) allows to perform very fast calculations.

Form of Cell Functions

Figs. 1.22 and 1.23 show the cell functions defined by problems (1.103a) - (1.103b) for a square block with α=0.5 or α=0.25. Solution has been obtained using the finite element method.

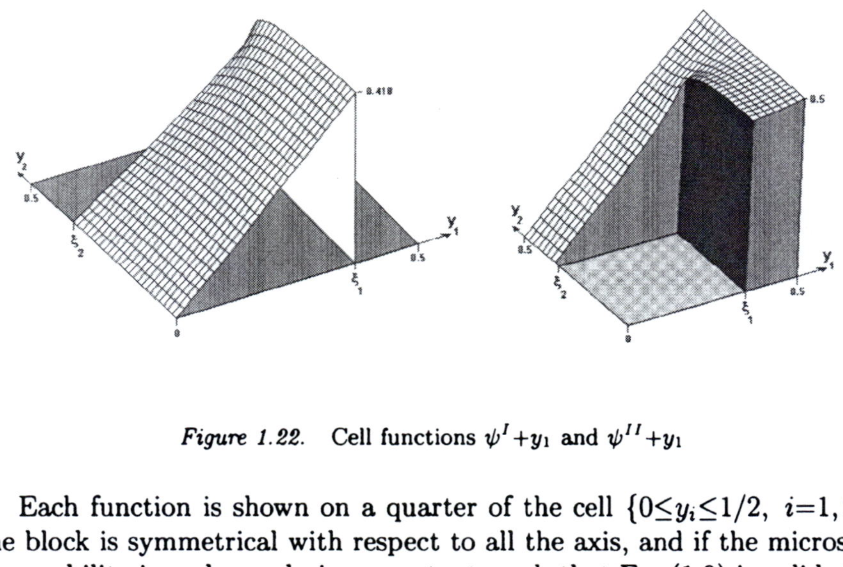

Figure 1.22. Cell functions ψ^I+y_1 and $\psi^{II}+y_1$

Each function is shown on a quarter of the cell $\{0\leq y_i\leq 1/2,\ i=1,2$. If the block is symmetrical with respect to all the axis, and if the microscale permeability is scalar and piece-constant, such that Eq. (1.9) is valid, then the cell functions can be easily reconstructed in other quarters of the cell Y using the following symmetry properties:

- for a fixed k, all three functions $\psi_k^{II}(y)$, $\psi_k^{I}(y)$ and $\psi_k^{II,I}(y)$ are odd with respect to y_k, and are even with respect to other variables.

Piecewise Constant Permeability Field

The system of three cell problems (1.103) can be considerably simplified in a particular case, when two conditions are true:

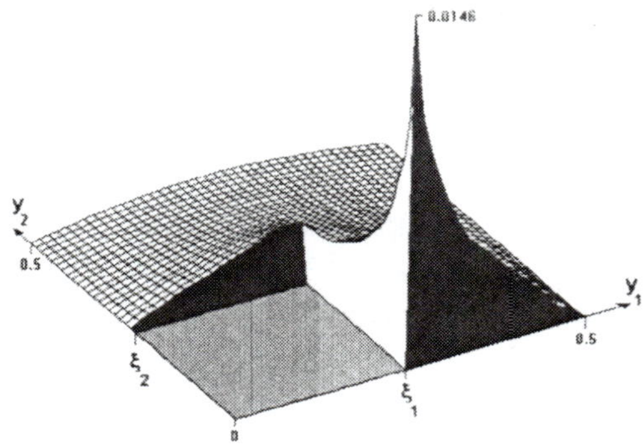

Figure 1.23. Cell function $\psi^{II,I}$

i) the microscale permeabilities are piecewise constant, such that $a_{ij}^{I}=const$ and $a_{ij}^{II}=const$, $i,j=1,...,n$;

ii) all the ratio $a_{ij}^{\omega_K,I}/a_{ij}^{\omega_K,II}=const=\omega_K$, $\forall i,j$; i.e., $a_{ij}^{I}=a_{ij}^{II}\equiv a_{ij}=const$, $\forall i,j$, where relation (1.7) of permeability $a_{ij}^{\omega_K}$ with rescaled permeability a_{ij} has been used.

Then we get from (1.101) using Gauss formulae and relation (1.102):

$$A_{ik}^{I}=\omega_K\left\langle a_{ij}\left(\frac{\partial\psi_k^I}{\partial y_j}+\delta_{jk}\right)\right\rangle_1=$$

$$=\omega_K a_{ij}\int_\Gamma \psi_k^I n_j^I dy+\omega_K a_{ij}\int_{\partial Y^I\cap\partial Y}\psi_k^I n_j^I dy+\left\langle a_{ij}\delta_{jk}\right\rangle_1=$$

$$=-\omega_K a_{ij}\int_\Gamma \psi_k^{II} n_j^{II} dy+\omega_K a_{ij}\int_{\partial Y^I\cap\partial Y}\psi_k^I n_j^I dy+\left\langle a_{ik}\right\rangle_1=$$

$$=\{\text{as the function } \psi(y) \text{ from (1.103c) is periodic}\}=$$

$$=-\omega_K a_{ij}\int_\Gamma \psi_k^{II} n_j^{II} dy-\omega_K a_{ij}\int_{\partial Y}\psi_k n_j dy+\omega_K a_{ij}\int_{\partial Y^I\cap\partial Y}\psi_k^I n_j^I dy+$$

$$+\left\langle a_{ik}\right\rangle_1=\ \{\text{from (1.103c)}\ \}\ =$$

$$=-\omega_K a_{ij}\int_\Gamma \psi_k^{II} n_j^{II} dy-\omega_K a_{ij}\int_{\partial Y^{II}\cap\partial Y}\psi_k^{II} n_j^{II} dy-$$

$$-\omega_K a_{ij}\int_{\partial Y^I\cap\partial Y}\psi_k^I n_j^I dy+\omega_K a_{ij}\int_{\partial Y^I\cap\partial Y}\psi_k^I n_j^I dy+\left\langle a_{ik}\right\rangle_1=$$

$$= -\omega_K a_{ij} \int_{\partial Y^{II}} \psi_k^{II} n_j^{II} dy + \left\langle a_{ij}\delta_{jk} \right\rangle_1 = -\omega_K \left\langle a_{ij} \frac{\partial \psi_k^{II}}{\partial y_j} \right\rangle_2 + \left\langle a_{ik} \right\rangle_1 =$$

$$= -\omega_K \left[A_{ik}^{II} - \left\langle a_{ik} \right\rangle_2 \right] + \left\langle a_{ik} \right\rangle_1$$

where n^i is the extern normal vector to the boundary ∂Y^i; $n^I = -n^{II}$, when $y \in \Gamma$.

Then, the following relation is true:

$$A_{ik}^I = a_{ik} \left[\alpha + \omega_K (1-\alpha)\right] - A_{ik}^{II} \omega_K$$

where α is the volume fraction of blocks.

This relation allows to accelerate the time of simulation for the effective permeability tensor.

Effective Permeability. Decomposition method

Eqs. (1.100) - (1.101) deduced for the effective permeability represent a method of decomposition, which consists in splitting the simulation in a heterogeneous domain into three components each of them is simulated in the homogeneous sub-domains of the cell.

To reach high accuracy, coarse grids are sufficient. Results of simulations were compared to that obtained using classical method to compute the effective permeability of a heterogeneous medium (see, for instance, [17]). Comparison has shown that the deviation of the results obtained by decomposition method constitutes 5% on the grids 70×70×10, with ω_K=0, 1, respectively to the results of [31] representing the solution of the full, non-splitted, problem in the overall cell by finite element method.

Figures 1.24, 1.24 image behavior of the effective permeability A (scalar case) as a function of α for ω_K=0.01, and as a function of ω_K for α=0.5. The dashed line corresponds to the classical method, while the solid line is obtained basing on the method of decomposition (1.100), (1.101). It may be seen that the decomposition method is satisfactory when ω_K<0.1.

1.10.2. ANALYTICAL APPROXIMATION

Method of Non-Smooth Approximation

This method has been proposed in [149], [127]. Let us examine the approximate solutions of the problems (1.103) in the form of the functions dependent on one variable y_1:

$$\psi + y_1 = \alpha_i y_1 + \beta_i, \; y \in i, \quad i = I, II, III \tag{1.125}$$

where $\psi \equiv \psi_1^{II}$ or $\psi \equiv \psi_1^{II,I}$ when $y \in II \cup III$; $\psi \equiv \psi^I$, when $y \in I$.

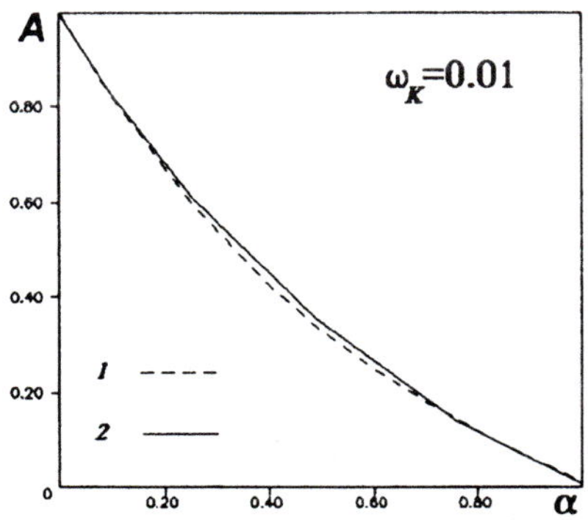

Figure 1.24. Effective permeability vs. volume fraction of blocks: decomposition method (solid line) and exact calculations (dashed line)

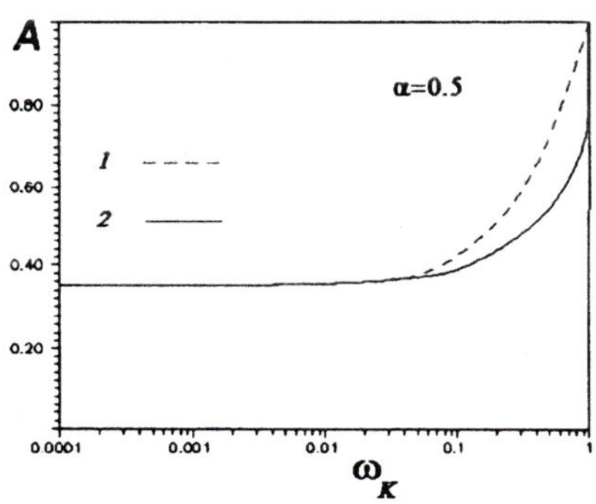

Figure 1.25. Effective permeability vs. heterogeneity degree: decomposition method (solid line) and exact calculations (dashed line)

The solution is assumed to be continuous, but with derivative discontinuity at the line BD (Fig. 1.26).

In this case, all boundary-value conditions in problems (1.103) are satisfied exactly, while the equation $\Delta\psi=0$ is not satisfied only on the line BD. Thus, the approximate solution (1.125) appears to be close to the exact solution only in some integral sense.

It is easy to obtain the explicit expressions for the effective parameters from (1.103) and (1.125). The expression determining the effective permeability in 2D case has the form:

$$A = 1-2\xi+\omega_K \tag{1.126}$$

It is possible to obtain more exact expression, by solving the initial non-decomposed problem in all Y in the similar way. This generalized problem

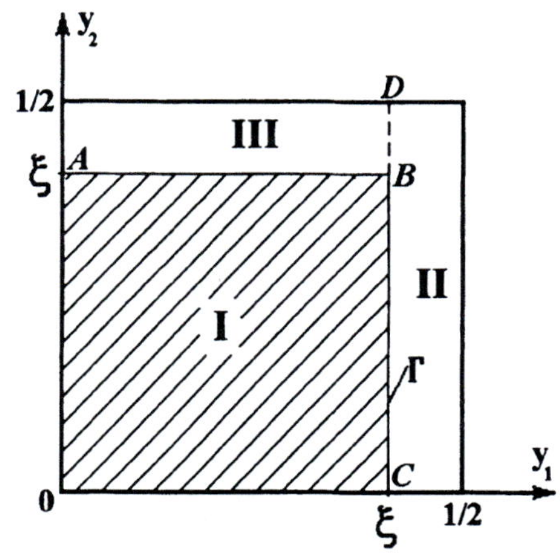

Figure 1.26. Decomposition of the cell into the areas

may be deduced in frameworks of usual homogenization method, excluding asymptotic expansions with respect to ω_K. It is formulated in the following way:

$$\begin{cases} \dfrac{\partial\left(a(y)\frac{\partial\psi}{\partial y_i}\right)}{\partial y_i} = -\dfrac{\partial a}{\partial y_1}, \; y \in Y; \quad \langle\psi\rangle = 0, \\ \psi \text{ is } y\text{-periodic} \end{cases} \tag{1.127}$$

where $a(y)$ is of structure (1.7) with any ω.

Then

$$A = \frac{2\xi(1-2\xi) + \omega_K(1-2\xi+4\xi^2)}{2\xi + \omega(1-2\xi)} \tag{1.128}$$

Relation (1.128) results from (1.127) when $\omega \to 0$.

Comparison with Exact Solutions

Exact data on the effective permeability are available, for example, in paper [31]. Moreover, fine numerical simulations have been performed in [149].

Comparison of the approximate solution (1.128) obtained by the decomposition method with the exact numerical solution is illustrated in Fig. 1.27, 1.28 for two magnitudes of the parameter ω_K.

The difference is small enough, when $\omega_K \leq 0,2$. The highest error is reached for small concentration of blocks. The maximal error of relation (1.128) is within 9−16%, when $\omega = 0,01$.

To compare, a solution obtained in [22] for very thin fractures is shown in Fig. 1.29.

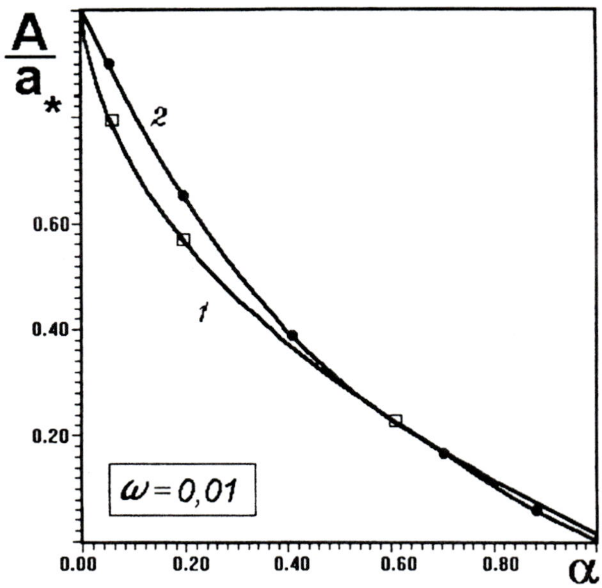

Figure 1.27. Effective permeability of double porosity medium versus block concentration for ω_K=0,01 :

1 - decomposition method (1.128) or (1.126);

2 - finite-difference method

1.10.3. EFFECTIVE PERMEABILITY FOR $\omega_K \leq \varepsilon^2$ (SOURCE MEDIA) AND SMALL BLOCKS

Basic Cell Problem when $\omega_K \leq \varepsilon^2$

In a source-type medium the effective permeability coincides with the effective permeability of the highly permeable matrix A_{ij}^{II} (1.58), and is determined by behavior of n similar functions ψ_k^{II} (n is the space dimension). In isotropic case (1.62), only one of these functions is sufficient to be calculated.

Examine the systems, where each block Y^I is uniformly symmetrical with respect to all the axes. Then it is enough to calculate one function $\psi(y) \equiv \psi_1^{II}(y)$, because the functions ψ_k^{II} have the equivalent structure and are converting one into the other after an appropriate replacement of the indexes "k" [16]. The problem for the function $\psi(y)$ will be called as the basic cell problem:

$$\Delta\psi=0, \ y \in Y^{II}; \quad \frac{\partial}{\partial n}(\psi+y_1)\big|_\Gamma=0, \qquad (1.129)$$

$$\langle\psi\rangle_2=0, \quad \psi - \text{y-periodic}$$

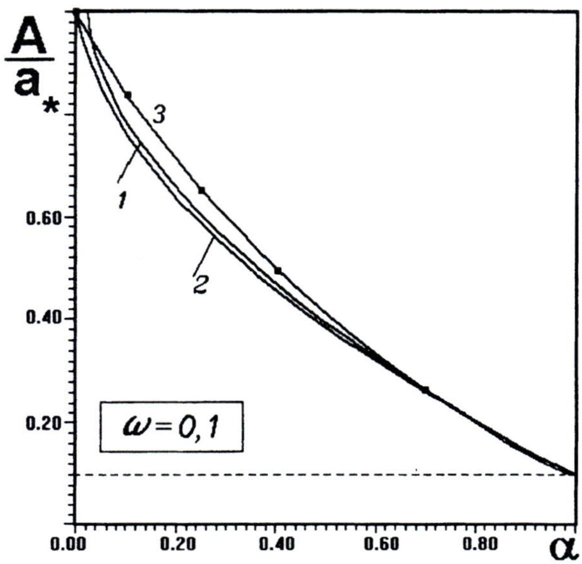

Figure 1.28. Effective permeability of double porosity medium versus block concentration for ω=0, 01 :

1 - decomposition method, version (1.126);

2 - decomposition method, version (1.128);

3 - finite-difference method

with a^{II}=1 according to (1.9). This problem can be reduced to a classical boundary-value problem [16], which is easily solved numerically.

In limiting cases, when the volume of blocks is small ($\alpha \to 0$) and the high permeable zones are thin ($\alpha \to 1$), it is possible to derive the approximate analytical relations.

Cell Problem

When the volumes of blocks are small, their form becomes insignificant, therefore it is convenient to use a spherical approximation for the blocks (Fig. 1.30. The approximate solution is deduced in the following way.

Similar to [16], the function ψ may be shown to be odd respectively to y_1, and even respectively to y_2 and y_3, when the boundary Γ has central symmetry. Then the problem (1.129) is reduced to the boundary-value in the segment Y_*^{II}, which is equal to a 1/8 of Y^{II}:

$$\Delta\psi=0, \quad y \in Y_*^{II}$$

$$\psi\big|_{y_1=1/2}=0, \quad \psi\big|_{y_1=0}=0$$

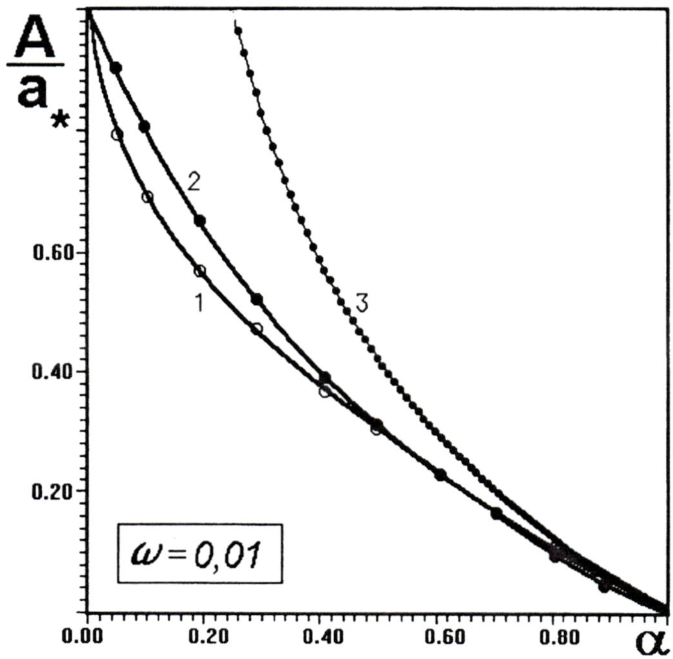

Figure 1.29. Effective permeability of double porosity medium versus block concentration:

1 - decomposition method (1.128);
2 - finite-difference method;
3 - solution of [22]

$$\left.\frac{\partial\psi}{\partial y_i}\right|_{y_i=1/2}=0, \quad \left.\frac{\partial\psi}{\partial y_i}\right|_{y_i=0}=0, \quad i=2,3$$
$$\left.\frac{\partial}{\partial n}(\psi+y_i)\right|_{\Gamma}=0,$$

The shape of the domain Y_*^{II} for spherical blocks is displayed in Fig. 1.30.

The condition $<\psi>_2=0$ is satisfied automatically for an odd function respectively one of arguments.

After introducing the function $\Psi(y)\equiv\psi(y)+y_1$ the basic cell problem

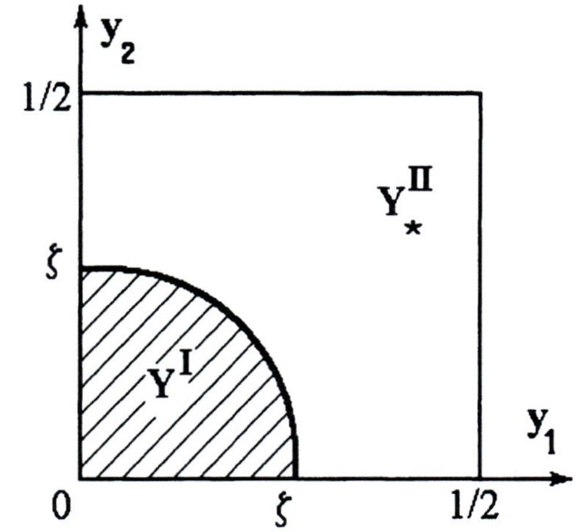

Figure 1.30. Cell structure with small block concentration α

gets the form:

$$\Delta\Psi=0, \quad y\in Y_*^{II}$$

$$\Psi\Big|_{y_1=0}=0, \quad \Psi\Big|_{y_1=1/2}=1/2$$

$$\frac{\partial\Psi}{\partial y_i}\Bigg|_{y_i=1/2}=0, \quad \frac{\partial\Psi}{\partial y_i}\Bigg|_{y_i=0}=0, \quad i=2,3 \tag{1.130}$$

$$\frac{\partial\Psi}{\partial n}\Bigg|_{\Gamma}=0$$

Approximate Solution for the Cell Problem

According to physical meaning, the function $\Psi(y)\equiv \psi(y)+y_1$ is a potential of the velocity field under the flow around a rigid impenetrable body (block), with a surface Γ located inside the cube Y, with impenetrable edges $y_2=\pm1/2$, $y_3=\pm1/2$ and with constant values of the potential at the remaining opposite edges: $\Psi=1/2$ when $Y^I=1/2$, $\Psi=-1/2$ when $y_1=-1/2$.

Let's use the function being a solution of the problem of potential flow around a rigid sphere by an infinite stream of incompressible fluid [75], [137]:

$$\Psi(y)=\frac{C_0y_1}{r^3}+C_1y_1, \quad r\equiv\sqrt{{y_1}^2+{y_2}^2+{y_3}^2}$$

where ξ is a radius of the sphere Γ.

This function satisfies exactly Laplace's equation and necessary symmetry conditions: it is odd respectively to y_1 and is even respectively to y_2, y_3. Moreover, it satisfies by identity the first kind conditions on the sites $y_1=0$, a condition of the second kind on the planes $y_2=0$, $y_3=0$, and a condition of impermeable sphere surface, if $C_1=2C_0/\xi^3$. The conditions of the first kind on the faces $y_1=1/2$, $y_2=1/2$, $y_3=1/2$ are not satisfied.

The remaining constant C_0 is selected in such a way, that a condition of the first kind on the outside faces $y_1=1/2$, should be satisfied in the average:

$$\frac{1}{(1/4)}\int\limits_0^{1/2}\int\limits_0^{1/2}\Psi(1/2,y_2,y_3)dy_2dy_3=1/2 \tag{1.131}$$

that leads to the following result:

$$C_0=\frac{\xi^3}{2+4\xi^3 I}, \qquad I\equiv\int\limits_0^{1/2}\int\limits_0^{1/2}\frac{dy_2dy_3}{(1/4+y_2^2+y_3^2)^{3/2}}$$

The double integral I can be calculated in quadratures. First, it can be reduced to a simple integral by once integration over y_3:

$$\frac{1}{2}\int_0^{1/2}\frac{dy_2}{(y_2^2+1/4)\sqrt{y^2+1/2}}$$

The last integral can be taken explicitly by the change of variables $u^2=(y_2^2+1/4)^2-1/16$:

$$\frac{1}{4}\int_{u(0)}^{u(1/2)}\frac{du}{u^2+\frac{1}{16}}=\operatorname{arctg}(4u)\Big|_{u(0)}^{u(1/2)}=\frac{\pi}{3}$$

Because $4\pi\xi^3/3=\alpha$, where α is the volume fraction of blocks, we obtain:

$$C_0=\frac{\xi^3}{2+\alpha}$$

At last, the solution takes the form:

$$\Psi(y)\simeq\frac{y_1}{2+\alpha}\left[2+\left(\frac{\xi}{r}\right)^3\right] \tag{1.132}$$

Effective Permeability
Let the block has a symmetrical form and be located in the cell center. Then we deduce for the effective permeability:

$$A^{II}=a\langle\partial\Psi/\partial y_1\rangle_2$$

The function Ψ is even respectively to all arguments, then:

$$A^{(2)}=8a_*\left\{\iiint_0^{1/2}\frac{\partial\Psi}{\partial y_1}dy_1dy_2dy_3-\int_0^{\xi}\int_0^{\sqrt{\xi^2-y_2^2}}\int_0^{\sqrt{\xi^2-y_2^2-y_3^2}}\frac{\partial\Psi}{\partial y_1}dy_1dy_2dy_3\right\}$$

Is easy to obtain from (1.130) at $y_1=0$ and (1.131):

$$A^{II}=8\left\{\frac{1}{8}-\int_0^{\xi}\int_0^{\sqrt{\xi^2-y_2^2}}\Psi\left(\sqrt{\xi^2-y_2^2-y_3^2},y_2,y_3\right)dy_2dy_3\right\}=$$

$$a\left\{1-\frac{24}{2+\alpha}\int_0^{\xi}\left(\int_0^{\sqrt{\xi^2-y_2^2}}\sqrt{\xi^2-y_2^2-y_3^2}\,dy_3\right)dy_2\right\}=$$

$$\left[1-\frac{4\pi\xi^3}{2+\alpha}\right]$$

Definitely:

$$A^{II}=\frac{(1-\alpha)}{(1+\alpha/2)}\sim(1-3\alpha/2)\quad\text{when}\quad\alpha\to0 \tag{1.133}$$

Relation (1.133) is more exact for small values of α as shown in Fig. 1.31, where solution (1.133) (curve 1) is compared to the exact solution (dashed line 3) obtained by means of numerical simulations of the cell problem using a fine 3D grid.

1.10.4. EFFECTIVE PERMEABILITY WHEN $\omega_K\leq\varepsilon^2$ AND $\alpha\to1$ (FISSURED MEDIA)

When the blocks are very large, the matrix has a form of thin bands-fractures, as shown in Fig. 1.32.

If the thickness of the bands is of order of one pore diameter, then, strictly speaking, Darcy's equations inside a fracture becomes invalid. The double porosity medium goes over in a *"fractured-porous medium"*. The flow

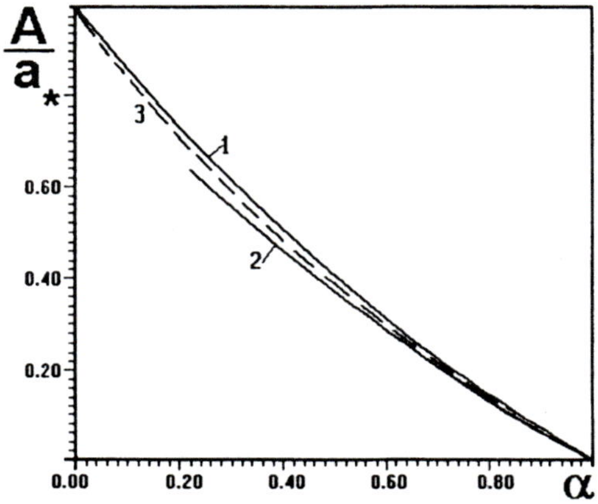

Figure 1.31. Source-type media. Comparison of exact (dashed line 3) and approximate dependences of the dimensionless effective permeability on the volume fraction of blocks α, when $\alpha \to 0$ (curve 1) and $\alpha \to 1$ (curve 2)

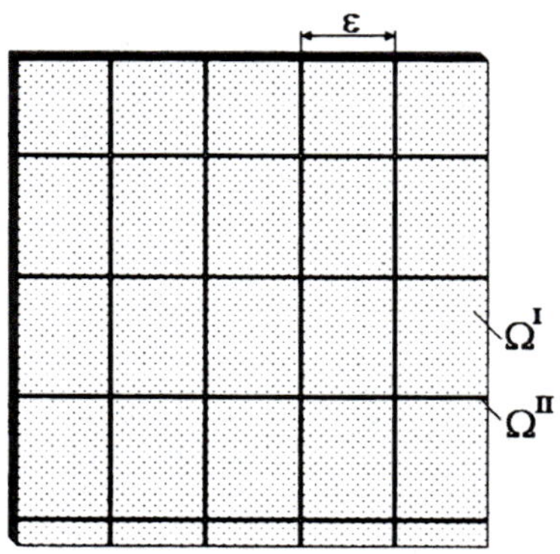

Figure 1.32. Transition of a double porosity medium into a fractured-porous medium when $\alpha \to 1$

inside each thin band should be described by Stokes or Navier-Stokes equations. As a result, a transient boundary layer arises near the block border, when Darcy's flow in porous blocks is matched with Stokes flow inside thin bands. To define the transient flow structures in the boundary layers, the total liquid motion should be described in more details, than using Darcy's equation which is already averaged over a number of pores. Then the liquid motion inside the blocks and inside the fractures should be examined on the same scale, and be described by uniform equations like Stokes's equations. Thus the considered system occurs to be more complex than in case of double porosity media. It includes three characteristic scales: the scale of one pore, the scale of a block, and the macroscale. Homogenization of such a system can be made by two steps.

Homogenization over the one pore scale leads to Darcy's law in blocks and Poiseuille-like equation in fractures, as well as to the *"macroscale contact conditions"*, which allow to conjugate Darcy's and Poiseuille's or Stokes's flows at the interface. In the literature, this procedure is known as a "contact problem", which was widely studied in [64], [65], [95].

The second step of homogenization is made over the scale equivalent to one block. Due to the procedure explained above, resulting equations are not intended to be similar to those derived for double porosity media.

In this book, we consider the case, when the matrix bands "are small, but not very small", in such a manner that the thickness of each band is more large than the size of one pore. Hence, Darcy's law remains valid to describe the flow in the overall medium, therefore the results obtained above are valid for this case too.

Decomposition of the Effective Operator

The magnitude of A^{II} for large value of α may be calculated using a principle of decomposition of the averaged operator [112], which has been introduced for frame constructions. In 2D case, it has the following meaning. If a subdomain Y^{II} consists of narrow bands Y_j^{II}, as shown in Fig. 1.33, then averaging over Y^{II} can be presented in the following asymptotic form, when $\xi \to 1/2$, where $(1-2\xi)$ is a thickness of a band:

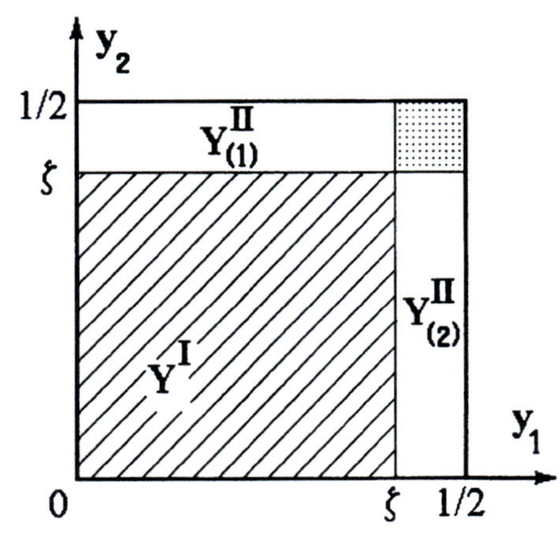

Figure 1.33. Cell structure for a fissured medium

$$\left\langle \frac{\partial \Psi}{\partial y_i} \right\rangle_2 \simeq \int_{Y_i^{II}} \frac{\partial \Psi}{\partial y_i} dy, \quad \forall i$$

In Fig. 1.33, the shaded area can be added to Y_i^{II} not influencing the asymptotics, as the integration over it gives a value of small order $\sim(1-2\xi)^2$. In 3D case, the technique is similar.

Let the bloc Y^I be a cube with a site ξ. The thickness of a "fractures" is equal to $(1-2\xi)$. Let the fractures are thin:

$$\xi \rightarrow 1/2, \qquad \text{or} \qquad \alpha \rightarrow 1$$

If the domain Y^{II} constitutes a sum of thin parallelepipeds Y^{k+}:

$$Y_k^{II+} = \{\xi \le y_k \le 1/2, -1/2 \le y_i \le 1/2, \forall i \neq k\},$$

$$Y_k^{II-} = \{-1/2 \le y_k \le -\xi, -1/2 \le y_i \le 1/2, \forall i \neq k\}$$

then

$$\left\langle \frac{\partial \Psi}{\partial y_j} \right\rangle_2 \sim \sum_{\substack{k=1 \\ k \neq j}}^{3} \left[\int\limits_{Y_k^{II+}} \frac{\partial \Psi}{\partial y_j} dy + \int\limits_{Y_k^{II-}} \frac{\partial \Psi}{\partial y_j} dy \right]$$

The effective permeability tensor is:

$$A^{II} \sim a_* \sum_{k=2}^{3} \left[\int\limits_{Y_k^{II+}} \frac{\partial \Psi}{\partial y_1} dy + \int\limits_{Y_k^{II-}} \frac{\partial \Psi}{\partial y_1} dy \right] \tag{1.134}$$

The 3D shape of integration in (1.134) is displayed in Fig. 1.34. It is the parallelepiped with a "hole".

In an element of such a domain, problem (1.130) has an exact solution: $\Psi(y) = y_1$.

Then: $A^{(2)} \simeq a_*(1-4\xi^2)$. Because $8\xi^3 = \alpha$, then definitely:

$$A^{II} \simeq a_*(1-\alpha^{2/3}), \quad \alpha \rightarrow 1 \tag{1.135}$$

Comparison of (1.135) with the exact solution is presented in Fig. 1.31. The "exact" solution has been obtained numerically, by finite-difference method for a medium with cubic impenetrable blocks.

It is necessary to underline, that in the source-type media, the effective permeability of blocks is always equal to zero. Because of this, on the plots the permeability is shown to be tending to zero when the fraction of matrix tends to zero too.

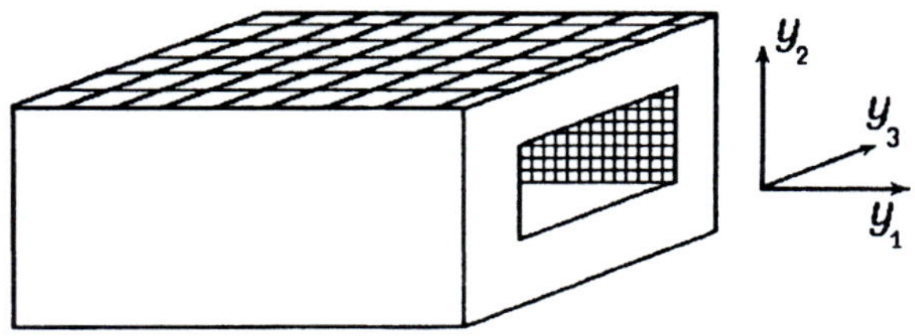

Figure 1.34. Scheme of a 3D cell with large blocks for the Source-type medium

CONCLUSIONS

1. Numerical method of permeability decomposition assures high accuracy and low CPU-time to calculate the effective permeability of highly heterogeneous media, when $\omega_K \leq 0,1$.

2. Approximate formulas (1.126) and (1.128) give practically identical results at $\omega_K \leq 0,1$.

3. Analytical relations (1.126) and (1.128) practically coincide with the exact solution when the blocks are large, i.e., at $\alpha \to 1$.

1.11. DEPLETION OF HIGHLY HETEROGENEOUS RESERVOIRS. SOLUTION OF MODEL PROBLEMS

1.11.1. DEPLETION OF TRANSLATION-SOURCE RESERVOIRS. A GALLERY OF WELLS

Consider a problem of elastic liquid flow between two galleries of wells in frameworks of model (1.83):

$$\begin{aligned}
&\tau^S\frac{\partial^2 P^{II}}{\partial\tau^2}-B\frac{\partial P^{II}}{\partial\tau}+A\frac{\partial^2 P^{II}}{\partial x^2}=0,\quad \tau>0,\ x\in(0,L)\\
&P^{II}(x,0)=P^0,\quad |P^{II}|<M\exp\sigma\tau,\ \forall\tau\geq 0,\quad M>0, \forall\sigma>0 \qquad (1.136)\\
&P^{II}(0,\tau)=P_A,\quad P^{II}(L,\tau)=P_B
\end{aligned}$$

The solution may be constructed by means of Fourier transform. It has the following form:

$$\begin{aligned}
P^{II}(x,\tau)=P_A-\frac{x\Delta P}{L}+2\sum_{k\geq 1}\Big[(P^0-P_A)(1+(-1)^k)+\Delta P\Big]\times\\
\exp\{-\lambda_1\tau(D_k-1)\}\frac{\sin(Hkx/L)}{Hk} \qquad (1.137)\\
\Delta P\equiv P_A-P_B,\quad \lambda_1\equiv\frac{B}{2\tau^S},\quad \lambda_2\equiv\frac{2\tau^S A\pi^2}{B^2L^2},\quad D_k\equiv\sqrt{1+\tau^S k^2}
\end{aligned}$$

An explicit relation can be derived for the block pressure from (1.137) by using formula (1.83a): $P^I\equiv P^{II}-\tau_*\partial P^{II}/\partial\tau$:

$$\begin{aligned}
P^{I}(x,\tau)=P_A-\frac{x\Delta P}{L}+2\sum_{k\geq 1}\Big[(P^0-P_A)(1+(-1)^k)+\Delta P\Big]\times\\
\exp\{-\lambda_1\tau(D_k-1)\}\frac{\sin(Hkx/L)}{Hk}\,[1+\tau_*\lambda_1(D_k-1)] \qquad (1.138)
\end{aligned}$$

When the boundary pressures are constant, the process is stabilizing in time, with the block and matrix pressure difference ($\sim\partial P^{II}/\partial\tau$) tending to zero. Due to the additive flow from blocks, the characteristic stabilization time increases relatively to a homogeneous reservoir, and is of order

$$\tau_{st}\sim\left(\lambda_1\sqrt{1+2\lambda_2}\right)^{-1} \qquad (1.139)$$

1.11.2. DEPLETION OF TRANSLATION-SOURCE RESERVOIRS.

RADIAL FLOW TOWARDS A WELL

Model Problem of Reservoir Depletion by a Single Well

Let a cylindrical reservoir of radius R_* with impenetrable boundary be developed by a central well of zero radius and fixed flow rate q:

$$\tau^S \frac{\partial^2 P^{II}}{\partial \tau^2} - B\frac{\partial P^{II}}{\partial \tau} + \frac{A}{r}\frac{\partial}{\partial r}\left(r\frac{\partial P^{II}}{\partial r}\right) = 0, \ \tau>0, \ r \in (0, R_*)$$

$$P^{II}(r,0) = P^0, \quad |P^{II}| < M \exp C\tau, \ \forall \tau \geq 0 \tag{1.140}$$

$$\frac{\partial P^{II}}{\partial r}\Big|_{r=R_*} = 0, \quad r\frac{\partial P^{II}}{\partial r}\Big|_{r=0} = \frac{q}{2}$$

The parameter τ^S is determined according to (1.89) or (1.124) and characterizes the intensity of exchange flow.

Solution for a Constant Flow Rate

The solution may be also constructed by expansions over eigen-functions of the operator $-(1/r)\partial(r\partial/\partial r)/\partial r$ and takes the form:

$$P^{II}(r,\tau) = P^0 - \frac{q}{4}\left(r^2 - 2\ln r - \frac{3}{2}\right) - \frac{q\tau}{B} + q\sum_{k\geq 1} \frac{J_0(\mu_k r)}{\mu_k^2 J_0^2(\mu_k)} \exp\{-\lambda_1 \tau (D_k^* - 1)\} \tag{1.141}$$

where J_i is a Bessel function of the first kind and i-th index, μ_k, $(k=1,2,...)$ are the zeros of a function $J_1(\mu)$, $D_k^* \equiv \sqrt{1+4\tau^S A\mu^2/B}$, $\lambda_1 = B/2\tau^S$.

Note that

$$\frac{1}{4}(r^2 - 2\ln r) = \frac{3}{8} + \sum_{k\geq 1} \frac{J_0(\mu_k r)}{\mu_k^2 J_0^2(\mu_k)},$$

therefore $P^{II} = P^0$ when $\tau = 0$ in (1.141).

The block pressure is determined as

$$P^I(r,\tau) = P^0 - \frac{q}{4}\left(r^2 - 2\ln r - \frac{3}{2}\right) - \frac{q(\tau - \tau_*)}{B} + q\sum_{k\geq 1} \frac{J_0(\mu_k r)}{\mu_k^2 J_0^2(\mu_k)} \left[\tau_* \lambda_1 (D_k^* - 1) + 1\right] \exp\{-\lambda_1 \tau (D_k^* - 1)\} \tag{1.142}$$

that follows from (1.83a) and (1.141).

Solution when the Flow Rate is Variable
If the flow rate is not constant: $q{=}q(\tau)$, then the solution of problem (1.140) is deduced from (1.141) with using the Duhamel principle [67]:

$$P^{II}(r,\tau){=}P^0-\int\limits_0^\tau \frac{\partial q(\theta)}{\partial\theta}\Big[\frac{1}{4}(r^2-2\ln r-3/2)+\frac{\tau-\theta}{B}-$$
$$\sum_{k\geq 1}\frac{J_0(\mu_k r)}{\mu_k^2 J_0^2(\mu_k)}\exp\{-\lambda_1(D_k^*-1)(\tau-\theta)\}\Big]d\theta \tag{1.143}$$

Behavior of the Difference between Block and Matrix Pressure
It follows for $P^I-P^{II}{=}-\tau_*\partial P^{II}/\partial\tau$ from (1.141) and (1.142), when q is constant :

$$P^I-P^{II}{=}\frac{q\tau_*}{B}+q\lambda_1\tau_*\sum_{k\geq 1}\frac{J_0(\mu_k r)(D_k^*-1)}{\mu_k^2 J_0^2(\mu_k)}e^{-\lambda_1\tau(D_k^*-1)} \tag{1.144}$$

The difference P^I-P^{II} appears to be exponentially decreasing in time to a non-zero limit when the flow rate is constant:

$$\lim_{\tau\to\infty}(P^I-P^{II}) = \frac{q\tau_*}{B} = \text{const} \tag{1.145}$$

Thus, the full exhaustion of blocks is impossible when the flow rate is constant. Therefore the depletion process may be controlled by the flow rate.

Regulating Oil Recovery by Depletion Rate
Let $q{=}q(\tau)$ with $q(0){=}0$. Then P^I-P^{II} is determined from (1.144):

$$P^I-P^{II}{=}\frac{q(\tau)\tau_*}{B}+\tau_*\int\limits_0^\tau\frac{\partial q(\theta)}{\partial\theta}\sum_{k\geq 1}\Phi_k(r)e^{-C_k(\tau-\theta)}d\theta, \tag{1.146}$$
$$\Phi_k(r){\equiv}C_k J_0(\mu_k r)\left[\mu_k^2 J_0^2(\mu_k)\right]^{-1},\quad C_k{\equiv}\lambda_1(D_k^*-1)$$

The value of the pressure difference P^I-P^{II} is a criterion of the oil recovery degree. Oil recovery is higher when this difference is small.

Let the well productivity be exponentially increasing (q_+), or exponentially decreasing (q_-):

$$q_+(\tau){=}q_0(1-e^{-\gamma\tau}),\qquad q_-(\tau){=}q_0\ e^{-\gamma\tau}\ 1(\tau)$$

where $1(\tau)$ is Hewiside's function. Then

$$(P^I - P^{II})_+ = \frac{q_0\tau_*}{B}\left[1 + B\gamma\sum_{k\geq 1}\frac{\Phi_k}{C_k-\gamma}(e^{-\gamma\tau} - e^{-C_k\tau}) - e^{-\gamma\tau}\right], \quad (1.147)$$

$$(P^I - P^{II})_- = \frac{q_0\tau_*}{B}\left[e^{-\gamma\tau}\left(1 - B\gamma\sum_{k\geq 1}\frac{\Phi_k}{C_k-\gamma}\right) + \right.$$

$$\left. + B\sum_{k\geq 1}\frac{\Phi_1 C_k}{C_k-\gamma}e^{-C_k\tau}\right] \quad (1.148)$$

It follows, that the pressure difference $P^I - P^{II}$ increases in time up to a fixed limit $q_0\tau_*/B$ when the depletion rate (q_+) grows, and decreases until zero when the rate (q_-) diminishes, independently of the ratio between the relaxation intensity of flow rate (γ) and pressure (C_1).

Type of the real dependence of $(P^I - P^{II})$ upon τ is shown in Fig. 1.35. The dashed curve corresponds to (1.148) and is fictitious for small time values, because the examined model is valid for the time $\tau > \varepsilon_p$. Due to the same raison, this model does not allow to calculate the extremum point $\tau_{\max}$.

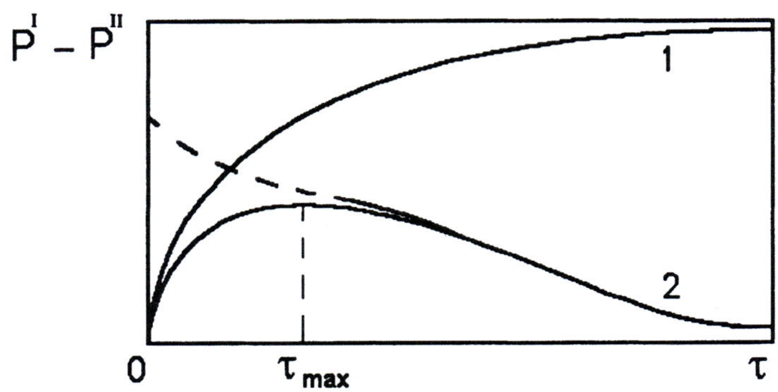

Figure 1.35. Variations of the difference between block and matrix pressures in Source-type media, with increasing (1) or decreasing (2) depletion rate

Thus, displacing the location of the extremum point by varying the depletion rate, one can reach various degrees of the oil extraction over the finite time. It is clear the depletion rate should not exceed the rate of exchange between blocks and matrix. Then, after both the pressures become equal, strict upper limitations are imposed on the depletion rate.

Otherwise, the blocks begin to be delaying, and the degree of oil recovery decreases.

At the first stage, when the pressure is equalized throughout all the volume with its own rate, a high depletion rate may be supported.

1.11.3. DEPLETION OF SOURCE-TYPE RESERVOIRS

Problem of Depletion by a Gallery of Wells

Consider the model (1.58), (1.59), which may be written in 1D case in the following form for P^{II}:

$$B^{II}\frac{\partial P^{II}}{\partial\tau}-A^{II}\frac{\partial^2 P^{II}}{\partial x^2}+B^I\int\limits_0^\tau\frac{\partial P^{II}}{\partial\theta}K_*(\tau-\theta)d\theta=0 \tag{1.149}$$

where $K_*(\tau)\equiv-dK/d\tau$, $K(\tau)$ is the exchange kernel studied in section 1.9.

Examine the depletion of a reservoir of the length L with an impenetrable boundary $x=0$, by a gallery of wells with the flow rate q:

$$P^{II}(x,0)=P^0,\quad \frac{\partial P^{II}}{\partial x}|_{x=0}=0,\quad \frac{\partial P^{II}}{\partial x}|_{x=L}=-q. \tag{1.150}$$

Solution Using Laplace Transform. Constant Flow Rate

Problem (1.149) and (1.150) with an arbitrary kernel K_* does not allow the analytical solution. Applying the Laplace's transform

$$\tilde{f}(s,x)\equiv\int_0^\infty f(x,\tau)e^{-s\tau}d\tau,$$

we get:

$$B^{II}s\tilde{U}-A^{II}\partial^2\tilde{U}/\partial x^2+B^I s\tilde{U}\tilde{K}_*(s)=0,$$
$$\frac{\partial\tilde{U}}{\partial x}|_{x=0}=0,\quad \frac{\partial\tilde{U}}{\partial x}|_{x=L}=-q/s,$$

where $\tilde{U}\equiv\tilde{P}^{II}-P^0$.

The image of solution takes the form:

$$\tilde{U}(s,x)=\frac{-q(e^{\omega x}+e^{-\omega x})}{s\omega(e^{\omega L}+e^{-\omega L})}, \tag{1.151}$$

where $\omega(s)\equiv\sqrt{[B^{II}+B^I\tilde{K}_*(s)]/A^{II}}$.

The image $\tilde{K}_*(s)$ of the derivative for the kernel $K(\tau)$ in the form (1.109) is as follows:

$$\tilde{K}_*(s)=\frac{6}{\lambda_0}\sum_{k\geq1}(s+\pi^2k^2/\lambda_0)^{-1}. \tag{1.152}$$

the parameter λ_0 is defined in (1.110).

If we use the asymptotic relations for $K(\tau)$ when $\tau \to 0$ and $\tau \to \infty$, then we obtain:

$$\tilde{K}_*(s) \sim 3/\sqrt{\lambda_0 s} \quad \text{when } s \to \infty \text{ (or } \tau \to 0) \tag{1.153}$$

and from (1.152):

$$\tilde{K}_*(s) \sim 1 - \lambda_0 s/15, \quad \text{when } s \to 0 \text{ (or } \tau \to \infty) \tag{1.154}$$

Asymptotics for the image (1.151) when $s \to \infty$:

$$\tilde{U}(s,x) = -\frac{q \exp\left(\alpha_2(x-L)\right)}{\alpha_1 s^{3/2}} \times \\ \exp\left[-\alpha_1 (L-x)\sqrt{s} + O\left(s^{-5/2}\exp\left(-\sqrt{s}\right)\right)\right],$$

where $\alpha_1 \equiv \sqrt{B^{II}/A^{II}}$, $\alpha_2 \equiv \frac{3}{2} B^I / \sqrt{\lambda_0 B^{II} A^{II}}$

Then the asymptotics of the solution for problem (1.149), (1.150) when $\tau \to 0$ has a form:

$$P^{II}(x,\tau) \sim P^0 - \frac{q \exp\left(-\alpha_2(L-x)\right)}{\alpha_1} \left\{ 2\sqrt{\frac{\tau}{\pi}} \exp\left\{-\frac{\alpha_1^2 (L-x)^2}{4\tau}\right\} - \\ \alpha_1 (L-x) \operatorname{erfc}\left\{\frac{\alpha_1 (L-x)}{2\sqrt{\tau}}\right\}\right\}, \tag{1.155}$$

where

$$\operatorname{erfc}(z) \equiv 1 - \operatorname{erf}(z) = 1 - \frac{2}{\sqrt{\pi}} \int_0^z e^{-t^2} dt$$

The asymptotics when $\tau \to \infty$ is developed without Laplace's transform, as the kernel $K(\tau)$ is now degenerating into a δ-function, and equation (1.149) is reduced to a model of Translation-Source flow with the effective permeability equal to the effective matrix permeability only. The solution of (1.150) for it is constructed by Fourier method, in the way explained in section 1.11.1.

Pressure in Blocks

Relationship for the block pressure may be derived from (1.58a):

$$P^I = P^{II} - \int_0^\tau \frac{\partial P^{II}}{\partial \theta} K(\tau - \theta) d\theta$$

For small τ, relation (1.155) yields first:

$$\frac{\partial P^{II}}{\partial \tau} = \frac{q}{\alpha_1 \sqrt{\pi \tau}} \exp\left[\frac{-\alpha_1^2 (L-x)^2}{4\tau} - \alpha_2 (L-x)\right]$$

because

$$\frac{\mathrm{derfc}(z)}{dz} = -\frac{2}{\sqrt{\pi}}e^{-z^2}$$

Then

$$P^I \sim P^0 + \frac{6q\exp[-\alpha_2(L-x)]}{\pi\alpha_1\sqrt{\lambda_0}}\int_0^\tau \exp\left\{\frac{-\alpha_1^2(L-x)^2}{4\theta}\right\}\sqrt{\frac{\tau-\theta}{\theta}}d\theta, \quad \tau \ll 1 \tag{1.156}$$

Asymptotic relation (1.116) for $K(\tau)$ at small τ has been used to deduce the last equation.

Furthermore the simplifications may be made, however the form (1.156) is sufficient to calculate the block pressure.

Solution for Variable Flow Rate

If the depletion rate is variable in time: $q{=}q(\tau)$, the solution results from (1.155) by a convolution, similarly to that in section 1.11.1:

$$P^{II}(x,\tau) \sim P^0 - \int_0^\tau \frac{\partial q}{\partial\theta}\frac{\exp\left(-\alpha_2(L-x)\right)}{\alpha_1}\left\{2\sqrt{\frac{\tau-\theta}{\pi}}\exp\left\{-\frac{\alpha_1^2(L-x)^2}{4(\tau-\theta)}\right\} - \alpha_1(L-x)\mathrm{erfc}\left\{\frac{\alpha_1(L-x)}{2\sqrt{\tau-\theta}}\right\}\right\}d\theta, \quad \tau \ll 1 \tag{1.157}$$

1.11.4. MEMORY EFFECTS

The memory effects described by relaxation-like term or integral operator differ quantitatively. Qualitative difference becomes obvious for the processes with inversion of the course direction.

Consider two systems. The first one, (1.83b), describes a process with a short memory, that corresponds to Translation-Source media. Remind, this model is close to Barenblatt-Zheltov-Kochina model, as shown in section 1.8.4, where the exchange term was proportional to the difference between the block and the matrix pressures, (1.106). The second formula, 1.59), describes a system with long-term memory, which corresponds to the Source-type media. The exchange flow can not be described by a pressure difference.

Examine the process with inversion of the coarse direction. Let the sign of depletion rate $q(\tau)$ is changing in time. Let the fluid be extracted from the reservoir up to the moment τ_1 with a constant flow rate q. Injection of the same fluid in the reservoir with a flow rate $-q$ then follows. Behavior of the matrix pressure P^{II} is shown in Fig. 1.36. The inversion of flowing is marked by a point of the minimum for P^{II}.

The analysis may be made basing on the formulas like (1.141) for a short memory system, and (1.157) for long memory case.

Let the minimum for block pressure P^I be designated as (τ_2). In this point the exchange flow changes its direction. At last, let the third point τ_3 be introduced, the block and the fracture pressures become equivalent in.

If the exchange was caused by the block and matrix pressure difference, as in the Barenblatt-Zheltov-Kochina model [19], then the point of the exchange flow inversion (τ_2) would coincide with the point where both the pressures become equivalent (τ_3), as shown in Fig. 1.36.

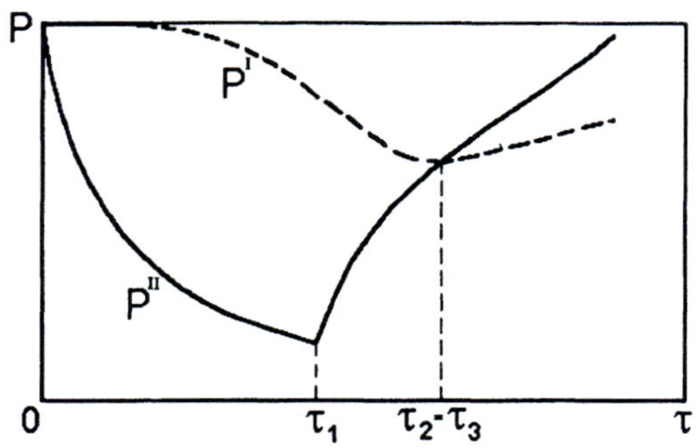

Figure 1.36. Behavior of pressure in blocks P^I and in matrix P^{II} in a system with short memory.
τ_1 is the point of flow rate inversion;
τ_2 is the point of exchange flow inversion;
τ_3 is the point where both the pressures become equivalent

In case of a long-memory system, these two points, τ_2 and τ_3, do not coincide, as shown in Fig. 1.37.

Inversion of exchange flow happens before two pressures become equivalent. Hence, in the interval (τ_2, τ_3) the exchange flow is directed from the matrix into the block, although the matrix pressure is lower than the pressure in blocks. This effect is not allowed in the model [19].

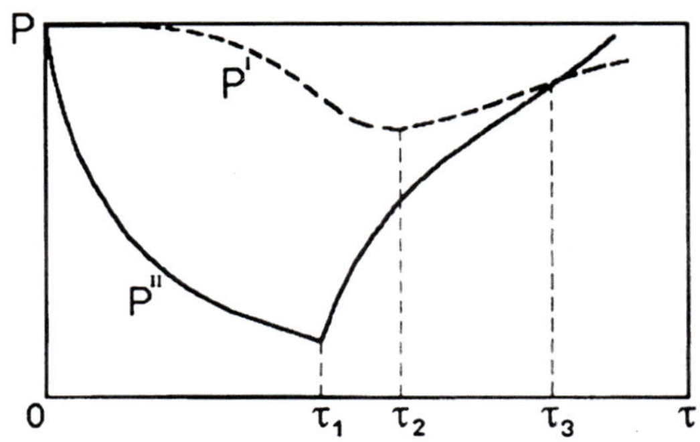

Figure 1.37. Behavior of pressure in blocks P^I and in matrix P^{II} for a long-memory system

1.12. HOMOGENIZATION OF TWO-SCALE DISORDERED MEDIA

Let us consider a fast oscillating non-periodic function $a(x)$, $x \in R^n$, $n=1,2,3$.

To describe an oscillating function the amplitude-spectral characteristic may be used, when the spectra is that of Fourier expansion. When the medium is not periodic, multiple characteristic scales determine them. Therefore, the amplitude-spectral characteristic may be rather wide. In this most general situation it is not very clear, what is the sense of averaging and whether averaging can lead to some universal results.

The problem of universality becomes basic, when we begin to work with disordered media. In fact, the success of the averaging theory for periodic media consists in that the results are independent of the used method, of the heterogeneity scale, e.t.c. Such a universality becomes not obvious in disordered media, where even the heterogeneity scale is not strictly defined.

1.12.1. TWO TYPES OF DISORDERED MEDIA

In the scientific literature two approaches are in development.

Multiscale Media

The first deals with so-called *multiscale media* [7] characterized by many scales ε_1, ε_2, ..., $\varepsilon ... N$. The spectrum ε_i of the function $a(x)$ is assumed to be very wide, with any domination of some frequencies. Moreover, the spectrum of heterogeneity is not separated from the macroscale, such that the macroscale enters in the heterogeneity spectrum. In other words, the value ε_N corresponds to the macroscale. In media of such a type, averaging (upscaling) is performed step-by-step, from the smaller scale ε_1 towards the maximal scale of the spectrum ε_N. As N is assumed to be large, averaging consists in performing an infinite set of scale-up procedures when $N \rightarrow \infty$.

It is clear that the effective results can be obtained when the limit of such a procedure is determined. If the limit does not exist, or exists but can not be defined, averaging has not any sense. In general case, any theory is not developed now in this field of knowledge. In particular cases, some constructive results can be, however, obtained.

A sufficient condition to the limit of upscaling would exist is a requirement of scale invariance of the equation type with respect to the change of the scale. In this case the type of the averaged equation is known, as it coincides with the original microscale equations. However, scale invariance of an equation depends on the heterogeneity structure. For example, diffusion equation is scale invariable in periodic media, but in disordered media this property may be lost. What are the classes of media, where the examined equation remains scale invariable, is an open problem. Usually scale invariance is postulated as an approximate property of the system.

Other problem consists in the following. Even the type of the averaged model is known, it does not mean that the effective parameters of this model are constructed. To develop some constructive procedure to define the effective parameters as the limit of an infinite set of scale-up transitions, some very strong requirement would be assigned to the heterogeneity field. A sufficient requirement is the self-similarity of the function $a(x)$. (Note, that self-similarity of the heterogeneity field does not lead to scale invariance of the process equations). The self similar fields were studied in theoretical physics, as a model of phase transitions near critical point, or in mathematics as a model of fractal sets. Averaging of self-similar fields can be performed basing on the renormalization method developed in theoretical physics.

It is clear, that this approach seems to be very fruitful, but it deals with some special case of media and is based on multiple hypothesis, many of them are not well proved. From this point of view this approach should be considered as approximate, however some brilliant results illustrate this method as very perspective.

Two-Scale Media

The second approach is based on a hypothesis that the spectrum of heterogeneity scales is separated from the macroscale of the system. The amplitude-spectral characteristic has the form shown in Fig. 1.38, where the macroscale is of order 1.

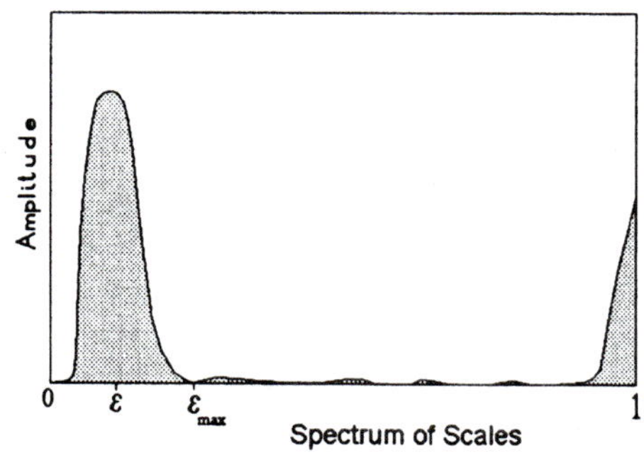

Figure 1.38. Amplitude-spectral characteristic of the oscillating two-scale function

In other words, all frequencies of the spectrum of the function $a(x)$ are much smaller than the macroscale. Let ε_{max} be the maximal scale of the spectrum, then:

$$\varepsilon_{max} \ll 1$$

Then within the scale 1, the width of the spectrum is small, and we can consider the spectrum as concentrated round some mean value ε which is of order ε_{max} of smaller. In any case, the relation $\varepsilon \ll 1$ remains true.

Existence of these two scales, ε and 1, is the basic property of those media.

This approach is more corespondent to describe natural porous media, where the spectrum ε_i is formed by the pore sizes. Despite of the sizes of various pores may be considerably distinct, the macroscale (in equations of mechanics) is determined in such a way that the corresponding volume contains a lot of pores. Thus, the macroscale is chosen as much greater than the maximal pore size ε_{max}. To underline existence of the small scale ε, one write the function of heterogeneity as $a(x, x/\varepsilon)$.

In two-scale media, averaging can be performed basing on homogenization technique. Due to an obvious analogy with periodic case (the same, two-scale, character of heterogeneity, the same homogenization method, e.t.c.), it is natural to expect the homogenization of disordered media will leads to some close results as in periodic case. If this is true, the universality of homogenization results will be considerably enlarged, and the periodic media can be examined as a simplified model of disordered media.

In this book the two-scale approach is examined.

The two-scale approach was suggested first in papers of [78], [79] and is now in development in papers [32] for highly heterogeneous media. In the mentioned papers the heterogeneous field is assumed to be a realization of a random field. In this book other ideas are used, without any randomization.

1.12.2. DECOMPOSITION OF A TWO-SCALE FIELD

Examine the two-scale non-periodic function (or the field) $a(x, x/\varepsilon)$ in a domain $\Omega \subset R^n$, $n=1,2,3$, determined as described above. I.e., the Fourier spectrum of the function a is assumed to be close to that shown in Fig. 1.38, with the characteristic scale $\varepsilon \ll 1$. For the present, the scale ε is not defined strictly, but it will be defined later as the result of some restrictions to the heterogeneity field.

Any function $a(x, x/\varepsilon)$ can be presented in the following form:

$$a(x, x/\varepsilon) = a^{(0)}(x) + a'(x, x/\varepsilon) \tag{1.158}$$

where

- $a^{(0)}(x)$ is the non-oscillating component of the function, which depends on the slow variable x only;

- $a'(x, x/\varepsilon)$ is the oscillating component dependent on the fast variable x/ε too.

The term "non-oscillating" refers to the scale ε, i.e., it means that the scale of probable oscillations is much larger than ε.

Decomposition Over Realizations

Let us cover the overall field a by a periodic grid with the characteristic scale of a cell equal to ε. The domain Ω will contain N cells $\{\Omega^{(k)}\}_{k=1}^{N}$.

Then the function a can be presented in the following form:

$$a(x, x/\varepsilon) = a^{(0)}(x) + \sum_{k=1}^{N} a^{(k)}(x, x/\varepsilon)\delta^{(k)}(x) \tag{1.159a}$$

or in (1.158):

$$a'(x, x/\varepsilon) = \sum_{k=1}^{N} a^{(k)}(x, x/\varepsilon)\delta^{(k)}(x) \tag{1.159b}$$

The function $a^{(0)}$ is always the non-oscillating component of the function a. Each function $a^{(k)}(x, x/\varepsilon) \in \Omega$ is periodic, but discontinuous in general case. It is constructed as a periodic prolongation to the overall domain Ω of the segment of the function $a - a^{(0)}$ given in the cell $\Omega^{(k)}$.

The function $\delta^{(k)}(x) = \delta(\Omega^{(k)})$ is the indicator function of the cell $(\Omega^{(k)})$:

$$\delta^{(k)}(x) = \begin{cases} 1, & x \in \Omega^{(k)} \\ 0, & x \notin \Omega^{(k)} \end{cases} \tag{1.160}$$

Not the following property is true:

$$\sum_{k=1}^{N} \delta^{(k)}(x) \equiv 1 \tag{1.161}$$

The set of periodic discontinuous (in general case) functions $\{a^{(k)}\}_{k=1}^{N}$ is called as the *ensemble of realizations* of the oscillating component a'. Decomposition (1.159) is called as *decomposition over realizations.*

The realizations satisfy some requirements. Let for any fixed cell $\Omega^{(k)}$ the numbers of the nearest neighboring cells be designated as k'. Let the boundary between two neighboring cells $\Omega^{(k)}$ and $\Omega^{(k')}$ be designated as

$S^{(k'k)}$. When the function a is continuous and differentiable, then the following contact conditions are true:

$$\begin{cases} a^{(k)}|_{S_{k'k}}=a^{(k')}|_{S_{k'k}}, \\ \frac{\partial a^{(k)}}{\partial x_i}\Big|_{S_{k'k}}=\frac{\partial a^{(k')}}{\partial x_i}\Big|_{S_{k'k}}, \quad \forall i \\ \text{for all nearest neighbors} \end{cases} \tag{1.162}$$

Introduction of a Single Cell

As each realization $a^{(k)}$ is periodic, let us introduce an independent cell Ω_ε where the local coordinate system $\tilde{x}=\{\tilde{x}_1,...\tilde{x}_n\}$ be determined in such a way that its origin is located in the geometrical center of the cell. The map $x\rightarrow\tilde{x}$ is defined as:

$$\tilde{x}=x-x_c^k, \quad \text{when } x\in\Omega^{(k)} \tag{1.163}$$

where x_c^k is the origin of the cell $\Omega^{(k)}$.

Then the decomposition (1.159) gets the form:

$$a(x,x/\varepsilon)=a^{(0)}(x)+\sum_{k=1}^{N}a^{(k)}(\tilde{x},\tilde{x}/\varepsilon)\delta^{(k)}(x)\equiv\tilde{a}(x,\tilde{x},\tilde{x}/\varepsilon) \tag{1.164a}$$

or in (1.158):

$$a'(x,x/\varepsilon)=\sum_{k=1}^{N}a^{(k)}(\tilde{x},\tilde{x}/\varepsilon)\delta^{(k)}(x)\equiv\tilde{a}'(x,\tilde{x},\tilde{x}/\varepsilon) \tag{1.164b}$$

1.12.3. AVERAGING OVER REALIZATIONS, OVER SPACE, AND DOUBLE AVERAGING

Let us introduce three operations of averaging for the oscillating field $a(x,x/\varepsilon)$:

Space Averaging over Cells

Space averaging over cells is equivalent to averaging over each cell. It is defined as:

$$\langle\cdot\rangle\equiv\frac{1}{|\Omega_\varepsilon|}\int_{\Omega_\varepsilon}(\cdot)d\tilde{x} \tag{1.165}$$

where $|\Omega_\varepsilon|$ is the volume of the cell.

For the function (1.164) we get:

$$\langle a\rangle=\langle\tilde{a}\rangle=a^{(0)}(x)+\sum_{k=1}^{N}\langle a^{(k)}\rangle\delta^{(k)}(x) \tag{1.166}$$

This, averaging over the cells leads to an oscillating piecewise constant function.

In Fig. 1.39 the original oscillating function $a(x,/x\varepsilon)$ and the function averaged over space correspond to the curves 1 and 2.

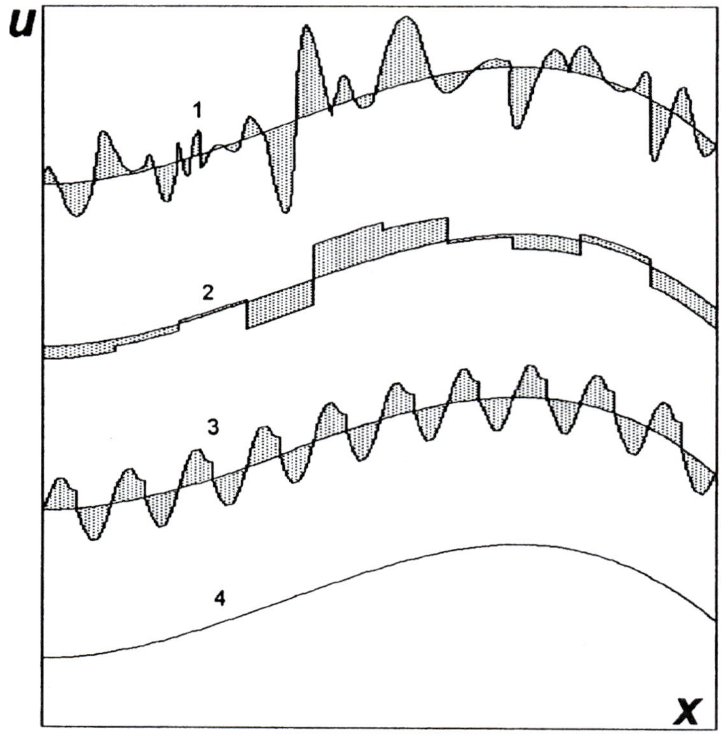

Figure 1.39. Disordered oscillating functions (1); the same function averaged over space (2); the same function averaged over realizations (3); the same function after double averaging (4)

Averaging over Realizations

Averaging over realizations for a function like (1.159) is defined as:

$$\overline{a}(x,x/\varepsilon)=a^{(0)}(x)+\frac{1}{N}\sum_{k=1}^{N}a^{(k)}(x,x/\varepsilon) \tag{1.167}$$

This operation replaces each individual component $a^{(k)}$ in the sum (1.159) by a single arithmetic mean function. In fact:

$$\bar{a}(x,y)-a^{(0)}(x)=\frac{1}{N}\sum_{k=1}^{N}a^{(k)}=$$
$$=\frac{1}{N}\sum_{k=1}^{N}a^{(k)}\sum_{m=1}^{N}\delta^{(m)}(x)=\sum_{m=1}^{N}\left(\frac{1}{N}\sum_{k=1}^{N}a^{(k)}\right)\delta^{(m)}(x)$$

Herein the property (1.161) has been used.

It follows from (1.158) that averaging over realization transforms the oscillating part a' of the function a into a periodic function:

$$\overline{a'}=\frac{1}{N}\sum_{k=1}^{N}a^{(k)} \tag{1.168}$$

Note, averaging over realizations can be defined even for a single realization $a^{(k)}(x,y)$:

$$\overline{a'}\equiv\bar{a}^{(k)}=\frac{1}{N}\sum_{m=1}^{N}a^{(m)},\quad \forall k$$

Thus, averaging over realizations converts the disordered function into a periodic function, as shown in Fig. 1.39, curve 3. The last is sectionally continuous in general case.

Double Averaging

Consequent averaging over realizations and over space give us double averaging:

$$\langle\bar{a}\rangle=a^{(0)}(x)+C_0,\quad C_0=\frac{1}{N}\sum_{k=1}^{N}\langle a^{(k)}\rangle \tag{1.169}$$

as spatial averaging of a periodic function yields a constant value C_0.

It is clear that only double averaging converts the original function into non-oscillating.

The non-oscillating component can be always chosen in such a way that the constant value $C_0=0$, then:

$$\langle\bar{a}\rangle=a^{(0)}(x) \tag{1.170}$$

Then the value $a^{(0)}$ in (1.158) is the double average of the function a. Note, the operations $\langle\cdot\rangle$ and $\overline{(\cdot)}$ commute.

1.12.4. PROBLEM OF AVERAGING THE TRANSPORT IN DISORDERED MEDIA

Let us consider the diffusion equation with disordered oscillating coefficients $a(x,x/\varepsilon)$ and $b(x,x/\varepsilon)$:

$$b\left(x,\frac{x}{\varepsilon}\right)\frac{\partial u}{\partial t}-\frac{\partial}{\partial x_i}\left(a\left(x,\frac{x}{\varepsilon}\right)\frac{\partial u}{\partial x_i}\right)=f\left(x,\frac{x}{\varepsilon},t\right),\quad x\in\Omega,\quad t>0 \qquad (1.171)$$

The oscillating functions a and b are assumed to be of two-scale type, as defined in the previous section, with ε some characteristic heterogeneity scale.

Formal Averaging over Realizations. Fluctuation Correlator

As averaging over realizations gives a periodic function, it is natural to expect that the problem of averaging the transport in disordered medium can be solved in two steps. At first, averaging over realizations being applied leads to a periodic system, which can be averaged afterwards over space using standard homogenization. However, averaging over realization is not easy to be realized in general case, in particular when two oscillating functions enter into the equation as a product.

For the operator

$$L\equiv\frac{\partial}{\partial x_i}\left(a(x,y)\frac{\partial u}{\partial x_i}\right)$$

Let us apply averaging over realizations taking into account the representation (1.158) and using the same representation for the solution $u(x,y)$: $u(x,y)=u^{(0)}+u'$, where $\overline{u'}\equiv 0$. As

$$\overline{\frac{\partial}{\partial x_i}\left(a\frac{\partial u}{\partial x_i}\right)}=$$

$$=\frac{\partial}{\partial x_i}\left(a^{(0)}\frac{\partial u^{(0)}}{\partial x_i}\right)+\frac{\partial}{\partial x_i}\left(\overline{a'}\frac{\partial u^{(0)}}{\partial x_i}\right)+\frac{\partial}{\partial x_i}\left(a^{(0)}\frac{\partial \overline{u'}}{\partial x_i}\right)+\frac{\partial}{\partial x_i}\left(\frac{\overline{a'\partial u'}}{\partial x_i}\right)=$$

$$=\frac{\partial}{\partial x_i}\left(\overline{a}\frac{\partial \overline{u}}{\partial x_i}\right)+\frac{\partial}{\partial x_i}\left(\overline{a'\frac{\partial u'}{\partial x_i}}\right)$$

then Eq. (1.171) yields after averaging over realizations:

$$\overline{b}\frac{\partial \overline{u}}{\partial t}-\frac{\partial}{\partial x_i}\left(\overline{a}\frac{\partial \overline{u}}{\partial x_i}\right)-\overline{f}=\frac{\partial}{\partial x_i}\left(\overline{a'\frac{\partial u'}{\partial x_i}}\right)-\overline{b'\frac{\partial u'}{\partial t}} \qquad (1.172)$$

Two terms in the right-hand part:

$$\overline{a'\frac{\partial u'}{\partial x_i}},\quad\text{and}\quad \overline{b'\frac{\partial u'}{\partial t}} \qquad (1.173)$$

are the *fluctuation correlators.*

It both the fluctuation correlators were equal to zero, then averaging over realizations being applied to the diffusion equation would lead to an equation of the same type written respectively to the periodic function $\overline{u}$ with the periodic coefficients $\overline{a}$ and $\overline{b}$. The problem of averaging would by solved then.

Unfortunately, the fluctuation correlator is not defined in general case. Equation (1.172) can be only defined respectively to the periodic function $\overline{u}$, when either the fluctuation u' is a linear function of some linear operators of $\overline{u}$, or each fluctuation correlator is small and can be neglected.

This simple analysis allows to make the following conclusions.

Corollaries

1. Averaging over realization does not allow to transform the problem of averaging the equation with disordered coefficients to the periodic problem, in general case. The fluctuation correlators remain undefined in the equation averaged over realizations. Definition of each correlator consists in establishing the relation between u' and $\overline{u}$, which will be called as *closure relation of first order.* (The closure problem of the second order will appears on the stage of space averaging over cells).

2. The form of closure relation between u' and $\overline{u}$ depends on the structure of the medium. Various types of heterogeneity lead to various closure relations. Thus, the same microscale equation is transformed into various macroscale models for various heterogeneity types.

3. The success of the homogenization of periodic media was related, in particular, with automatic solution of the closure problem, without any additional assumptions, except for the assumption about small heterogeneity scale. The same automatic solution of the closure problem would be a very wished property of the method of averaging designated to disordered media.

4. In general case of heterogeneity, the equation type is not kept after averaging due to arising of the fluctuation correlator. From this point of view, the procedures of up-scaling customary used in reservoir engineering have not a good theoretical basis.

1.12.5. EXCLUDING THE FLUCTUATION CORRELATORS. CONDITIONS ON THE SPATIAL AVERAGE

The fluctuation correlators can be excluded or neglected, if the heterogeneity field satisfies some basic conditions, which are discussed below in this section. In the same time, these conditions can be examined as the limitations for the heterogeneity scale ε which is not yet defined strictly.

For such a type of media averaging over realizations gives the periodic problem of the same type as the original one.

Let us assume such a scale ε exists that the following, additional, properties are satisfied for the parameters of the equation (1.171).

Weak Oscillation of Spatial Average

First, let us assume space averaging over cells, as defined in (1.166), leads to a weakly oscillating function:

$$\langle a\rangle = a^{(0)}(x) + \sum_{k=1}^{N} \langle a^{(k)}\rangle \delta^{(k)}(x), \quad \langle a^{(k)}\rangle = O(\varepsilon), \ \forall k$$

$$\langle b\rangle = b^{(0)}(x) + \sum_{k=1}^{N} \langle b^{(k)}\rangle \delta^{(k)}(x), \quad \langle b^{(k)}\rangle = O(\varepsilon), \ \forall k \tag{1.174a}$$

The best situation is when space averaging over cells leads to a non-oscillating function, as in periodic case. But this requirement is not necessary. The average over cells can oscillate, but the amplitude of these oscillations should be rather small, i.e., of order of the heterogeneity scale.

This requirement is very weak. As shown in the previous section, the non-oscillating function can be obtained only after double averaging. Then it is natural to require that one step of averaging would lead to a *"semi-averaged"* function. This heuristic notion of the semi-average is expressed above in the qualitative form (1.174). Otherwise, the spatial averaging would lead to a function with a large amplitude of the oscillations, that is not more simple than the original function. So, spatial averaging becomes useless. Thus, requirement (1.174) is natural and equilibrate the role of two averaging procedures.

Practically in any two-scale disordered field a rather small scale ε can be found out, such that the spatially averaged heterogeneity becomes weakly oscillating.

Then the function $a - \langle a\rangle$ is close to be a constant value equal to zero.

Invariance of Spatial Average with Respect to the Shift

Both the functions a and b should be stochastically homogeneous at the scale ε, in order to the results of averaging would be invariable respectively to the shift of the grid $\{\Omega^{(k)}\}$.

This condition can be formulated in the following way:

let at the scale ε :

1) the non-oscillating components $a^{(0)}$ and $b^{(0)}$, which are the double averages, be invariable respectively to any spatial shift;
2) condition (1.174a) for the oscillating components remains true for any spatial shift

(1.174b)

Relation of Conditions on the Space Average with the Heterogeneity Scale

Two basic requirements (1.174) are the limitations for the parameter ε. They limit the freedom of the probable variations of this parameter. Numerical experiments show that step-by-step increasing of the parameter ε leads to some critical value. Since this value conditions (1.174) are satisfied. This critical value may be chosen as the "heterogeneity scale".

For any two-scale field such scale exists. In fact if the Fourier spectrum is such as shown in Fig. 1.38, then spatial averaging over the maximal scale of high-frequency harmonics, ε_{max}, leads to smoothing of all the fast oscillations. The corresponding space average is non-oscillating function, then condition (1.174a) is satisfied at the scale ε_{max}.

Thus, for a two-scale field the heterogeneity scale can be determined as the maximal period of high-frequency harmonics in the Fourier spectrum.

Conditions (1.174) are basic to develop the effective homogenization method.

Smallness of the Fluctuation Correlator when the Spatial Average is Weakly Oscillating

Let us prove the fluctuation correlators in (1.172) are small and can be neglected.

The solution of (1.171) can be found in the two-scale asymptotic form, as usual in homogenization theory:

$$u(x,y)=u_0(x,y)+\varepsilon u_1(x,y)+\varepsilon^2 u_2(x,y)+\varepsilon^3 ..., \quad y=x/\varepsilon$$

$$u_i(x,y)=u_i^{(0)}(x)+u_i'(x,y), \quad i\geq 1 \tag{1.175}$$

where $u_i^{(0)}(x)$ is the non-oscillating part of the ith term (independent of y), while $u_i'(x,x/\varepsilon)$ is the oscillating part of each term (dependent on y).

The extended function $u(x,y)$, which coincides with the solution $u(x,x/\varepsilon)$ when $y=x/\varepsilon$, will be assumed to satisfy the same conditions (1.174) as the parameters of equation (1.171).

Let us substitute the last expansion into the two-scale variant of (1.171). Then we get for the zero term:

$$\frac{\partial}{\partial y_i}\left(a\frac{\partial u_0}{\partial y_i}\right)=0$$

Then u_0 does not depend on the fast variable y. Then u_0 is the spatially averaged value over each cell, namely:

$$u_0 = u_0^{(0)} + \langle u_0' \rangle$$

where $\langle u_0' \rangle$ is its oscillating component.

But the value averaged over the cells is weakly oscillating, according to (1.174a), or:

$$\langle u_0' \rangle \sim \varepsilon$$

Therefore:

$$u_0 = u_0^{(0)} + O(\varepsilon)$$

Then we get:

$$\begin{aligned} u &= u^{(0)} + u' = u_0^{(0)} + O(\varepsilon) + \varepsilon u_1^{(0)} + \varepsilon u_1' + \varepsilon^2 \ldots = \\ &= u_0^{(0)} + \varepsilon\left(u_1^{(0)} + u_1'\right) + O(\varepsilon) == u^{(0)} + O(\varepsilon) \end{aligned}$$

Thus,

$$u' = O(\varepsilon) \tag{1.176}$$

Then the fluctuation correlators similar to (1.173) are of order ε. In fact:

$$\overline{\left(a'\frac{\partial u'}{\partial x_i}\right)} = O(\varepsilon), \qquad \overline{\left(b'\frac{\partial u'}{\partial t}\right)} = O(\varepsilon) \tag{1.177}$$

Thus in media, satisfying the basic condition (1.174), the fluctuation correlators can be ignored.

1.12.6. ALGORITHM OF HOMOGENIZATION

First Step: Establishing the Heterogeneity Scale

The first step consists in founding such a scale ε when the conditions (1.174a) and (1.174b) are satisfied.

For two-scale field the heterogeneity scale can be determined as the maximal period of high-frequency harmonics in the Fourier spectrum.

In the next section, an alternative method will be explained, which is not related to Fourier transform.

Second Step: Averaging Over Realizations
Averaging over realizations of the equation (1.171) leads to: the following "semi-averaged" relationship:

$$\overline{b}\left(x,\frac{x}{\varepsilon}\right)\frac{\partial\overline{u}}{\partial t}-\frac{\partial}{\partial x_i}\left(\overline{a}\left(x,\frac{x}{\varepsilon}\right)\frac{\partial\overline{u}}{\partial x_i}\right)=\overline{f}+O(\varepsilon),\ x\in\Omega,\ t>0 \quad (1.178)$$

where (1.177) has been taken into account.

The functions $\overline{a}$, $\overline{b}$ and $\overline{f}$ are periodic with respect to the second argument, and piecewise continuous. Therefore the equation (1.178) should be meant in generalized sense.

We seek for the periodic and continuous solution $\overline{u}$.

Thus we obtain the periodic model, which should be averaged over space.

Third Step: Averaging Over Space
The next algorithm based on homogenization technique is standard. The results of such double averaging are the following:

Let $U(x,t)\equiv\langle\overline{u}\rangle$, then:

$$B\frac{\partial U}{\partial t}-\frac{\partial}{\partial x_i}\left(\widehat{A}_{ik}\frac{\partial U}{\partial x_k}\right)=F \quad (1.179)$$

where:

$$B=\langle\overline{b}\rangle,\quad \widehat{A}_{ik}=\left\langle\overline{a}\left(\frac{\partial\psi_k(y)}{\partial y_i}+\delta_{ik}\right)\right\rangle,\quad i,k=1,...,n$$

The periodic continuous functions $\psi_k(y)$, $k=1,..,n$ are solutions of the following cell problems for each k:

$$\frac{\partial}{\partial y_i}\left(\overline{a}\frac{\partial\psi_k}{\partial y_i}\right)=-\frac{\partial\overline{a}}{\partial y_k},\quad y\in Y$$

$$\langle\psi_k\rangle=0$$

The problem is solved.

Basic Conditions are not True
What should one do when the basic condition (1.174a) is not satisfied ? In this case, averaging over realizations becomes useless, as it leads to arising of fluctuation correlators which are not small.

It only remains for us to apply the procedure of averaging over space directly, at first step. However, it should be noted, that homogenized equations result from a necessary condition of solution periodicity with respect to the fast variable, in periodic case. In random case, homogenized model results from the necessary conditions of stochastic homogeneity with respect to the fast variable and ergodicity of the solution. In any case, the transition to a two-scale formulation requires inserting some additional conditions with respect to the fast variable.

In our case, these conditions are undefined. Thus the strict spatial homogenization in not possible, and we have to induce some additional hypothesis, for example, about the scale invariance of the analyzed equations.

1.12.7. ITERATING DEFINITION OF THE HETEROGENEITY SCALE

Developed method is based on definition of such a scale ε which ensures the conditions on the space average (1.174). On the other hand, these two conditions may be used as the definition of the scale ε. In other words, the *heterogeneity scale* ε can be defined as such a scale, on which the spatial cell average of the functions a and b yields two weakly oscillating functions, $\langle a \rangle$ and $\langle b \rangle$.

It was noted that for a two-scale field the heterogeneity scale can be determined as the maximal period of high-frequency harmonics in the Fourier spectrum. Another method which is not related to the Fourier transform, but is based directly on the property (1.174a) may be useful in practice.

Set of Grids of Various Scales

To define the heterogeneity scale for any two-scale function $a(x)$, the following algorithm can be suggested.

Let the function $a(x, x/\varepsilon)$ be a disordered two-scale function in the domain Ω, with rather non-strictly defined heterogeneity scale ε. It is only known, that $\varepsilon \ll 1$, where 1 is the characteristic macroscale.

Let the medium be covered by a sequence of the regular grids of the scale l_n, $n=0, 1, 2, ...$, as shown in Fig. 1.40.

On this figure, the length is measured in terms of the numbers of the smallest cells. In practice, the original heterogeneous field is always discrete with some finite small size of discrete cells. This size may be chosen as the smallest scale l_0, and it can be taken as a unit of measuring. Then the length l can be measured in terms of the number of smallest cells, such that the size of the scale l_n is equal to n.

Standard Deviation of the Averages over the Cells

The function averaged over realizations, $\bar{b}$, can be easily determined according to the definition (1.167). Then the double average, $\langle \bar{a} \rangle = a^{(0)}$, can be calculated according to (1.169). As the double average depends on the grid scale l, we will designate the double average of the function a corresponding to the grid scale l as $\langle \bar{a} \rangle_l$, or $a_l^{(0)}$.

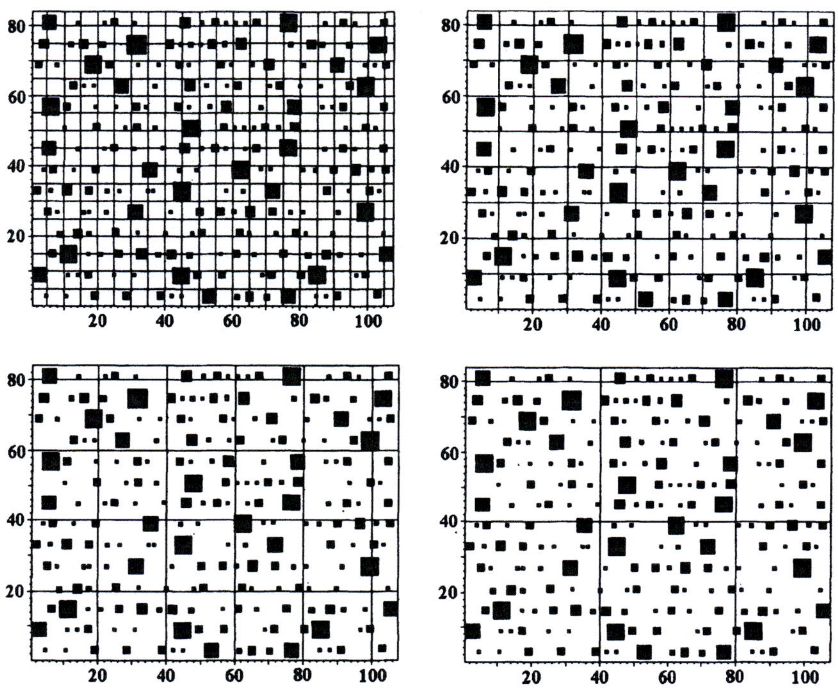

Figure 1.40. Set of grids covering the disordered heterogeneity field

Let us construct the space fluctuation function, $a_l'=a-a_l^{(0)}$, and the periodic (discontinuous) realizations of fluctuations $\left\{a_l^{(k)}\right\}_{k=1}^{N}$, according to (1.159):

$$a_l'=\sum_{k=1}^{N} a_l^{(k)}\delta^{(k)}(x)$$

The following value is the average over the kth cell of the fluctuation function a_l':

$$Z_l^{(k)}\equiv\langle a_l'\rangle^{(k)}\equiv \int\limits_{\Omega^{(k)}} a_l' dx=\langle a^{(k)}\rangle_l$$

The set of the values $\left\{Z_l^{(k)}\right\}_{k=1}^{N}$ at the fixed scale l forms the ensemble of realizations of a random value Z_l. Let the standard deviation of this random value is σ_l.

Standard Deviation as the Function of Grid Scale
If the basic assumption about weak oscillation of the space average over cells (1.174) is satisfied, then we obtain for the standard deviation:

$$\sigma_l \sim l \ll 1 \tag{1.180}$$

i.e., the difference between various values $Z_l^{(k)}$ (when l is fixed) becomes small.

Then the standard deviation σ_l is expected to be a decreasing function of the scale l. In Fig. 1.41, the form of the function σ_l is presented (the scale l is measured in terms of the number of smallest cells, n), which was calculated for the field shown in Fig. 1.40:

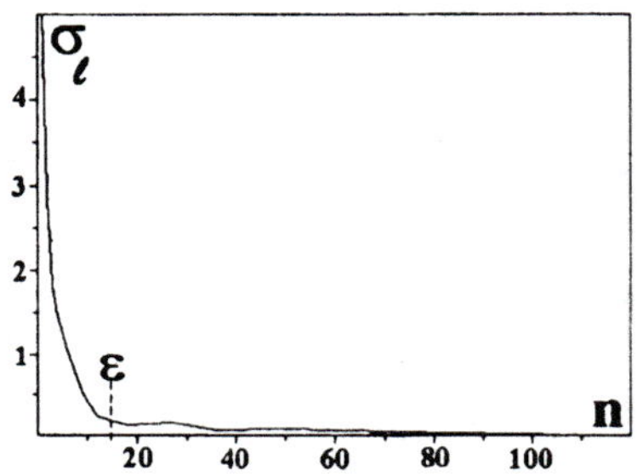

Figure 1.41. Dependence of the standard deviation of the random value Z_l on the grid scale

Then, the heterogeneity scale ε can be chosen as equal to the value l_n, when σ becomes vanishing small. In the presented example, $\varepsilon \approx l_{15}$.

Definition of the Unite ε for Two Oscillating Functions
The same algorithm been applied to both the fields of permeability $a(x)$ and porosity $b(x)$ leads to two dependences of the standard deviations σ_l^a and σ_l^b on the grid scale l.

The unit heterogeneity scale is defined as the maximal value from those obtained for each function, a and b. For the presented example, the unite heterogeneity scale is $\varepsilon \sim l_{15}$.

In all this sections, numerical experiments have been performed for a 2D bleak/white function b shown in Fig. 1.40, which is defined as the indicator function of the white areas:

$$b(x) = \begin{cases} 1, x \in \text{ white areas,} \\ 0, x \in \text{ bleak areas} \end{cases}$$

and for the piecewise permeability defined as:

$$a(x) = \begin{cases} 1, x \in \text{ white areas,} \\ a_{min}, x \in \text{ bleak areas} \end{cases}$$

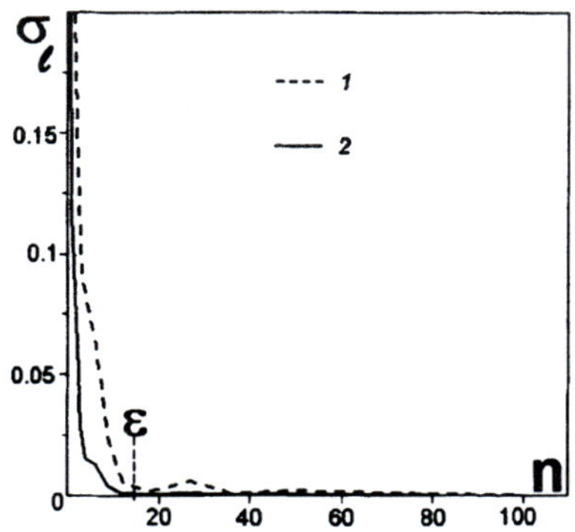

Figure 1.42. Dependence of the standard deviations of the random value Z_l for porosity, σ_l^b (curve 1), and for permeability, σ_l^a, (curve 2) on the grid scale

where a_{min}=0.2 for the shown example.

References

1. Adler, P.M. (1992) *Porous media: Geometry and transports*. Butterworth-Heinemann, N.Y., USA.
2. Aharony, A., Hinrichsen, E.L. and Hansen, A. (1991) Effective renormalization group algorithm for transport in oil reservoirs, *Physica A.*, **177**, no. 1-3, pp. 260–266.
3. Ahiezer, A.I. and Peletinski, S.V. (1977) *Methods of Statistical Physics*. Nauka, Moscow.
4. Ahmadi, G. (1982) A continuum theory for two phase media, *Acta mech.*, **44**, no. 3-4, pp. 299–317.
5. Allaire, G. (1992) Homogenization and two-scale convergence, *SIAM J. Math. Anal.*, **23**, no.6, pp. 1482–1518.
6. Allaire, G. (1997) Mathematical approaches and methods, *In: Homogenization and Porous Media*, Ed. U.Hornung, Springer, pp. 225–246.
7. An, L., Glimm, J., Sharp, D.H., and Zhang, Q. (1995) Scale-Up Flow in Porous Media, In: *Mathematical Modeling of Flow Through Porous Media*, Eds. A. Bourgeat et al., World Scientific Publishing, Singapore, 1995, pp. 26–43.
8. Arbogast, T. (1989) Analysis of the simulation of single phase flow through a naturally fractured reservoir, *SIAM J. Math. Anal.*, **26**, pp. 12–29.
9. Arbogast, T. and Douglas, J. (1988) Two-phase immiscible flow in naturally fractured reservoirs, *Numer. simul. oil recov.: Proc. Symp., Munneapolis, Minn., Dec. 1-12, 1986. N. Y. etc.*, 1988. pp. 47–66.
10. Arbogast, T., Douglas, J. and Hornung, U. (1990) Derivation of the double porosity model of single phase flow via homogenization theory, *SIAM J. Math. Anal.*, **21**, pp. 823-836.
11. Axell, J. (1991) Conduction in two and three dimensions: effective medium theory, *Phys. Lett.*, **153**, no. 1, pp. 43–48.
12. Bachmat, Y. and Bear, J. (1986) Macroscopic modelling of transport phenomena in porous media. 1. The continuum approach, *Transport in Porous Media*, no. 1, pp. 213–240.
13. Bakhvalov, N.S. (1977) *Averaging PDE With Fast Oscillating Coefficients*. Nauka, Moscow.
14. Bakhvalov, N.S. (1975) Averaging of partial derivation equations with quiclky oscillating coefficiennts, *Doklady Academii Nauk SSSR*, **221**, no. 3, 1975, pp. 516–519.
15. Bakhvalov, N.S. (1974) Averaged properties of bodies wiht periodic structure, *Doklady Academii Nauk SSSR*, **218**, no. 5, pp. 1046–1048.
16. Bakhvalov, N.S. and Panasenko G.P. (1984) *Homogenization of Processes In Periodic Media*. Nauka, Moscow.
17. Bakhvalov, N.S. and Panasenko, G.P. (1989) *Homogenization : Averaging Processes in Periodic Media*. Kluwer Academic Publishers, Dordrecht.
18. Barenblatt, G.I., Entov, V.M. and Ryzhik, V.M. (1972) *Theory of Non-Stationary Flow of Liquids and Gases Through Porous Media*. Nedra, Moscow.
19. Barenblatt, G.I., Zheltov, Yu.P. and Kochina, I.N. (1960) On main representations in the theory of flow of homogeneous liquids in fractured rocks, *Prikladnaya Mathematika i Mechanika*, **24**, no. 5, pp. 852–864.
20. Barker, J.A. (1985) Block-geometry functions characterizing transport in densely fissured media, *J. Hydrol.*, **77**, pp. 263–279.
21. Barker, J.A. (1985) Modelling the effects of matrix diffusion on transport in densely fissured media, *Mem. of XVIII Congr. of the Intern. Assoc. of Hydrogeol.*, Cambridge, pp. 250–269.
22. Basniev, K.S., Bedrikovetsky, P.G. and Dedinets, E.N. (1988) Detrmination of effective permeability of the porous-fractured medium, *Inzhenerno-Phyzicheskii Journal*, **55**, no. 6, pp. 940–948.

23. Bazarov, I.P., Gevorkian, E.V. and Nikolaev, P.N. (1989) *Nonequilibrium Thermodynamics and Physical Kinetics*. Ed. Moscow University, Moscow.
24. Berdichevski, V.L. (1983) *Variational Principles of Continuum Mechanics*. Nauka, Moscow.
25. Berdichevski, V.L. (1977) *On averaging of periodic structures, Prikladnaya Mathematika i Mechanika*, **41**, no. 6, pp. 993–1006.
26. Bensoussan, A., Lions, J.-L. and Papanicolaou, G. (1978) *Asymptotic analysis for periodic structures*, Amsterdam: North-Holland.
27. Bernasconi, J. (1978) Real space renormalization of bond-disordered conductance lattices, *Phys. Rev. B.*, **18**, pp. 2185–2191.
28. Bleher, P.M. and Surgailis, D. (1983) Self-similar random fields, *in : Issues of Science and Thechnology. Probability theory. Mathematical Statistics. Theoretical cibenetics*, Ed. VINITI, Moscow, **20**, pp. 3–51.
29. Böhm, M. and Showalter, R.E. (1985) Diffusion in fissured media, *SIAM J. Math. Anal.*, **16**, no. 3, pp. 500–509.
30. Bolotin, V.V. and Moscalenko, V.N. (1969) To the calculation of strongly isotrope composite materials, *Izvestiya Academii Nauk SSSR, Ser. Mechanika Tverdogo Tela*, no. 3.
31. Bourgat, J.F. (1978) Numerical experiments of the homogenization method for operators with periodic coefficients, Rapport de Recherche IRI, Rocquencourt, no. 277, pp. 212–245.
32. Bourgeat, A., Mikelić, A., and Piatnitski, A. (1998) Modeèle de double porosité aléatoire, *C.R. Acad. Sci. Paris*, Série 1, 327, pp. 99–104.
33. Bruggeman, D.A.G. (1935) Berechnung verschiedenier physikalisher Konstanten von heterogenen Substanzen, *I. Dielektrizitäts Konstanten und Leitfähigkeiten der Mischkörper aus isotropen Substanzen, Ann. Phys.*, **5**, pp. 636–664.
34. Buevich, Y.A. and Markov, V.G. (1973) Continuum mechanics of monodisperse suspension: reological equations of state for suspensions of moderate concentration, *Prikladnaya Mathematika i Mechanika*, **37**, no. 6, pp. 1059–1077.
35. Buzinov, S.N. and Umrikhin, I.D. (1973) *Gydrodynamic methods of well and field investigation*, Nedra, Moscow.
36. Charlaix, E., Hulin, J.P. and Plona, T.J. (1987) Experimental study of tracer dispersion in sintered glass porous materials of variable compaction, *Phys. Fluids*, **30**, pp. 1690–1698.
37. Chen, Z.-X. (1990) Analytical solutions for double-porosity, double-permeability and layered systems, *J. Petrol. Sci. and Eng.*, **5**, no. 1, pp. 1–24.
38. Coats, K.H. and Smith, B.D. (1964) Dead-end pore volume and dispersion in porous media, *Trans. Soc. Petol. Eng.*, **231**, no. 3, pp. 73–84.
39. Crapiste, G.H., Rotstein, E. and Whitaker, S. (1986) A general closure scheme for the method of volume averaging, *Chem. Eng. Sci.*, **41**, pp. 227–235.
40. Danian, F. (1986) *Interprétation des essais de puits: Les méthodes nouvelles*. Technip, Paris.
41. De Genne, P. *Ideas of Scaling in Polymer Physics*. Mir, Moscow.
42. De Swaan, A.O. (1976) Analytic solutions for determining naturally fractured reservoir properties by well testing, *SPE J.*, no. 6, pp. 117–122.
43. Dean, R.H. and Lo, L.L. (1988) Simulations of naturally fractured reservoirs, *SPE Reservoir Eng.*, **3**, no. 2, pp. 638–648.
44. Deans, H.A. (1963) A mathematical model for dispersion in the direction of flow in porous media, *Trans. Soc. Petrol. Eng.*, **288**, pp. 49–52.
45. Deich, M.E. and Philippov, G.A. (1968) *Gas Dynamics of Two-Phase Media*. Energyia, Moscow.
46. Desbarats, A.J. (1978) Numerical estimation of effective permeability in sand-shale formations, *Water Resour. Res.*, **23**, no. 2, pp. 273–286.
47. Douglas, J., Paes Leme, P.J., Arbogast, T. and Schmitt T. (1987) Simulation of flow in naturally fractured reserviors, *Proc. IX SPE symp. on reservoir simula-*

tion, Soc. of Petrol Eng. Dallas (Tex.), Paper SPE 16019, pp. 271–279.
48. Duguid, I.O. and Lee, P.C. (1977) Flow in fractured porous media, *Water Resour. Res.*, **13**, no. 3, pp. 558–566.
49. Dykhne, A.M. (1970) Conductivity of a two-dimentional two-phase system, *Soviet Physics JETP* **59**, no. 1(7), pp. 111–115.
50. Emets, Y.P. (1986) *Electric Properties of Composite Materials With Regular Structure*. Naukova Dumka, Kiev.
51. Evans, R.D. (1983) A proposed model for multiphase flow through naturally fractured reservoir, *Sos. Petr. Eng. J.*, **22**, no. 5, pp. 669–680.
52. Fllgelman, H., Cinco-Ley, H., Ramey H.J., et al. (1989) Pressure drawdown test analysis of a gas well-application of new correlations, *SPE Formation Evaluation*, Sept. pp. 406–453.
53. Geld, P.V. and Mitushov, E.A. (1990) Generalized method of self-similar field to determine elastic properties of heterogeneous materials, *J. Prikladnoi Mathematiki i Technicheskoi Physiki*, no. 1, pp. 96–100.
54. Geld, P.V., Sachkov, I.N., Gofman, A.G. and Sidorenko, F.A. (1990) Conductivity of heterogeneous systems: finit element method, *Doklady Academii Nauk SSSR*, **315**, no. 3, pp. 604–607.
55. Golf-Rakht, T.D. (1986) *Basis of oil geology and development of fractured reservoirs*, Nedra, Moscow.
56. Happel, G. and Brenner, G. (1976) *Hydrodynamics at low Reynolds Numbers*. Mir, Moscow.
57. Hashin, Z. and Shtikmen, S. (1962) On some variational priciples inanisotropic and homogeneous elasticity, *J. Mech. Phys. Solids.*, **10**, no. 4, pp. 343.
58. Hill, R. (1963) Elastic properties of reinforced solids: some theoretical principles, *J. Mech. Phys. Solids.*, **11**, no. 5.
59. Hinch, E.J. (1977) An averaged-equation approach to particle interactions in a fluid suspension, *J. Fluid Mech.*, **83**, pt. 4, pp. 695–720.
60. Hornung, U. and Jäger, W. (1987) A model for chemical reactions in porous media, *In: Complex chemical reactions, modelling and simulation*, B., Springer, New-York.
61. Hornung, U. and Showalter, R.E. (1990) Diffusion models for fractured media, *J. Math. Anal. and Appl.*, **147**, no. 1, pp. 69–80.
62. Hornung, U., Showalter, R.E. and Walkington, N.J. (1991) Microstructure models of diffuion in fissured media, *J. Math. Anal. and Appl.*, **155**, no. 1, pp. 1–20.
63. Ibraguimov, A.I. (1984) *Some problems of qualitative theory of elliptic and parabolic equations*. Dr. Thesis, Steklov Mathematical Institute, oscow.
64. Jäger, W. and Mikelić, A. (1994) On the flow conditions at the boundary between a porous medium and an imprevious solid, *In: Progress in partial Differential Equations*, vol. 314 of *Pitman Research Notes in Mathematics*, Eds. M. Chipot, J. Saint Jean Paulin and I. Shafrir, Longman Scientific and Technical, London.
65. Jäger, W. and Mikelić, A. (1995) On the boundary conditions at the contact interface between a porous medium and a free fluid, *Annali della Scuola Normale Supriore di Piza, Classe di Scienze.*
66. Kalinin, V.A. and Baiuk, N.O. (1990) Energy limitations for the effective modulus of anisotropic microheterogeneous media, *Doklady Academii Nauk SSSR*, **313**, no. 5, pp. 1090–1094.
67. Kartashov, E.M. (1979) *Analytical methods in thermo-conductivity of the solid bodies*. Ed. Superior School, Moscow.
68. Kazemi, H., Seth and Thomas, G.W. (1969) The interpretation of interference tests in naturally fractured reservoirs with uniform fracture distribution, *SPE J.* no. 12, pp. 463–472.
69. Kent, T.L., Dixon, T.N. and Pierson, R.G. (1983) Fractured reservoirs simulation, Ibid. **23**, no. 1, pp. 42–54.
70. Khrouslov, E.Ya. (1988) Hommohenized model of non-stationary flow in fractured

porous media, *Preprint Ukrainian Academy of Sciences*, Kiev, no. 50.
71. Khrouslov, E.Ya. (1990) Hommohenized model of highly heterogeneous medium with memory, *Uspekhi Mathematicheskikh Nauk*, **45**, no. 1(297), pp. 197–198.
72. King P.R. (1988) The use of renormalization for calculating effective permeability, *Transport in Porous Media*, p. 643
73. King P.R., Muggeridge, A.H. and Price, W.G. (1993) Renormalization calculations of immiscible flow, *Transport in Porous Media*, bf 12, pp. 237–260.
74. Kirkpatrick, S. (1973) Percolation and conduction, *Rev. Mod. phys.*, **45**, no. 4, pp. 574–588.
75. Kochin, N.E., Kibel, I.A. and Rose, N.V. (1963) *Theoretical Hydromechanics. Part 2*. Fizmatgiz, Moscow.
76. Kohring, G.A. (1991) Calculation of the permeability of porous media using hydrodynamic cellular automata, *J. Stat.Phys.*, **63**, no. 1/2, pp. 411–418.
77. Kolesnichenko, V.V. (1978) Main equations of hydro-thermodynamics of the multiphase multicomponent continuum with chemical reactions: (phenomenological theory), *Institut Prikladnoi Mathematiki Academii Nauk SSSR*, Preprint, no. 125.
78. Kozlov, S.M. (1978) Homogenization of differential operators with closely periodic quickly oscillating coefficients, *Mathematicheskii Sbornik. Novaya Seriia*, **107(149)**, no. 2(10), pp. 199–217.
79. Kozlov, S.M. (1978) Homogenization of random structures, *Doklady Academii Nauk SSSR*, **241**, no. 5, pp. 1016–1019.
80. Kozlov, S.M. (1979) Homogenization of random operators, *Mathematicheskii Sbornik. Novaya Seriia*, **109(151)**, no. 2(6), pp. 188–202.
81. Kozlov, S.M. (1979) Conductivity of two-dimentional random media, *Uspekhi Mathematicheskikh Nauk*, **34**, no. 4(208), pp. 193–194.
82. Kozlov, S.M. (1985) Homogenization method and random walk in heterogeneous media, *Uspekhi Mathematicheskikh Nauk*, **40**, no. 2(242), pp. 61–120.
83. Kozlov, S.M. (1989) Geometric aspects of homogenization, *Uspekhi Mathematicheskikh Nauk*, **44**, no. 2, pp. 79–120.
84. Landau, L.D. and Lifhitz, E.M. (1960), *Electodynamics of Continuum Media*. Pergamon Press, Oxford.
85. Landau, L.D. and Lifshits, E.M. *Continuum Mechanics*. Fizmatgiz, Moscow.
86. Landauer, R. (1952) The electrical resistance of binary metallic mixtures, *J. Appl. Phys.*, **23**, pp. 779–784.
87. Levy, T. (1990) Filtration in a porous fissured rock: influence of the fissures connexity, *Europ. J. Mech. B.*, **9**, no. 4, pp. 309–327.
88. Livshits, I.M. and Rozentsveig, L.N. (1946) To the theory of elastic properties of liquid cristals, *Russian J. Experimental and Theoretical Physics*, **16**, no. 11, p. 967.
89. Lifshits, E.M. and Pitaevsky, L.P. (1979) *Physical Kinetics*. Nauka, Moscow.
90. Lin, S.H. (1992) Transient conduction in heterogeneous media, *Intern. Commun. Heat and Mass Transfer.*, **19**, no. 2, pp. 165–174.
91. Lindstrom, F.T. and Narasimham, N.L. (1973) Mathematical theory of a kinetic model for dispersion of previously distributed chemicals in a sorbing porous medium, *SIAM J. Appl. Math.*, **24**, pp. 496–510.
92. Lions, J.-L. (1978) Notes on somes computational aspects of homogenization in composite materials, *Numerical methods in mathematical physics and optimal control*, Ed. Marchuk G.N. and J.-L.Lions, Nauka, Novosibirsk, pp. 5–19.
93. Marinbakh, M.A. and Lyusternik, V.E. (1985) Effect of macroheterogeneity on effective permeability of porous media, *Dynamika mnogofaznykh sred*, Nauka, Novosibirsk, pp. 123–127.
94. Marles, C.M. (1982) On macroscopic equations governing multiphase flow with difuusion and chemical reactions in porous media, *Int. J. Eng. Sci.*, **20**, pp. 643–662.
95. Mikelić, A. (1997) On the transmission conditions at the contact interface between

a porous medium and a free fluid, *In: Homogenization and Porous Media*, Ed. U.Hornung, Springer, pp. 69–76.
96. Mohanty, S. and Sharma, M.M. (1991) A Monte Carlo RSRG method for the percolation/conduction properties of correlated lattices, *Phys. Lett.*, **154**, no. 9, pp. 475–481.
97. Monin, A.Kh. and Yaglom, F.M. (1975) *Statistical Theoty of Turbulence*. Nauka, Moscow.
98. Neale, G. and Jacob, H.N. (1975) Flow perpendicular to mats of randomly arranged cilindrical fibers: (Importance of cell models), *AIChE J.*, **21**, no. 4, pp. 805–807.
99. Nigmatulin, R.I. (1978) *Basis for Mechanics of Heterogeneous Media*. Nauka, Moscow.
100. Nigmatulin, R.I. (1987) *Dynamics of Multiphase Media*. Nauka, Moscow.
101. Nikolaevski, V.N. (1996) *Geomechanics and Fluidodynamics With Applications to Reservoir Engineering*. Kluwer Academic Publishers, Dordrecht.
102. Nikolaevskii, V.N., Bondarev, E.A., Mirkin M.I. et al, (1968) *Motion of Hydrocarbon Mixtures in Porous Media*. Nedra, Moscow.
103. Nguetseng, G. (1989) A general converegence result for a functional related to the theory of homogenization, *SIAM J. Math. Anal.*, **20**, pp. 608–629.
104. Odeh, A.S. (1965) Unsteady-state behaviour of fractured reservoirs, *SPE J.*, no. 1, pp. 60–66.
105. Olarewaju, J.S. and Lee, W.J. (1989) New pressure-transient analysis model for dual-porosuty reservoir, *SPE Form. Evaluation*, no. 9, pp. 384–390.
106. Ootani, I., Ohashi, Y.H., Ohashi, K. and Fukuchi M. (1992) Breakdown properties of random systems with distributed conductances, *J. Phys. Soc. Jap.*, **61**, no. 4, pp. 1399–1407.
107. Palatnik, B.. (1991) Numerical simulation of multiphase flow through highly heterogeneous periodic media, *In: Proc. 7-th All Russian Congress on Theoretics and Applied Mechanics*, Moscow, Aug. 15-21.
108. Panasenko, G.P. (1983) Avaeraging the fields in composite materials with highly modulus structure, *Vestnik MGU, Ser.15, Vychislitelnaya matematika i kibernetika*, no. 2, pp. 20–27.
109. Panasenko, G.P. (1988) Homogenization of processes in strongly heterogeneous media, *Doklady AN SSSR*, **298**, no. 1, pp. 76–79 (translated in: Soviet Phys. Dokl., 1988.)
110. Panasenko, G.P. (1988) Averaging the carcas constructions with random structures, *Doklady AN SSSR*, **244**, no. 2. pp. 335–337 (translated in: Soviet Phys. Dokl., 1988.)
111. Panasenko, G.P. (1990) Numerical-Analytical Method of Multicomponent Homogenization for Equations with Contrast Parameters, *J. Vychislitelnoi Mathematiki i Mathematicheskoi Physiki*, **30**, no. 2, pp. 243–253 (translated on English).
112. Panasenko, G.P. (1992) Principle of decomposition for the homogenized operator for nonlinear system of equations in periodic and carcas constructions, *Doklady AN SSSR*, **263**, no. 1, pp. 35–40.
113. Panasenko, G.P. and Sysuev, V.V. (1986) Averaging the parameters of mathematic models of heat, water and mass transport in an aggregated random-periodic porous medium, *Mathematical methods of rock description and calculation of their effective propeties*, Moscow, MOIP, pp. 33–42.
114. Panfilov, M. (1995) Averaged Models of Convection-Diffusion Transfer Through Highly Heterogeneous Porous Media, *in : Math. Modelling of Flow Through Porous Media*, Proc. Int. Conf., World Scientific Pub., Singapore, pp. 276–300.
115. Panfilov, M.B. (1995) Homogenised Models of Convection-Diffusion Transfer in Periodic Network Media, *Preprint Oil & Gas Research Institute, Russian Acad. Sci.*, no. 26.
116. Panfilov, M. (1994) Averaged Model-Type Transition In Flows Through Multiple

Heterogeneous Porous Media, *C.R. Acad. Sci. Paris*, Ser.II, **318**, pp. 1437–1444.
117. Panfilov, M. (1993) Structural Averaging of Porous Flow Processes in Heterogeneous Media, *Fluid Dynamics*, no. 3, pp. 112–120.
118. Panfilov, M.B. (1991) Averaging transport through heterogeneous porous media, *In: Proc. 7-th All Russian Congress on Theoretics and Applied Mechanics*, Moscow, Aug. 15-21.
119. Panfilov, M. (1990) Mean mode of porous flow in highly inhomogeneous media, *Soviet Physics Doklady*, **35**, pp. 225–227.
120. Panfilov, M. (1990) Irregular averaging of filtration transfer processes in heterogeneous media, *Proc. 2nd Europ. Conf. Mathematics of Oil Recovery*, Arles, pp. 347-350.
121. Panfilov, M.B. (1990) Averaged irreditary models of flow through fractured-porous media, *Proc. Int. Symp. on Development Oil Fields with Fractured Rocks, Varna, Oct. 22-25*, Varna, book 3, pp. 52–57.
122. Panfilov, M.B. (1990) Averaged irreditary models of flow with highly heterogeneous interior structure, *Proc. Int. Conf. Development of Gas-Condensate Reservoirs, Krasnodar, May 29 - June 2*, Section 6, pp. 83–87.
123. Panfilov, M.B. (1989) Unlocal models and hydrodynamic irrevirsibility of flow through porous media in oil and gas reservoirs, *In: Proc. 3rd Republican Conf. Ukrainian Acad. Sci. "Integral Equations in Applied Modeling", Odessa, Nov. 14-16*, Ed. Naukova Dumka, Kiev, pp. 102–103.
124. Panfilov, M.B. (1987) Regularized asymptotics and averaged models of flow in periodic heterogeneous media, *In: Thechnology and Technique of Enhanced Oil, Gas and Condensate Recovery*, Ed. Moscow Oil & Gas Institute, Moscow, no. 199.
125. Panfilov, M.B. (1986) Flow through porous media with fast oscillating properties, *Proc. VI All Russian, Congress on Theor. and Appl. Mechanics, Tashkent, Sept. 24-30*, Ed. Russian Acad. Sci., Tashkent, pp. 503–504.
126. Panfilov, M.B. (1986) Quasi-equilibrium asymptotic of underground depletion in oil and gas reservoirs, Soviet Physics Doklady, **31(5)**, pp. 225–227.
127. Panfilov, M. and Panfilova, I. (1996) *Averaged models of flows with heterogeneous internal structure*. Nauka, Moscow.
128. Prues, K. and Narasimhan, T.N. (1985) A practical method for modelling fluid and heat flow in fractured porous media, *SPE J.*, no. 2, pp. 14–26.
129. Raats, P.A.C. (1968) Transport in soils: The balance of momentum, *Soil Sci. Sor. Amer. Proc.*, **32**, pp. 452–456.
130. Rakhmatulin, Kh.A. (1956) Basis of gas dynamics for coexisting motions of compressible media, *Prikladnaya Mathematika and Mekhanika*, **20**, no. 2, pp. 184–195.
131. Reynolds, P.J., Klein, W. and Stanley, H.E. (1977) A real space renormalization group for site and bond percolation, *J. Phys. C.*, **10**, pp. L167–L172.
132. Rojdestvensky, B.L. and Yanenko, N.N. (1978) *Systems of Quasilinear Equations*. Nauka, Moscow.
133. Ross, B. (1986) Dispersion in fractal networks, *Water Resour. Res.*, **22**, no. 5, pp. 823–827.
134. Sahimi, M., Hughes, B.D., Scriven, L.E. and Davis, H.T. (1983) Real-space renormalization and effective medium approximation to the percolation conduction problem, *Chem. Eng. Sci.*, **28**, pp. 307–311.
135. Salamatin, A.N. (1987) *Mathematicak Models of Disperse Flows*. Ed. Kazan University, Kazan.
136. Sanchez-Palensia, E. (1980) *Non-homogeneous media and vibration theory*. Volume 129 of *Lecture Notes in Physics*, Springer Verlag, Berlin.
137. Sedov, L.I. (1976) *Continuum Mechanics. Part 1*. Nauka, Moscow.
138. Sedov, L.I. (1965) Mathematical methods to construct the models of continuum media, *Uspekhi matematicheskih nauk*, **20**, no. 3, pp. 331–335.
139. Selyakov, V.I. (1985) Effective permeability of heterogeneous medium, *Dynamics*

of multiophase media, Nauka, Novosibirsk, pp. 199–203.
140. Shah, N. and Ottino, J.M. (1986) Effective transport properties of disordered multiphase composites: Application of real-space renormalization group theory, *Chem. Eng. Sci.*, **41**, pp. 283–296.
141. Sheidegger, A. (1960) *Physics of flow through porous media*. Gostoptekhizdat, Moscow.
142. Shermegor, T.D. (1977) *Elasticity Theory for Micro-heterogeneous Media*. Nauka, Moscow.
143. Shvidler, M.I. (1985) *Statistical Hydrodynamics of Porous Media*. Nedra, Moscow.
144. Showalter, R.E. and Walkington, N.J. (1991) Micro-structure models of diffusion in fissured media, *J. Math. Anal. and Appl.*, **155**, no. 1, pp. 1–20.
145. Stinchcombe, R.B. (1977) Conductivity and spin-move stiffness in disordered systems: An exactly soluble model, *J. Phys. C.*, **7**, pp. 179–203.
146. Toffoli, T. and Marolus, N. (1991) *Cellular Automata Machines*. Mir, Moscow.
147. Tsybulsky, G.P. (1977) Equations of nonequilibrium two-phase flow through porous media, *Numerical solution in problems of multiphase noncompressible liquid flow through porous media*, Ed. Siberian Branch of Acad. Sci., Novosibirsk, pp. 203–213.
148. Tsybulsky, G.P. (1985) Equations of macro-nonequilibrium flow through porous media, *Numerical methods in continuum mechanics*, Ed. Siberian Branch of Acad. Sci., Novosibirsk, **16**, no. 5, pp. 133–140.
149. Tuvaeva, I.V. and Panfilov, M.B. (1990) Convective dispersion of the front n heterogeneous medium and the problem of its description, *Preprint, Oil & Gas Research Institute, Moscow*, no. 6.
150. Van Dyke, M. (1964) *Perturbation methods in fluid mechanics*. Academic Press, New York - London.
151. Vasiliev, L.L. and Tanaeva, S.A. (1971) *Thermophysic properties of Porous Materials*. Nauka i Technika, Minsk.
152. Vladimirov, V.S. (1983) *Equations of Mathematical Physics*. Nauka, Moscow.
153. Volkov, S.D. and Stavrov, V.P. (1978) *Statistical Mechanics of Composite Materials*. Ed. Belorussian University, Minsk.
154. Warren, J.E. and Root, J. (1963) The behavior of naturally fractured reservoirs, *SPE J*, **3**, no. 9, pp. 245–255.
155. Wilson, W.G. and Loidlaw, W.G. (1992) Microscopic-based flow invasion simulations, *J. Stat. Phys.*, **66**, no. 3/4, pp. 1165–1176.
156. Wu, Y.-S. and Prues, K. (1988) A multiple-porosity method for simulation of naturally fractured petroleum reservoirs, *SPE Res. Eng.*, **3**, no. 1, pp. 327–336.
157. Yu Boming and Yao, K.L. (1991) Computation of heat conduction in self-similar porous structures, *Phys. Rev. A.*, **44**, no. 6, pp. 3664–3668.
158. Yurinsky, V.V. (1980) On averading of an elliptic boundary-value problem with random coefficients, *Syberian Math. J.*, **21**, no. 3, pp. 209–223.
159. Yurinsky, V.V. (1982) On averaging og the non-divergent equations of the second order with random coefficients, *Syberian Math. J.*, **23**, no. 2, pp. 176–188.
160. Zhikov, V.V. (1991) On the estimations for the averaged matrix and averaged tensor, *Uspekhi Mathematicheskikh Nauk*, **46**, no. 3(279), pp. 49–109.
161. Zhikov, V.V., Kozlov, S.M. and Oleinik, O.A. (1991) *Homogenization of Differential Operators*. Nauka, Moscow.
162. Zhikov, V.V., Kozlov, S.M., Oleinik, O.A. and T'en Ngoan, K. (1979) Averaging and G-convergence of differential operators, *Russian Math. Surveys*, **34**, no. 5, pp. 65–133.
163. Zubarev, D.N. (1977) *Nonequilibrium Statistical Thermodynamics*. Nauka, Moscow.

CHEMICAL OR HEAT CONVECTION-DIFFUSION TRANSPORT THROUGH HIGHLY HETEROGENEOUS POROUS MEDIA

2.1. INTRODUCTORY NOTES

The next step of the developed theory is related with the flow of miscible mixtures. The linear convection-diffusion equation written relatively to the concentration of a chemical component is a simplest model of mixture flow. In this chapter a more general model is studied when the field of convection transport velocity is not given and should be found as the solution of a parabolic equation for the velocity potential (the pressure).

This model corresponds to the miscible displacement in porous media.

Within the frameworks of this approach the fluid constitutes a homogeneous mixture of two or more components which are entirely miscible. Any chemical reactions are not taken into account. The total flow of mixture is governed by Darcy's law. Transport velocity of each component within the mixture differs from the total velocity by diffusion component which is proportional to the concentration gradient. The diffusion tensor is assumed to be not dependent on the concentration.

Both the weak compressibility of fluid and the small deformations of porous medium are taken into account, that leads to a linear parabolic equation for the fluid pressure.

A general type of double-porosity media is examined, which is composed by a connected high conductive matrix and periodically located tight blocks. Both the permeabilities and the diffusion factors of blocks and matrix are contrasting.

The governing system of equations is as follows:

$$m\left(\frac{x}{\varepsilon},\omega_m\right)\frac{\partial c}{\partial t}-\frac{K_{ij}}{\mu}\left(\frac{x}{\varepsilon},\omega_K\right)\frac{\partial p}{\partial x_i}\frac{\partial c}{\partial x_j}-\frac{\partial}{\partial x_i}\left(D_{ij}\left(\frac{x}{\varepsilon},\omega_D\right)\frac{\partial c}{\partial x_j}\right)=0 \tag{2.1a}$$

$$\beta m\left(\frac{x}{\varepsilon},\omega_m\right)\frac{\partial p}{\partial t}-\frac{\partial}{\partial x_i}\left(\frac{K_{ij}(\frac{x}{\varepsilon},\omega_K)}{\mu}\frac{\partial p}{\partial x_j}\right)=f(x,t,\omega_K) \qquad (2.1b)$$

where p is the fluid pressure, c is the concentration of a component in the mixture, K is the permeability tensor, m is the porosity, D is the diffusion tensor.

Four main parameters determine the process, namely:
- ε, the period of heterogeneity;
- ω_K, the ratio of permeabilities of blocks and matrix;
- ω_D, the ratio of diffusion parameters of block and matrix;
- ω_m, the ratio of blocks and matrix porosities.

Figure 2.1 shows the structure of the medium and its parameters.

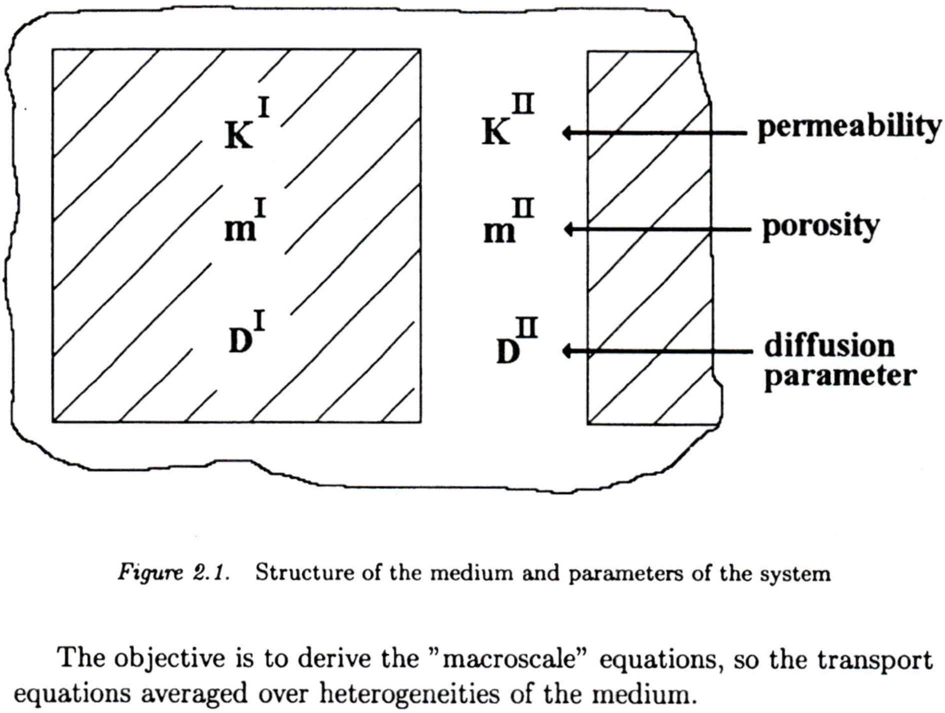

Figure 2.1. Structure of the medium and parameters of the system

The objective is to derive the "macroscale" equations, so the transport equations averaged over heterogeneities of the medium.

2.1.1. FEATURES OF THE PROBLEM

It is easy to see that the second equation (2.1b) is closed respectively to the pressure p, therefore the pressure does not depend on the concentration, and all results obtained in chapter 1 for diffusion-like equation hold true for the variable p. Hence the problem is to homogenize the first convection-diffusion equation in highly heterogeneous medium (2.1a), that can be also written in the equivalent form:

$$m(\frac{x}{\varepsilon},\omega_m)\frac{\partial c}{\partial t}+V_i(x,\frac{x}{\varepsilon},\omega_K)\frac{\partial c}{\partial x_i}=\frac{\partial}{\partial x_i}\left(D_{ij}(\frac{x}{\varepsilon},\omega_D)\frac{\partial c}{\partial x_j}\right) \tag{2.2a}$$

$$V(x,\frac{x}{\varepsilon})=-\frac{K}{\mu}gradp \tag{2.2b}$$

where V is the vector of convective transport velocity which is defined by Darcy's law.

More exactly, we attempt to understand what consequences may be produced by convective terms being introduced in the parabolic equation with quickly and highly oscillating coefficients.

A considerable feature of equation (2.1a) consists in that the velocity of convective transport is not periodic.

Five main phenomena caused by introducing the convection terms may be forecasted a-priori.

i) The system of elementary flows described in chapter 1 is kept unchanged for the total mixture. However, this system becomes insufficient for the flow of each component of the mixture, as the diffusion transport is added. Hence a more general classification of elementary flows and more general set of averaged models can be expected to arise.

ii) The transport equation becomes hyperbolic of first order when diffusion is vanishing. It is known that hyperbolic equation is a bad object for the homogenization theory, even when the heterogeneity degree not very high. In highly heterogeneous case, the physics of travelling convective waves over block and matrix interfaces is not clear yet.

iii) The principle of superposition for elementary flows proved in chapter 1 holds true for the flow of total mixture, but becomes invalid for concentration transport of each component. This principle says that the total flow is a superposition of the source-type and the translation-type flows. Due to this fact the convection transport velocity in (2.2a) may be presented as a sum of two terms corresponding to the translation flow (V_t) and to the source flow (V_s):

$$V=V_t+V_s$$

However equation (2.2a), being nonlinear relatively the assemblage "velocity and concentration", has a solution c which can not be presented as a sum of two functions c_t and c_s, each of them corresponds to a pure translation flow or to a pure source flow.

iv) The interaction between convection and diffusion is not clear, especially at the heterogeneity interface. As the flow through interfaces produces the exchange terms in averaged equations, non-trivial exchange phenomena are expected to appear on the macroscale. More strictly speaking, the problem is the following. If any process on the blocs and the matrix has the same physical origin (convection flow, for instance) but is shifted in time due to some delaying, the block-matrix interaction can be displayed in the form of a time convolution with a self-correlation function as a kernel. Examples are shown in chapter 1. If the processes on blocs and matrix have various origins (diffusion and convection, or vice versa), the interaction is not clear. Arising of some convolution operator can be assumed, but the form of correlation functions, or kernels, can not be forecasted.

v) It can be expected that the convective term, being prevailing over diffusion, would lead to a coupling between the macro-scale and the micro-scale variables, which can not be splitted. In statistical physics it is known [59] that the first order convection equation with random velocity may be effectively averaged for ergodic velocity field. When the medium heterogeneity is periodic effective macroscale model can be constructed for periodic velocity field [50], [52]. Otherwise, the macro-scale model may be formally constructed, however it contains coupled variable appropriated to various scale. Then, the homogenized description occurs non effective for practical use, though it may be feasible. The averaged model is found to be not more simple than the original micro-scale equations.

Thus, simple insertion of convective terms in diffusion equation determines five complex problems at least, which remain still unresolved.

In a number of papers the problems of convection-diffusion flow in highly heterogeneous media were studied fore reduced systems where one of transport components predominated. However, this does not resolve the problem. Indeed, if convection predominates in the whole system, then the corresponding hyperbolic equation is a bad object for homogenization theory. If diffusion predominates, then convection is suppressed by the second derivatives.

More effective way consists in considering the cases where diffusion dominates not in the whole medium, but in one of the medium components only. Even in this case only some sub-classes may be analyzed up to closed macroscale equations, when convection and diffusion effects are decomposed.

Thus, we examine the flow when convection and diffusion are of the same order in the matrix, whereas they can be of distinct order in the blocks. In other words, the matrix Peclet number is assumed to be equal to one, while the block Peclet number is various.

2.1.2. OVERVIEW OF EXISTING RESULTS

Main results of this chapter was obtained in the series of papers [46],[48], [50], [51].

It is convenient to introduce the local (microscale) Peclet number on the blocks, Pe^I, and on the matrix, Pe^{II}. The local Peclet number is defined as a ratio of convection transport microscale velocity over diffusion transport microscale velocity.

The heterogeneity scale is designated as ε and is supposed to be small.

Both Peclet Numbers are Small

The most number of papers of convection-diffusion transport in highly heterogeneous media is devoted to the case of small Peclet numbers in the whole medium. It is a pure diffusion process examined in the previous chapter of this book.

"Models with quasi equilibrium exchange" were obtained at the physical level in [11],[12]. The exchange term is written as the difference between concentrations in blocks and matrix, that corresponds to the quasi equilibrium exchange. There are variants, where the blocks are considered as capacitor element only. In [22] such model was named as "kinetic model of first order". Such models are used to describe the diffusion with sorption in porous beds having stagnant zones, which play the role of low permeable blocks [21]. The analogous models were offered later in [16],[20],[25],[28],[29], [30],[57].

"Quasi uniform models". Assumption of quasi equilibrium exchange was checked up in [33] where it was shown that stabilization is practically instantaneous and the assumption was justified. Other results where obtained in [40] where it has been shown, basing of fine-grid simulations, that the exchange is not governed by equations [12]. In [39] it was shown quantitatively that the exchange process is not only fast stabilizing but is always vanishing. Therefore the uniform diffusion equation may be used for the whole medium without any exchange term. This conclusion is consistent to the results obtained in [44] by homogenization method, where the exchange term has been shown to be of the order ε in some cases.

"The models with memory" were offered in [23] for the first time. The exchange function was admitted as a convolution of solution gradient at the block border. Solution of diffusion equation in a single block under constant boundary condition gives closure relations. Basing on homogenization methods, this model was derived in [8], [9], [24], [31], [45], [46] for a particular ratio between the medium parameters. The models with non-stationary exchange were considered in [13], [14], [19], where the solution of the cell problem was brought up only to some relation for the Laplace image.

The following results were obtained in papers [34],[35]. If the conductivity is oscillating and the ellipticity condition is infringed in some areas Q, the averaged model is non trivial. Its kind depends on the shape of the areas Q. In particular, it displays the memory effects.
Media Classification. In papers [46], [47], [48], [49] three sub-classes of similar systems were found out. They differ in the heterogeneity degree respectively to the permeability and the porosity fields. The system of *"translation-type"* or ("through-type") has a uniform macroscale equation for the whole medium, which is of the same type as the microscale equations. The system of *"source-type"* corresponds to De Swaan's model describing the exchange process by an integro-differential operator. The system of *"translation-source type"* is described by a decomposition of the exchange process in such a way that the source component of the exchange stipulates a second order time derivative in the matrix transport equation, while the translation component gives a supplement to the effective permeability tensor of the medium.

Both Peclet Numbers Are of Order One
The case of total balance between convection and diffusion ($Pe^I \sim 1$, $Pe^{II} \sim 1$) was considered in [2], [17], [18], [50]. The microscale equations were treated as the model of saturation transport by two-phase porous flow. However a linearized case has been studied, thereto the fields of pressure and saturation were independent in. This case is equivalent to the concentration transport within a single-phase flow. The homogenized equation occurs to be of the same type as the initial one and is uniform for the whole medium.

Peclet Number Is Small Only in Blocks
A system, in which the diffusion prevails on the blocks, whereas the diffusion and the convection are of the same order in the matrix, such that ($Pe^I \ll 1$, $Pe^{II} \sim 1$), was analyzed in [36], [55], [56]. The macroscale model of concentration transport constitutes a system of two equations, for the blocks and the matrix, related by an exchange term proportional to the difference between the averaged concentrations.

In this chapter several new sub-classes of such systems will be shown, which have not been described earlier and where the exchange process is governed by other laws.

Both Peclet Numbers Are Large
In all papers devoted to the case of convection dominating in the overall medium, only the one-dimensional flow is examined and only for moderately heterogeneous case [3], [4], [5], [6], [7]. Governing transport equation is

hyperbolic of first order. Basic problem is associated to the phenomena of interaction between solution discontinuity and discontinuity of permeability and porosity. Multiple desintegrations of the solution discontinuity passing through the block-matrix interface, reflection of the desintegrated waves, as well as the interaction of desintegrated and reflected waves form a complex picture of multiple small jumps moving in various directions, colliding with the interfaces and one with another, and desintegrating after each collision.

Due to collisions of jumps, their motion is mutually correlated. The correlation scale is a new spatial scale arising in the system. It is equal to the time between to neighbor collisions multiplied by the flow velocity. When the correlation scale is larger than the medium heterogeneity scale, the averaged equations will be unlocal. The problem is to define the kernels.

The situation is complicated by the fact that the solution discontinuity must satisfy the Hugoniot-like conservation conditions, as well as the entropy conditions of stability. Thus, the correlation scale is varying in time and can not be determined before the full solution be constructed.

For highly heterogeneous case a single, entirely approximate, attempt to propose some averaged equations for prevailing convective transport in the overall medium is known, [15]. It is based on the assumption that the convection transport should induce an exchange term which would has the form of the difference between two concentrations divided by some characteristic time, which is as usual unknown.

Peclet Number Is Large in the Matrix

In [26], [27] the case is investigated when convection dominates only in thin fractures (contrary, the diffusion dominates inside the blocks). The condition of thin fractures is in essence, because the hyperbolic equations of convection transport appear locally one-dimensional and may be studied via homogenization technique.

Numerical Models

In the same time the numerical methods with two scale grids are applied. The large-scale grid describes the averaged transport in the matrix; the small-scale grid describes the process in each block. The large-scale equation contains a function of exchange with the blocks. This function is computed from the numerical solution of the local problem in each block, using the small-scale grid. To be closed, it needs the boundary conditions, that conjugate both solutions of different scale. Two kinds of conjugate boundary conditions are used.

In [37], [54], [53], [58] it is admitted that the concentration at the block border is equal to the averaged concentration in the matrix. In [54] the model of similar type is named as *"micro structural conjugate model"*.

Other conditions at the interface are proposed in paper [54]. The flow rate at the block border is assumed to be proportional to the difference between the averaged concentration inside the matrix and the block border concentration. The corresponding model was named as *"regularized micro structural model"*. In [32] the correctness of Cauchy's problem was studied.

Methods of Averaging

Techniques named *"multicomponent homogenization"* was elaborated in [10], [41], [42], [43] for the media where high conductive inclusions are non connected. The results of homogenization have been shown to be splitted onto equations appropriated to each component of the medium.

In the same time, for those media where the connected sub-domain is always high conductive, the formal techniques of *"nonuniform homogenization"* was developed in [44], [45], [46], [47]. The nonuniform averaged value can be obtained by the integration over a period with small weighting parameter in the blocks. The idea of method consists in developing the asymptotic expansions when the zero term constitutes a nonuniform averaged value.

The more strict mathematical approach is based on a two-scale convergence method [38], [1].

2.2. SETTING OF THE PROBLEM

2.2.1. PHYSICAL MODEL

Model of the Fluid

Assume the one-phase flow of a fluid consisting of N various chemical components. All components are entirely miscible. Any chemical reaction is not allowed. Let c_i be the weight concentration of $i-th$ component in the mixture. A component differs from another by its diffusion parameter.

To keep linearity of the equations, we assume the fluid viscosity μ is constant and the fluid volume density ρ is not dependent on the concentration. However the density may be dependent on the pressure (compressible fluid).

Elastic Properties of the Fluid and the Medium

We consider a weakly compressible fluid and a weakly deformable porous rock, such that the porosity and the fluid density are linear functions of pressure with small derivatives:

$$m{=}m_0{+}m'(p{-}p_0), \qquad \rho{=}\rho_0{+}\rho'(p{-}p_0)$$

Due to this assumption, a usual linear parabolic equation may be applied to describe the pressure field:

$$\gamma\frac{\partial p}{\partial t} - \frac{\partial}{\partial x_i}\left(\frac{K_{ij}}{\mu}\frac{\partial p}{\partial x_j}\right) = f(x,t) \tag{2.3}$$

where γ is the "elastic capacity" of the system determined as:

$$\gamma{=}m_0\rho'{+}m'\rho_0$$

where the prime denotes derivation with respect to the pressure.

Let the following splitting of variables takes place: $m(p,x){=}\ \varphi(p)m_0(x)$. Then we obtain for γ:

$$\gamma = m_0(x)\Big[\rho'{+}\varphi'\rho_0\Big] = \beta m_0, \qquad \beta{\equiv}\rho'{+}\varphi'\rho_0{\equiv}const \tag{2.4}$$

Model of the Heterogeneous Field

Consider the medium in the form of a connected highly permeable matrix Ω^{II} and dense periodically located separated blocks Ω^{I}. Each of subareas Ω^{i} is characterized by the permeability tensor K^i, the porosity m^i, and the diffusion tensor D^i. The viscosity of fluid μ is assumed to be constant.

The volume fractions of blocks and matrix are of the same order, as usual in dual-porosity media.

Let us introduce the ratio of "ij"-th components of the permeability tensor as:

$$\omega_{K,ij}=\frac{K^I_{ij}}{K^{II}_{ij}}, \qquad \forall i,j \tag{2.5}$$

Assume for the sake of simplicity all $\omega_{K,ij}$ are of the same order:

$$\omega_{K,ij} \sim \omega_K, \qquad \forall i,j \tag{2.6}$$

More strictly speaking, the parameter ω_K may be determined as the mean value among all $\omega_{K,ij}$. Introducing the ratios of porosity and diffusion parameters in the same way, we obtain three parameters defining the degree of heterogeneity:

$$\omega_K=\frac{\overline{K}^I}{\overline{K}^{II}}, \qquad \omega_D=\frac{\overline{D}^I}{\overline{D}^{II}}, \qquad \omega_m = \frac{m^I}{m^{II}} \tag{2.7}$$

where the bar over symbol implies for a mean value.

To simplify the notations the following values will be introduced:

- the tensor $a=K/\mu$ named as the medium conductivity;
- the scalar $b=m_0$ equal to the main part of the porosity.

Note that the ratio $\omega_a = \overline{a}^I/\overline{a}^{II}$ is equal to the parameter ω_K, while $\omega_b=b^I/b^{II}$ is equal to ω_m:

$$\omega_a = \omega_K, \qquad \omega_b = \omega_m \tag{2.8}$$

2.2.2. MATHEMATICAL SETTING

Micro-Scale Transport Equations

Equations of concentration c transport and pressure p variation in such a medium take the following form:

$$L_\varepsilon^{\omega_a,\omega_b} p \equiv \frac{\partial}{\partial x_i}\left(a_{ij}(\frac{x}{\varepsilon},\omega_a)\frac{\partial p}{\partial x_j}\right) - \beta b(\frac{x}{\varepsilon},\omega_b)\frac{\partial p}{\partial t} = -f(x,t,\omega_a) \tag{2.9a}$$

$$L_\varepsilon^{\omega_D,\omega_a,\omega_b} c \equiv \frac{\partial}{\partial x_i}\left(D_{ij}(\frac{x}{\varepsilon},\omega_D)\frac{\partial c}{\partial x_j}\right)+a_{ij}(\frac{x}{\varepsilon},\omega_a)\frac{\partial p}{\partial x_i}\frac{\partial c}{\partial x_j} - b(\frac{x}{\varepsilon},\omega_b)\frac{\partial c}{\partial t}=0 \tag{2.9b}$$

$$t>0, \qquad x\in\Omega/\Gamma, \qquad \Omega\equiv R^3$$

$$[p]_\Gamma = 0, \qquad \left[a_{ij}\frac{\partial p}{\partial x_j n_i}\right]_\Gamma = 0 \tag{2.10}$$

$$0 = \left[D_{ij}\frac{\partial c}{\partial x_j}n_i\right]_\Gamma + a^I_{ij}\frac{\partial p}{\partial x_j}n_i\Big|_{\Gamma^I}[c]_\Gamma \tag{2.11}$$

$$[c]_\Gamma = 0 \tag{2.12}$$

$$\Omega = \Omega^I \cup \Omega^{II}; \quad \Omega^I = \cup\partial\Omega^I_{(i)}; \quad \Omega^I_{(i)} \cap \Omega^I_{(j)} = \emptyset, \ i{\neq}j; \quad \Gamma = \cup\partial\Omega^I_{(i)}$$

We consider the initial problem:

$$p(x,0) = p^0(x), \qquad c(x,0) = c^0(x) \tag{2.13}$$

Herein n is the external normal vector to the boundary; ε is the period of the functions a_{ij}, b, D_{ij} ; Γ is the interface between the block and the matrix; Γ^I or Γ^{II} means the appropriate side of the interface Γ. The symbol Γ which was used earlier for the interface between Ω^I and Ω^{II} will be kept for the image of this interface after rescaling $y{=}x/\varepsilon$, in order to reduce the volume of new notations.

The factor $b{>}0$, the matrixes a_{ij} and D_{ij} are positively defined. The constant β is defined according to (2.4).

Equations (2.9a) and (2.9b) are assumed to be written in the dimensionless form.

Note the problem for the pressure is entirely independent.

Conditions at the Interface

Conditions (2.10) mean continuity of both the pressure and the total flow rate through the interface between the block and the matrix. Condition (2.11) means continuity of the flow rate for a single component of the mixture.

As to condition (2.12), it is only needed when diffusion takes place everywhere in the medium. In this case condition (2.11) may be simplified to the following:

$$0 = \left[D_{ij}\frac{\partial c}{\partial x_j}n_i\right]_\Gamma \tag{2.14}$$

that means continuity of diffusion flux.

When diffusion takes place only in one of sub-domains the concentration may appear discontinuous, such that relation (2.12) becomes invalid. Hence, condition (2.11) should be applied instead of (2.14).

2.2.3. PARAMETER RANGE. COARSE CLASSIFICATION OF SYSTEMS

Parameters of the Process

Four parameters determine the process:

$$\varepsilon, \quad \omega_a, \quad \omega_D, \quad \omega_b \tag{2.15}$$

Furthermore two additional parameters, such as local Peclet numbers, will be used:

$$Pe^i = \frac{\overline{a}^i}{\overline{D}^i}, \quad i = I, II$$

Let the following be true:

$$\varepsilon \ll 1, \qquad \omega_a \ll 1, \qquad \omega_b \sim 1 \tag{2.16}$$

The first condition means the medium is fast oscillating. The second one is the condition of high heterogeneity. The third relation in (2.16) says that porosity variations are considerably smaller than variations of permeability, that happens in practice (see chapter 1).

Variations of the parameter ω_D are not limited, such that it can be of order one, or be tending to zero, either be tending to infinity. The meaning of all these cases is analyzed in the next section.

Respectively to section 2.1.1, convection and diffusion are assumed to be of the same order in the matrix:

$$Pe^{II} = \frac{\overline{a}^{II}}{\overline{D}^{II}} \sim 1 \tag{2.17}$$

Behavior of the model (2.9a)-(2.11) will be examined at $\varepsilon, \omega_a \to 0$, and at various magnitudes of the parameter ω_D.

Allowed Classes of Systems

The last condition in (2.16) allows to exclude the parameter ω_b from analysis. Due to condition (2.17) and the second limitation in (2.16) for ω_a, the set of probable systems becomes relatively small. Four pictures in the Figs. 2.2a, 2.2b, 2.2c and 2.2d show all probable classes.

Three cases differ in the order of the block Peclet number. In all cases the matrix Peclet number Pe^{II} is of order one, according to (2.17).

As the parameter ω_D is not limited, theoretically a situation is admissible when diffusion in blocks is more important than in matrix. This case is shown in Fig. 2.2d. It corresponds to the relation $Pe^I << 1$, therefore this is a particular case of the situation shown in Fig. 2.2c.

Explained classification is rather coarse, because it does not take into account various geometrical types of flow. A more fine classification will be presented in the next sections.

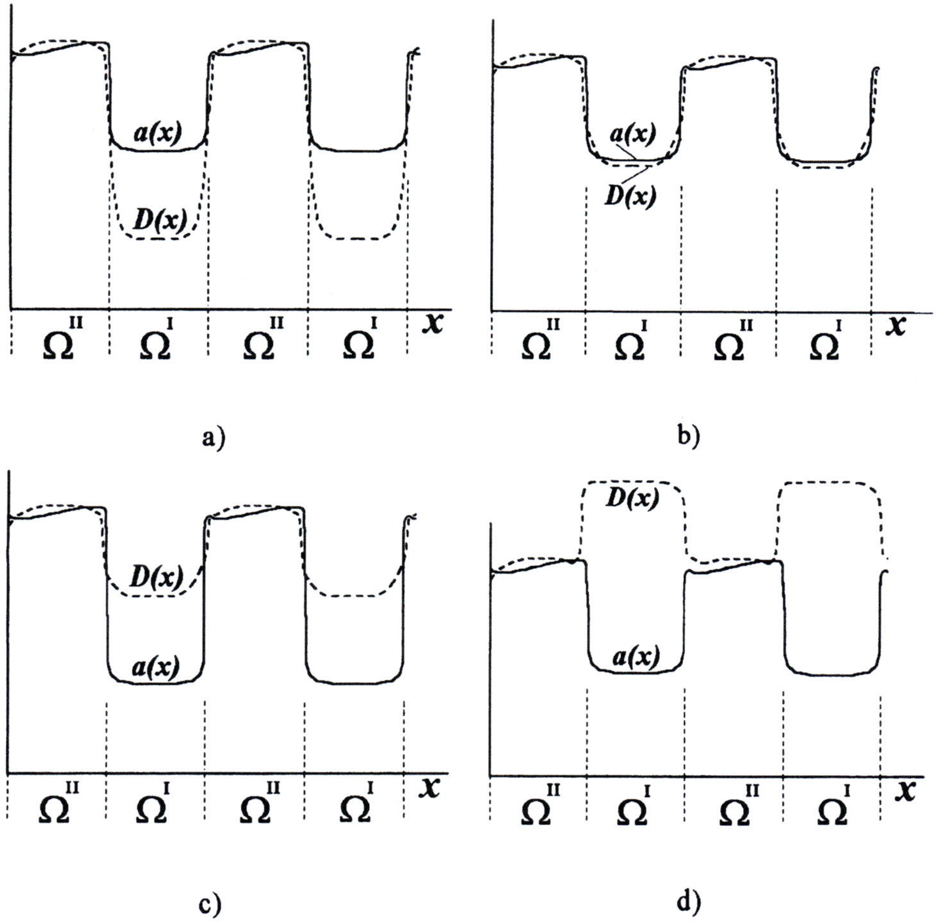

Figure 2.2. Spatial variations of conductivity $a(x)$ and diffusion parameter $D(x)$ in highly heterogeneous 1D-medium:
a) convection prevails in blocks, $Pe^I >> 1$,
b) convection and diffusion in blocks are of the same order, $Pe^I \sim 1$,
c) diffusion prevails in blocks, $Pe^I << 1$
d) diffusion in blocks prevails and is higher than in matrix

2.2.4. PHYSICAL CONTENT OF THE PROBLEM

The physical content of the problem examined corresponds to both the transport of chemical components and the heat transport by porous flow.

For these two cases the dimensionless variables and the variables with dimension are related in the following ways.

Chemical Transport by Porous Flow

$$\begin{cases} x=X/L, \quad p=P/P^0, \quad \tau=t/t^*, \\ t^*=\frac{L^2}{\langle K\rangle P^0/\langle m\rangle\mu^0}, \quad \beta=\frac{d}{dp}\left(\frac{\rho(p)}{\rho^0}\right), \\ a(x)=\frac{K(x)\mu^0}{\langle K\rangle\mu}, \quad b(x)=\frac{m}{\langle m\rangle}, \quad D(x)=\frac{\tilde{D}}{\langle\tilde{D}\rangle} \\ Pe=\frac{\langle K\rangle P^0}{\mu^0\langle\tilde{D}\rangle}, \quad \rho^0=\rho(p^0) \end{cases} \tag{2.18}$$

where K is the permeability; m is the porosity; P is the dimension pressure; $\tilde{D}$ is the dimensional dispersion (diffusion) parameter; ρ is the fluid density; μ is the viscosity; X, t are dimensional vector coordinate and time. The value μ^0 is some characteristic viscosity. If the viscosity is constant, then $a(x)$ is the dimensionless permeability of the medium. The symbol $\langle\rangle$ denotes averaging over the domain Ω.

Heat Transport in Saturated Porous Medium

$$\begin{cases} x=X/L, \quad p=P/P^0, \quad \tau=t/t^*, \quad t^*=\frac{L^2\langle m\rangle\mu^0}{\langle K\rangle P^0}, \\ \theta=T/T^0, \quad C_e=C_s+C_p m\rho, \quad a(x)=\frac{K(x)\mu^0}{\langle K\rangle\mu}, \quad b(x)=\frac{C_e}{C_p\langle m\rangle\rho^0}, \\ D(x)=\frac{\lambda_T}{\langle\lambda_T\rangle}, \quad Pe=\frac{\langle K\rangle P^0 C_p}{\mu^0\langle\lambda_T\rangle} \end{cases} \tag{2.19}$$

where T is the temperature; C_p and C_s are the heat capacities of fluid and rocks; λ_T is the heat conductivity of rocks. The value t^* is the characteristic time of conductive propagation for the heat wave.

The Peclet number Pe means the ratio of fluid flow velocity over conductive heat transport velocity: $Pe=\frac{\langle a\rangle}{\langle D\rangle}$

Heterogeneity Scales

The microscale transport equations, being written in the form with dissipative term, implicitly mean that the medium has also a submicroscopic scale of heterogeneity which corresponds to the size of a pore. Hence, "microscale" equations (2.9a)-(2.9b) should be considered as averaged over submicroscopic heterogeneity.

Thus, in this problem the microscopic level corresponds to the scale l including a rather large ensemble of pores, but in such a way that a medium element of size l would have uniform permeability, porosity and diffusion parameter. The macroscale corresponds to the length L along which the permeability and other parameters are varying many times, such that $l/L=\varepsilon$.

2.2.5. TWO-SCALE FORMULATION

Preparatory to do the asymptotic analysis, the problem will be reformulate in two-scale form.

Let the extended two-scale variables $\tilde{p}=p(x,y,t)$ and $\tilde{c}= c(x,y,t)$ be introduced, where y is a new independent argument. The original functions $p(x,t)$ and $c(x,t)$ are equal to $\tilde{p}(x,y,t)$ and $\tilde{c}(x,y,t)$, when $y=x/\varepsilon$.

Furthermore only two-scale variables will be used, such that the symbol "tilde" will be omitted anywhere.

Along with area Ω the image $y = x/\varepsilon$ of its separate periodic element on the cell $Y=\{-1/2<Y^i<1/2, \quad i=1,2,3\}$ will be considered. The surface $\Gamma=\partial Y^I$ breaks it on two subareas Y^I (block) and Y^{II} (matrix), $\Gamma\cap\partial Y=\emptyset$. The factors $a(y), b(y)$ undergo a discontinuity of first kind on Γ.

i) For pressure:

$$L^a p - \varepsilon_p \beta b^{II} \partial_t p = 0, \ \ y \in Y^I; \tag{2.20a}$$

$$L^a p - \varepsilon^2 \beta b^{II} \partial_t p = 0, \ \ y \in Y^{II} \tag{2.20b}$$

$$\frac{\varepsilon^2 \omega_b}{\varepsilon_p} a_{ij}^{II} \nabla_j^\varepsilon p^I n_i \Big|_\Gamma - a_{ij}^{II} \nabla_j^\varepsilon p^{(2)} n_i \Big|_\Gamma = 0 \tag{2.20c}$$

ii) For concentration:

$$L^D c + Pe{\cdot}a_{ij}^{II} \nabla_i^\varepsilon p \nabla_j^\varepsilon c - \varepsilon_c b^{II} \partial_t c = 0, \ \ y \in Y^I \tag{2.21a}$$

$$L^D c + a_{ij}^{II} \nabla_i^\varepsilon p \nabla_j^\varepsilon c - \varepsilon^2 b^{II} \partial_t c = 0, \ \ y \in Y^{II} \tag{2.21b}$$

$$\frac{\varepsilon^2 \omega_b}{\varepsilon_p Pe} D_{ij}^{II} \nabla_j^\varepsilon c^I n_i \Big|_\Gamma - D_{ij}^{II} \nabla_j^\varepsilon c^{II} n_i \Big|_\Gamma = -\frac{\varepsilon^2 \omega_b}{\varepsilon_p} a_{ij}^{II} \nabla_i^\varepsilon p^I n_j \Big|_\Gamma \cdot [c]_\Gamma, \tag{2.21c}$$

where $[c]_\Gamma \equiv c^I|_\Gamma - c^{II}|_\Gamma$, $c^i \equiv c(x,y,t)$ at $y \in Y^i$, $\vec{n}$ is the normal vector to the interface $\Gamma \equiv \partial Y^I$ external with respect to the area Y^I.

$$L^D \equiv L_0^D + \varepsilon(L_{11}^D + L_{12}^D) + \varepsilon^2 L_2^D, \quad \nabla_i^\varepsilon \equiv (\partial_i + \varepsilon \partial_{xi}), \quad L_0^D \equiv \partial_i (D_{ij}^{II} \partial_j),$$

$$L_{11}^D \equiv \partial_i (D_{ij}^{II} \partial_{xj}), \quad L_{12}^D \equiv \partial_{xi} (D_{ij}^{II} \partial_j), \quad L_2^D \equiv \partial_{xi} (D_{ij}^{II} \partial_{xj})$$

$$\partial_i \equiv \frac{\partial}{\partial y_i}, \quad \partial_{xi} \equiv \frac{\partial}{\partial x_i}, \quad \partial_t \equiv \frac{\partial}{\partial t},$$

L^a is determined in the same way as L^D , if one replace "D_{ij}^{II}" by "a_{ij}^{II}".

iii) Integral relations playing a role of necessary conditions for existence of the periodic solution of problem (2.20), (2.21), can be obtained by integrating Eqs. (2.20), (2.21) over the whole cell Y:

$$\frac{\varepsilon^2\omega_b}{\varepsilon_p}\langle L_{12}^a p + \varepsilon L_2^a p\rangle_1 + \langle L_{12}^a p + \varepsilon L_2^a p\rangle_2 - \varepsilon\langle b\partial_t p\rangle = 0 \qquad (2.22)$$

$$\frac{\varepsilon^3\omega_b}{\varepsilon_p Pe}\langle L_{12}^D c + \varepsilon L_2^D c\rangle_1 + \varepsilon\langle L_{12}^D c + \varepsilon L_2^D c\rangle_2 -$$
$$- \varepsilon^2\langle b\partial_t c\rangle + \frac{\varepsilon^2\omega_b}{\varepsilon_p}\left\langle a_{ij}^{II}\nabla_i^\varepsilon p\nabla_j^\varepsilon c\right\rangle_1 + \left\langle a_{ij}^{II}\nabla_i^\varepsilon p\nabla_j^\varepsilon c\right\rangle_2 -$$
$$- \frac{\varepsilon^2\omega_b}{\varepsilon_p}\int_\Gamma a_{ij}^{II}\nabla_j^\varepsilon p^I n_i[c]_\Gamma dy = 0 \qquad (2.23)$$

where the symbol $<>^i$ designates integration over Y^i, i=I, II.

Parameters of the Problem

Instead of the parameters (2.15) another system of parameters will be considered which is more convenient to be used in next deductions. They are:

$$\varepsilon, \quad \omega_b, \quad Pe \equiv \frac{\omega_a}{\omega_D}, \quad \varepsilon_p \equiv \frac{\varepsilon^2\omega_b}{\omega_a}, \quad \text{or} \quad \varepsilon_c \equiv \frac{\varepsilon^2\omega_b}{\omega_D} = \varepsilon_p Pe$$

Parameters ε_p and ε_c mean the ratio of the time of pressure (or concentration) wave propagation through a single block over the time of wave propagation through the overall matrix ; Pe=Pe^I/Pe^{II}~Pe^I is dimensionless Peclet number in blocks.

As ω_b is fixed according to (2.16), we have two parameters $(\varepsilon, \varepsilon_p)$ that determines the pressure field, and 3 parameters $(\varepsilon, \varepsilon_p, Pe)$, that determines the concentration field.

2.3. CLASSIFICATION OF FLOWS

2.3.1. ELEMENTARY FLOWS IN A CELL

It is necessary to distinguish the flow of the whole fluid which is entirely of convective type, and the flow of each component of fluid which can be additionally of diffusion type.

Difference between the convection and the diffusion flow consists in its physical origin. However, the flows cam also differ in the geometry of streamlines.

Elementary Flows with Various Streamline Geometry
Results of homogenization for the pressure field presented in chapter 1 show the existence of two main kinds of convection elementary flow in the block which are distinguished in the type of streamline geometry. They are:

- the convection source-type flow is directed out of the block center towards its boundary, or vice versa; it is caused by the difference between the averaged block and matrix pressures ;
- the convection translation-type flow is directed in the sense of the macroscale matrix pressure gradient.

The difference between these elementary flows is shown in Fig. 1.5 of section 1.2.3, chapter 1.

The total flow of the whole fluid is a superposition of two elementary:

$$grad_y p = \nabla_y^{CS} p + \nabla_y^{CT} p \tag{2.24}$$

where CS, CT correspond to convection-source and convection-translation flows. It was shown that:

$$\nabla_y^{CT} p \sim \omega_a, \qquad \nabla_y^{CS} p \sim \varepsilon_p \tag{2.25}$$

All Classes of Elementary Flows
These results allow to find all probable types of concentration transport through a cell. It is clear a priori that the geometry of streamlines does not depend on the physical origin of the flow. Therefore four classes of elementary flows are expected to arise.

Equation (2.21) can be written in the following way:

$$b^{II}\frac{\partial c}{\partial t} = \frac{1}{\varepsilon_p Pe}\Big[L_0^D c + \varepsilon(L_1^D + \varepsilon L_2^D)c + Pe{\cdot}a^{II}\nabla_y^{CS}p{\cdot}\nabla_y^{\varepsilon}c + $$

$$+ Pe{\cdot}a^{II}\nabla_y^{CT}p{\cdot}\nabla_y^{\varepsilon}c\Big], \qquad y{\in}Y^I \tag{2.26}$$

where relation (2.24) has been used.

Four terms in the right-hand part define four elementary types of concentration transport through a block , which differ in the streamline geometry (source or translation type), as well as in the prevailing physical component of transport (convection or diffusion):

D-S*ource flow* is caused by the difference between averaged block and matrix concentrations , that produces a diffusion flux inside a block with the source-type geometry of streamlines;

D-T*ranslation* is caused by the macroscale matrix concentration gradient , that produces a diffusion flux through a block with a translation type geometry of streamlines;

C-S*ource flow* is caused by the difference between averaged block and matrix pressure , that produces a convection transport inside a block, with a source type geometry of streamlines;

C-T*ranslation* is caused by the macroscale matrix pressure gradient producing a convective flow through a block, with translation type geometry of streamlines.

Two last flow types were defined above. Letters "C" and "D" correspond to "convection" and "diffusion".

2.3.2. CONSTRUCTION OF THE DIAGRAM OF ELEMENTARY FLOWS

Various parameter ratio determines predomination of various flow types. Analysis of (2.26) and (2.25) allows to construct the full classification of probable systems, before any solution.

Remind that such media are examined where the blocks are less permeable, such that:

$$\varepsilon_p > \varepsilon^2 \tag{2.27}$$

Basic Transport Equation in Blocks

The concentration field can be presented as the sum of a non-oscillating component $c_0(x,t)$ and an oscillating component dependent on the fast variable y:

$$c \equiv c^I \sim c_0(x,t) + \delta c_1(x,y,t), \qquad \delta \geq \varepsilon \tag{2.28}$$

The value δ can not be smaller than ε, because the asymptotic expansion of a homogenized problem always contains a term of order ε.

According to (2.24) and (2.25), the following notations are introduced:

$$a^{II} \nabla_y^{CT} p \sim \varepsilon v^{CT}, \qquad a^{II} \nabla_y^{CS} p \sim \varepsilon_p v^{CS} \tag{2.29}$$

where v^{CT} and v^{CS} are some flow velocities of order one.

Then transport equation (2.26) gets the form:

$$\underbrace{b^{II}\partial_t c_0}_{A} = \underbrace{\frac{\delta}{\varepsilon_p Pe} L_S^D c_1}_{DS} + \underbrace{\frac{\varepsilon}{\varepsilon_p Pe} L_T^D c_0}_{DT} + \underbrace{v_i^{CS}\left(\delta\partial_i c_1 + \varepsilon\partial_{xi} c_0\right)}_{CS} +$$

$$\underbrace{\frac{\varepsilon}{\varepsilon_p} v_i^{CT}\left(\delta\partial_i c_1 + \varepsilon\partial_{xi} c_0\right)}_{CT} + \{\text{ macro-flow}\} \; ... \qquad (2.30)$$

where the "macro-flow" means the terms describing the flow at the macro-scale; the operators are defined as:

$$L_S^D \equiv L_0^D \equiv \partial_i(D_{ij}\partial_j), \quad L_T^D \equiv \partial_i(D_{ij}\partial_{xj}), \quad \partial_i \equiv \frac{\partial}{\partial y_i}, \quad \partial_{xi} \equiv \frac{\partial}{\partial x_i}, \quad \partial_t \equiv \frac{\partial}{\partial t}$$

Let the symbol $\succ$ replaces the words "prevails over", such that a formula

$$A \succ B$$

means that the event A prevails over the event B.

Correspondence Between C-T and C-S Flows
It follows from (2.30) that the CT-flow $\succ$ CS-flow when $\varepsilon_p < \varepsilon$, and vice versa. This situation is imaged in Fig. 2.3.

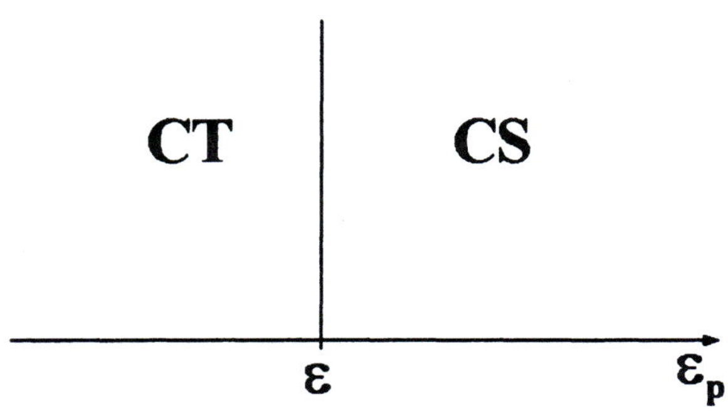

Figure 2.3. Correspondence between C-T and C-S flows

Domination of D-S Flow when $\varepsilon_p < \varepsilon$
Examine the case when D-S flow prevails. This is satisfied when equation (2.30) yields in zero approximation: $A = DS$. Then the necessary conditions

are the following:

$$\delta > \varepsilon, \qquad \frac{\delta}{\varepsilon_p Pe} > \frac{\varepsilon}{\varepsilon_p}\delta, \qquad \frac{\delta}{\varepsilon_p Pe} \sim 1$$

The first condition means that DS-term≻DT-term in (2.30), the second condition means that DS-term≻CT-term. The third condition implies that the DS-term is equilibrated by the time derivative A in the left-hand part. Then the flow is nonstationary by its definition.

As the result we get:

$$\frac{\varepsilon}{\varepsilon_p} < Pe < \frac{1}{\varepsilon}, \tag{2.31}$$

$$\delta = \varepsilon_p Pe > \varepsilon \tag{2.32}$$

This subdomain of D-S flow predomination is shown in Fig. 2.4.

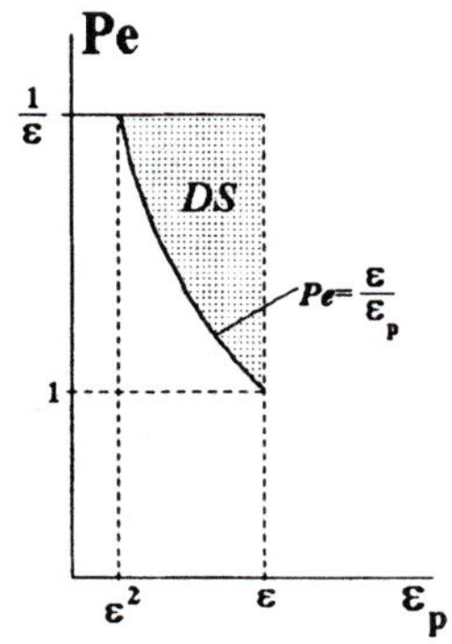

Figure 2.4. Predomination of D-S flow at $\varepsilon_p < \varepsilon$

Domination of D-S Flow when $\varepsilon_p > \varepsilon$

To D-S flow would prevail in this case, the following is necessary to be satisfied:

$$\delta > \varepsilon, \qquad \Longrightarrow \qquad DS \succ DT,$$

$$\frac{\delta}{\varepsilon_p Pe} > \delta, \qquad \Longrightarrow \qquad DS \succ CS,$$

$$\frac{\delta}{\varepsilon_p Pe} \sim 1, \qquad \Longrightarrow \qquad DS \sim A$$

Then we obtain:

$$\frac{\varepsilon}{\varepsilon_p} < Pe < \frac{1}{\varepsilon_p}, \tag{2.33}$$

$$\delta = \varepsilon_p Pe > \varepsilon \tag{2.34}$$

This subdomain of D-S flow predomination is shown in Fig. 2.5.

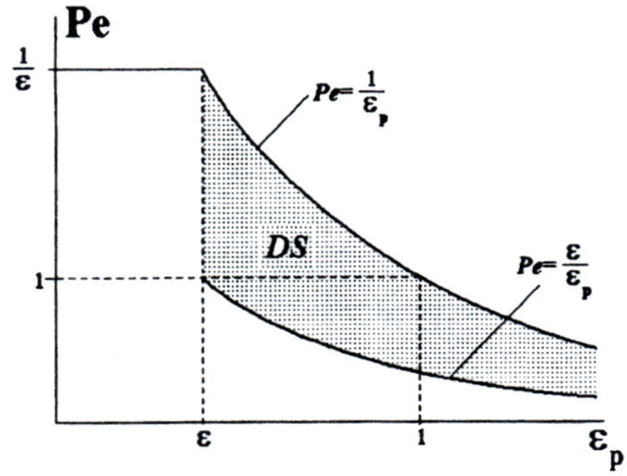

Figure 2.5. Predomination of D-S flow at $\varepsilon_p>\varepsilon$

Domination of D-T Flow when $\varepsilon_p<\varepsilon$
In general case, the necessary condition of D-T flow domination is $DT{=}DS$ in zero approximation of (2.30). Then the following is necessary to be satisfied:

$$\delta\sim\varepsilon, \qquad \Longrightarrow \qquad DT\sim DS,$$
$$\frac{\varepsilon}{\varepsilon_p Pe}>\frac{\varepsilon^2}{\varepsilon_p}, \qquad \Longrightarrow \qquad DT\succ CT,$$
$$\frac{\varepsilon}{\varepsilon_p Pe}>1, \qquad \Longrightarrow \qquad DT\succ A$$

The last condition means the D-T flow is quasi-stationary.
Then we obtain:

$$Pe<\frac{1}{\varepsilon}, \quad Pe<\frac{\varepsilon}{\varepsilon_p} \tag{2.35}$$
$$\delta=\varepsilon \tag{2.36}$$

The appropriate domain is shown in Fig. 2.6.

Domination of D-T Flow when $\varepsilon_p>\varepsilon$
To D-T flow prevails in this case the following is necessary:

$$\delta\sim\varepsilon, \qquad \Longrightarrow \qquad DT\sim DS,$$
$$\frac{\varepsilon}{\varepsilon_p Pe}>\varepsilon, \qquad \Longrightarrow \qquad DT\succ CS,$$
$$\frac{\varepsilon}{\varepsilon_p Pe}>1, \qquad \Longrightarrow \qquad DT\succ A$$

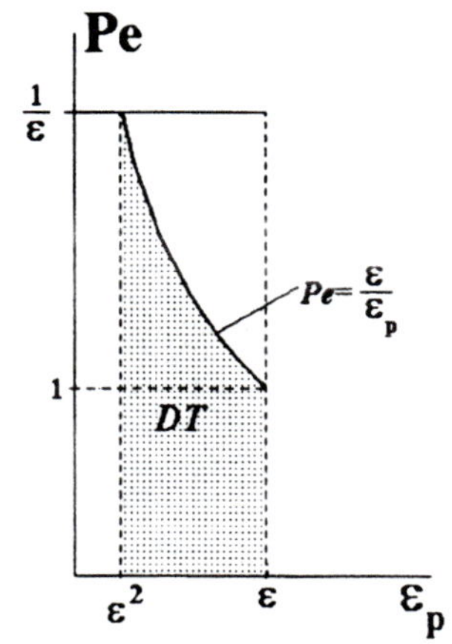

Figure 2.6. Predomination of D-T flow at $\varepsilon_p<\varepsilon$

Then it follows:

$$Pe<\frac{1}{\varepsilon_p}, \quad Pe<\frac{\varepsilon}{\varepsilon_p} \tag{2.37}$$

$$\delta=\varepsilon \tag{2.38}$$

The corresponding domain is shown in Fig. 2.7.

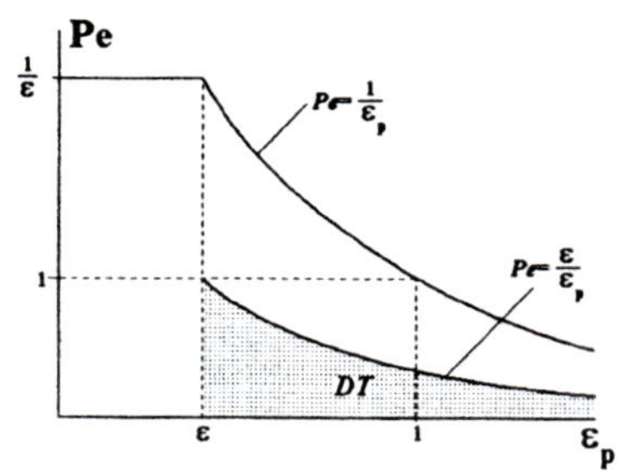

Figure 2.7. Predomination of D-T flow at $\varepsilon_p<\varepsilon$

2.3.3. DIAGRAM OF ELEMENTARY FLOWS

Collecting results (2.31), (2.33), (2.35) and (2.37), we obtain the total diagram of elementary flow predomination which is shown in Fig. 2.8.

The top and the bottom curve lines are the functions $Pe=1/\varepsilon_p$ and $Pe=\varepsilon/\varepsilon_p$.

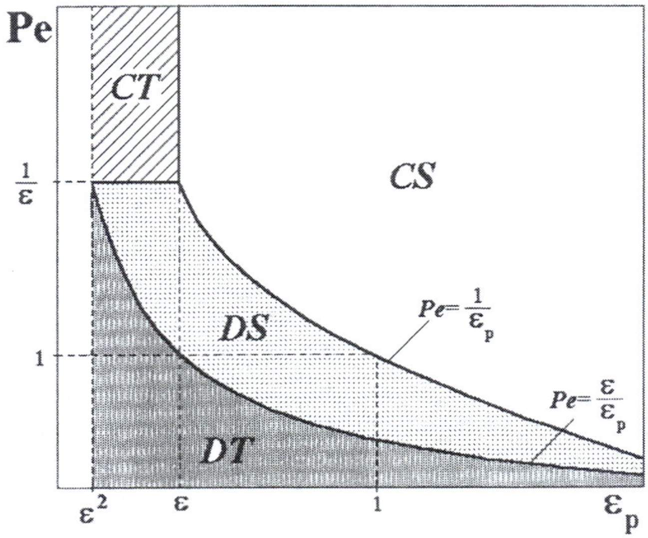

Figure 2.8. Diagram of Elementary Flow Predominance for Convection-Diffusion Transport

According to (2.28), (2.32), (2.34), (2.36) and (2.38), first terms of the asymptotic expansions for c^I in each zone may be written in the following way:

$$
\begin{aligned}
CS:&\quad c^I = c_0(x,y,t)+o(1)\ldots\\
CT:&\quad c^I = c_0(x,y,t)+\frac{\varepsilon_p}{\varepsilon}c_1(x,y,t)\ldots\\
DT:&\quad c^I = c_0(x,t)+\varepsilon c_1(x,y,t)\ldots\\
DS:&\quad c^I = c_0(x,t)+\varepsilon_p Pe\cdot c_1(x,y,t)\ldots
\end{aligned}
$$

In the diagram the interfaces mean the equilibrium between two corresponding flow types. The point ($\varepsilon_p=\varepsilon$, $Pe=1/\varepsilon$) corresponds to the full equilibrium between all four types of flow.

Note, the Peclet number can be higher than 1, however diffusion prevails over convection. It is not strange, because the value Pe is the macroscale parameter, while the microscale Peclet number, $Pe_{mic}=Pe\cdot\varepsilon$. Hence, the large values of Pe up to $1/\varepsilon$ correspond to the value of the microscale Peclet number lower than one.

In this chapter all the types of transport will be examined.

2.4. C-S SYSTEM: DEDUCTION OF HOMOGENIZED EQUATIONS

2.4.1. FEATURE OF THE SYSTEM

We will homogenize the sub-system from area II of the diagram in Fig. 2.8. In the blocks, the convective transport is prevailing, whereas the geometry of streamlines corresponds to the source type.

Basic problems which are expected to appear are caused by the fact that the concentration transport inside each block happens in the form of travelling waves propagating from the block border towards the block center or vice versa. The following features should be noted as corollaries.

1. The transport in each block is sufficiently non steady-state. Then the steady-state description of exchange phenomena between the blocks and the matrix is not valid, while the true description is not clear.

2. As the equation of traveling wave is not invariant respectively to the sign of the transport velocity, a stream penetrating into the block and a stream going out of the block are non symmetric. Thus, the macroscale behavior is expected to be dependent on the direction of the microscale flow within a cell.

3. In general case various blocks may be operating in various modes, either in the mode of filling or in the mode of draining. Thus, the macroscale transport is expected to be spatially unlocal with a large correlation length.

4. The following property of the travelling wave equation is known. Let the transport velocity v be constant and the concentration in the initial point $x{=}0$ be fixed as some function of time, $c_0(t)$, such that:

$$\frac{\partial c}{\partial t}+v\frac{\partial c}{\partial x}=0, \qquad c(0,t)=c_0(t)$$

Then concentration evolution in any other point x repeats the evolution of the boundary concentration $c_0(t)$ but with the delay time equal to $\Delta t{=}x/v$: $c(x,t){=}c_0(t-\Delta t)$. In the considered problem the value c is the concentration inside the block, while c_0 corresponds to the matrix concentration. Thus, the averaged block concentration is expected to be related with the averaged matrix concentration as a delay differential equation showing unlocal behavior in time.

5. A concentration discontinuity can arise at the block interface. However, as diffusion in blocks is assumed to be small but non zero, any discontinuity is not possible. It appears only as the result of neglecting the thin diffusion boundary layer. Appearance of such a layer generates a new

spatial scale in the system which is smaller than the microscale l. Within the diffusion boundary layer the asymptotic expansions must be corrected, such that the small diffusion terms becomes of zero order there.

6. When the flow direction inside a block is changed many times, several surfaces of concentration discontinuity can appear, that should be reflected on the averaged relations.

Let the parameter ratio be fixed as:

$$Pe \sim 1/\varepsilon, \qquad \varepsilon<\varepsilon_p<1 \tag{2.39}$$

According to chapter 1, the single phase flow with the parameter ε_p chosen in such a manner is intermediate between the source-type and translation-source. The first advantage of such a choose consists in that, the long memory equation for the pressure field, which is appropriate for a source-type system, will be replaced by more simple relaxation equations. Secondly, conditions (2.39) allow to take into account the effect of C-Translation flow in the first approximation.

Moreover, the effect of D-Source flow will be also taken into consideration, when the diffusion boundary layers are expected to arise.

Hence, below in the zero approximation the C-Source system is examined, while in the first approximation the C-Translation and the D-Source effects will be added.

2.4.2. RESULTS FOR THE PRESSURE FIELD

The pressure field may be homogenized separately. The following results obtained in chapter 1 take place:

$$p(x,y,t) = \begin{cases} p_{00}(x,t) + \varepsilon_p p_{01}^I(x,y,t) + \varepsilon p_{10}^I(x,y,t) + \ldots, & y \in Y^I \\ p_{00}(x,t) + \varepsilon_p p_{01}^{II}(x,t) + \varepsilon p_{10}^{II}(x,y,t) + \ldots, & y \in Y^{II} \end{cases} \tag{2.40}$$

$$p_{01}^I(x,y,t) = \partial_t p_{00}^I \mu(y) + p_{01}^{II}(x,t) \tag{2.41}$$

$$L_0^a \mu = b^{II}, \quad y \in Y^I; \qquad \mu|_\Gamma = 0 \tag{2.42}$$

$$p_{10}^{(i)}(x,y,t) = \psi_k^{(i)}(y)\partial_{xk} p_{00} + w_{10}(x,t), \qquad i = I, II \tag{2.43}$$

$$L_0^a \psi_k^I = -\partial_i a_{ki}, \quad y{\in}Y^I, \quad \psi_k^I|_\Gamma = \psi_k^{II}|_\Gamma, \quad k = 1,2,3 \tag{2.44a}$$

$$L_0^a \psi_k^I = -\partial_k a^{II}, \quad y{\in}Y^{II}; \quad (\partial_i \psi_k^{II} + \delta_{ki}) n_i|_\Gamma = 0, \quad \langle \psi_k^{II} \rangle = 0, \tag{2.44b}$$

where ψ_k^{II} is 1-periodic for any k, $L_0^a \equiv \partial_i (a_{ij} \partial_i ())$.

Let the block be a sphere of radius R, moreover $a_{ij}^{II}=0$, $i\neq j$; $a_{ii}^{II}\equiv a^{II}=\text{const}$, $i=1,2,3$; and $b^{II}=\text{const}$, then

$$\mu(y)=\frac{b^{II}}{6a^{II}}(r^2-R^2),\quad r^2=y_1^2+y_2^2+y_3^2 \tag{2.45}$$

Note that $L_0^a p_{01}^I=b^{II}\partial_t p_{00}$, $\quad L_0^a p_{10}^I+\varepsilon L_{11}^a p_{00}=0.$

2.4.3. ASYMPTOTIC EXPANSIONS FOR CONCENTRATION

Form of Expansions

We shall seek the solution of the two-scale problem (2.20)-(2.23) in the following form:

$$c(x,y,t)=\sum_{k=0}^{\infty}\sum_{m=-k}^{\infty}\varepsilon^k\varepsilon_p^m c_{km}(x,y,t) \tag{2.46}$$

$$c_{ok}^{II}=c_{ok}^{II}(x,t);\quad c_{k,-k}^{II}=c_{k,-k}^{II}(x,t),\quad \forall k\geq 0$$

The zero term is assumed to be corresponding to a following, non-uniformly averaged value:

$$c_{00}=\langle\chi(\omega_a)c\rangle\equiv\langle\omega_a c^I\rangle_1+\langle c^{II}\rangle_2$$

Thus, we obtain:

$$\langle\chi(\omega_a)c_{km}\rangle=0,\quad k,m>0 \tag{2.47}$$

It is easy to show by a simple substitution that this expansion is correctly determined, so the problem (2.20)-(2.23) gives strongly one independent relationship for each new term of the series.

To construct averaged equations of zero order, it is sufficient to determine the functions c_{00}^I in the blocks and c_{00}^{II}, c_{10}^{II} in the matrix. To take into account the first order terms, it is necessary to determine $c_{1,-1}^I$, $c_{2,-1}^{II}$, $c_{1,-1}^{II}$ too.

Expansion for the Micro-Scale Equations

Substituting series (2.46) into (2.20), (2.21) and collecting the terms of the same power-values of ε and ε_p, we obtain the following results for the first terms of the asymptotic expansions.

1) In Y^I:

$$\sim\varepsilon_p: \quad L^* c_{00}^I = 0, \quad y \in Y^I \tag{2.48a}$$

$$\sim\varepsilon: \quad L^* c_{1,-1}^I + z_i^{(10)} \partial_i c_{00}^I + (\varepsilon Pe)^{-1} L_0^D c_{00}^I = 0; \quad y \in Y^I \tag{2.48b}$$

where

$$L^* \equiv z_i^{(01)} \partial_i - b^{II} \partial_t,$$

$$z_i^{(01)} \equiv a_{ij}^{II} \partial_j p_{01}^I = a_{ij}^{II} \partial_t p_{00} \partial_j \mu, \quad z_i^{km} \equiv a_{ij}^{II} (\partial_j p_{km} + \partial_{\mathrm{x}j} p_{k-1,m}) \tag{2.49}$$

2) In Y^{II}:

$$\sim\varepsilon: \quad L_0^D c_{10}^{II} + L_{11}^D c_{00}^{II} = 0, \quad y \in Y^{II} \tag{2.50a}$$

$$\sim\frac{\varepsilon^2}{\varepsilon_p}: \quad L_0^D c_{2,-1}^{II} + L_{11}^D c_{1,-1}^{II} = 0, \quad y \in Y^{II} \tag{2.50b}$$

3) On Γ:

$$\sim\varepsilon: \quad D_{ij}^{II} \left(\partial_i c_{10}^{II} + \partial_{\mathrm{x}i} c_{00}^{II} \right) n_j|_\Gamma = 0; \tag{2.51a}$$

$$\sim\frac{\varepsilon^2}{\varepsilon_p}: \quad D_{ij}^{II} \left(\partial_i c_{2,-1}^{II} + \partial_{\mathrm{x}i} c_{1,-1}^{II} \right) n_j|_\Gamma = 0 \tag{2.51b}$$

4) Integral relationships:

$\sim\varepsilon^2$:

$$\langle L_{12}^D c_{10}^{II} + L_2^D c_{00}^{II} \rangle_2 - \langle b^{II} \partial_t c_{00}^{II} \rangle_2 - \langle b^{II} \omega_b \partial_t c_{00}^I \rangle_1 + \omega_b \langle z_i^{(01)} \partial_i c_{00}^I \rangle_1 + \left\langle z_i^{(10)} \left(\partial_i c_{10}^{II} + \partial_{\mathrm{x}i} c_{00}^{II} \right) \right\rangle_2 = \omega_b \int_\Gamma z_i^{(01)} n_i \left(c_{00}^I - c_{00}^{II} \right) dy \tag{2.52a}$$

$\sim\dfrac{\varepsilon^2}{\varepsilon_p}$:

$$\langle L_{12}^D c_{2,-1}^{II} + L_2^D c_{1,-1}^{II} \rangle_2 + \langle z_i^{(10)} \left(\partial_i c_{2,-1}^{II} + \partial_{\mathrm{x}i} c_{1,-1}^{II} \right) \rangle_2 - \langle b^{II} \partial_t c_{1,-1}^{II} \rangle_2 + \omega_b \langle z_i^{(01)} \partial_i c_{1,-1}^I + z_i^{(10)} \partial_i c_{00}^I \rangle_1 - \langle b^{II} \omega_b c_{1,-1}^I \rangle_I = \omega_b \int_\Gamma z_i^{(10)} n_i (c_{00}^I - c_{00}^{II}) dy + \omega_b \int_\Gamma z_i^{(01)} n_i (c_{1,-1}^I - c_{1,-1}^{II}) dy \tag{2.52b}$$

here the relation (2.41) has been taken into account.

Cell Problems in the Matrix

The following representations for c^{II}_{10}, $c^{II}_{2,-1}$ result from (2.50), (2.51), (2.43), (2.46):

$$c^{II}_{10} = \varphi^{II}_k(y)\partial_{\mathrm{x}k}c^{II}_{00} + c^{II*}_{10}(x,t), \quad c^{II}_{2,-1} = \varphi^{II}_k(y)\partial_{\mathrm{x}k}c^{II}_{1,-1} + c^{II*}_{2,-1}(x,t)$$

where $c^{II*}_{1,0}(x,t)$ and $c^{II*}_{2,-1}(x,t)$ are some functions which do not depend on the fast variables.

Each function $\varphi^{II}_k(y)$ for any k=1,2,3 is defined as the solution of the following problem

$$\begin{cases} L^D_0\varphi^{II}_k = -\partial_j D^{II}_{ik}, \quad y \in Y^{II}; \qquad <\varphi^{II}_k>^{II}=0; \\ D^{II}_{ij}(\partial_j\varphi^{II}_k + \delta_{kj})n_i|_\Gamma = 0; \qquad \varphi^{II}_k \text{ is 1-periodic} \end{cases} \tag{2.53}$$

Cell Problems in the Block

The following relations for c^I_{00} and $c^I_{1,-1}$ result from (2.48), (2.41):

$$L^*c^I_{00} = 0, \quad y{\in}Y^i; \quad c^I_{00}|_{t=0} = c^0(x); \quad c^I_{00}|_\Gamma = c^{II}_{00}(x,t)^{*)} \tag{2.54}$$

$$\begin{cases} L^*c^I_{1,-1} = z^{(10)}_i\partial_i c^I_{00} + (\varepsilon Pe)^{-1}L^D_0 c^I_{00}, \quad y{\in}Y^i \\ c^I_{1,-1}|_{t=0} = 0; \quad c^I_{1,-1}|_\Gamma = c^{II}_{1,-1}(x,t)^{*)} \end{cases} \tag{2.55}$$

Symbol $^{*)}$ means that this condition can not be satisfied, if the characteristics of equation (2.54) or (2.55) go out of Y^I.

The operator L^* describes the pure C-Source flow in the block. In the right-hand part of (2.55), the first and the second terms correspond to C-Translation and D-Source flow. Thus, the zero approximation c^I_{00} describes the C-Source flow. However, the first approximation $c^I_{1,-1}$ includes the effects of C-Translation and D-Source elementary flows.

Equations Averaged Over the Matrix

It can be obtained for c^{II}_{00} and $c^{II}_{1,-1}$ from (2.52):

$$\mathbf{D}^{II}_{km}\partial^2_{\mathrm{x}km}c^{II}_{00} - B^{II}\partial_t c^{II}_{00} + A^{II}_{km}\partial_{\mathrm{x}k}p_{00}\partial_{\mathrm{x}m}c^{II}_{00} =$$
$$-\omega_b\langle L^*c^I_{00}\rangle_1 + \omega_b\partial_t p_{00}\int_\Gamma a^{II}_{ij}(\partial_j\mu)n_i(c^I_{00} - c^{II}_{00})dy \equiv q \tag{2.56a}$$

$$\mathbf{D}^{II}_{km}\partial^2_{\mathrm{x}km}c^{II}_{1,-1} - B^{II}\partial_t c^{II}_{1,-1} + A^{II}_{km}\partial_{xk}p_{00}\partial_{xm}c^{II}_{1,-1} =$$
$$-\omega_b\langle L^*c^I_{1,-1}\rangle_1 - \omega_b\langle z^{(10)}_i\partial_i c^I_{00}\rangle_1 +$$

$$\omega_b \partial_t p_{00} \int_\Gamma a_{ij}^{II}(\partial_j \mu) n_i (c_{1,-1}^I - c_{1,-1}^{II}) dy +$$

$$\omega_b \partial_{xk} p_{00} \int_\Gamma a_{ij}^{II}(\partial_j \psi_k^{II} + \delta_{ki}) n_i (c_{00}^I - c_{00}^{II}) dy \equiv q_2 \qquad (2.56b)$$

$$\begin{cases} B^{II} = < b^{II} >^{II}, \quad B^I = < b^{II} \omega_b >^I = < b^I >^I, \\ \mathbf{D}_{km}^{II} \equiv \langle D_{mi}^{II}(\partial_i \varphi_k^{II} + \delta_{ki}) \rangle_2, \quad A_{km}^{II} \equiv \langle a_{mi}^{II}(\partial_i \psi_k^{II} + \delta_{kj}) \rangle_2 \end{cases} \qquad (2.57)$$

To close these equations, the properties of the operator $<>^I$ and the function c_{00}^I are needed to be studied.

2.4.4. AVERAGING OVER THE BLOCK

Surfaces of Discontinuity Inside the Block

Let us assume a closed surface $\partial\mathcal{B}$ exists inside the block, on which the function c^I and its derivatives $\partial c^I/\partial y_i$ are discontinuous. The domain bounded by this surface is $\mathcal{B} \subset Y^I$. Let the domain $\mathcal{B}$ does not intersect the block boundary, and can be varying in time, such that the surface $\partial\mathcal{B}$ is mobile.

Let $\alpha_\mathcal{B}$ be the ratio of the volume of the domain $\mathcal{B}$ to the volume of the whole block. Let $\beta_\mathcal{B}$ be the area of the surface $\partial\mathcal{B}$.

Concentration and its derivative may be discontinuous on different surfaces. Let the block concentration and its derivative undergo M jumps. The $i-$th surface of discontinuity is designated as $\partial\mathcal{B}_i$. All these surfaces are assumed to be not intersecting one other and be closed, as shown in Fig. 2.9.

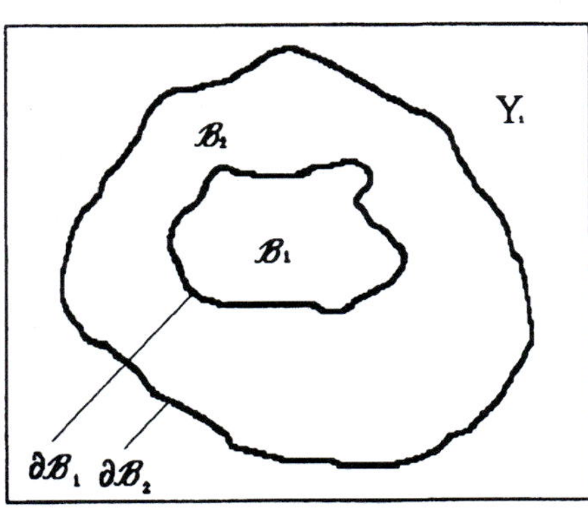

Figure 2.9. Surfaces of discontinuity for the concentration and its derivative inside the block

Then, $\mathcal{B}_i$ is a ring-like area located between the surfaces $\partial\mathcal{B}_i$ and $\partial\mathcal{B}_{i-1}$.

The problem how many discontinuity surfaces exist in the block is open. To solve it, we need to study the structures of flow inside the blocks.

Furthermore all the discontinuity surfaces will be designated by one symbol $\partial\mathcal{B}$.

Properties of Averaging Over a Block

Let $f(y)$ be some function defined in Y^I with the following properties.

i) $f = f^* = \text{const}, \quad y \in \Gamma$;

ii) Let $\partial\mathcal{B}$ be a closed surface dividing the area Y^I on two parts: internal $\mathcal{B}^I$ and external $\mathcal{B}^{II}$. Let $f(y)$ has a discontinuity on $\partial\mathcal{B}$ and $[f]_{\partial\mathcal{B}} \equiv f|_{\mathcal{B}^{II}} - f|_{\mathcal{B}^I} = \text{const}, \forall y \in \partial\mathcal{B}$.

Then the following properties are true.

$$\int_\Gamma a^{II}(\partial_i\mu)n_i dy = \langle b^{II}\rangle_1, \tag{2.58a}$$

$$\langle a^{II}(\partial_i\mu)(\partial_i f)\rangle_1 = f^*\langle b^{II}\rangle_1 - \langle f b^{II}\rangle_1 + [f]_{\partial\mathcal{B}}\langle b^{II}\rangle_{\mathcal{B}^I} \tag{2.58b}$$

$$\int_\Gamma a^{II}(\partial_i\psi_k^I + \delta_{ki})n_i dy = 0, \tag{2.58c}$$

$$\langle a^{II}(\partial_i\psi_k^I + \delta_{ki})\partial_i f\rangle_1 = 0, \quad \forall k \tag{2.58d}$$

All these relations result from (2.42), (2.44a) and can be easily proved.

Property of Functions c_{00}^I and $c_{1,-1}^I$ at the Block Boundary

Assume the block has a symmetric shape with respect to all axes. Then the following properties take place.

i) $c_{00}^I|_\Gamma = \text{const}(x,t)$ (does not depend on y)

ii) The function c_{00}^I may be discontinuous on some surface ($\partial\mathcal{B}$) symmetric about all axes.

iii) $[c_{00}^I]_{\partial\mathcal{B}} = \text{const}(x,t), \quad \forall y \in \partial\mathcal{B}$

As the function $\mu(y)$ determined in (2.42) is symmetric with respect to all axes, the equations of all characteristic lines for (2.54) are identical. Thus, the velocities of concentration propagation inside a block along each axes are equivalent. This fact proves the properties (2.4.4) and (2.4.4).

These properties are not valid for the function $c_{1,-1}^I$, because equation (2.55) has a non symmetric right-hand part. This asymmetry is caused by the convection translation phenomenon arising within the frameworks of such an approximation. It means that the rates of concentration waves along all axes are identical, but the shape of the wave becomes not symmetrical.

As the corollary of these properties and (2.4.4), (2.4.4), it follows that the integral in the right-hand part of (2.56a) is equal to zero. In the same time, the first integral in the right-hand part of (2.56b) differs from zero. This fact constitutes a basic difficulty to construct closed relations, when we take into account the convection translation phenomena.

Equations Averaged Over the Block
Let C^I and C^{II} be the concentrations averaged over the block and the matrix correspondingly:

$$C^I = \frac{1}{\alpha}\langle c\rangle^I = \frac{1}{\alpha}[\langle c^I_{00}\rangle_1 + \frac{\varepsilon}{\varepsilon_p}\langle c^I_{1,-1}\rangle_1] + O\Big(\varepsilon_p + \frac{\varepsilon^2}{\varepsilon_p^2}\Big),$$

$$C^{II} = \frac{1}{1-\alpha}\langle c\rangle_2 = c^{II}_{00} + \frac{\varepsilon}{\varepsilon_p}c^{II}_{1,-1} + O\Big(\varepsilon_p + \frac{\varepsilon^2}{\varepsilon_p^2}\Big)$$

Let us integrate equations (2.56), (2.57) and take into account relations (2.58), (2.4.4), (2.4.4):

$$-b^{II}\partial t\langle c^I_{00}\rangle_1 + b^{II}\partial_t p_{00}\Big[c^I_{00}|_\Gamma\alpha - \langle c^I_{00}\rangle_1\Big] + b^{II}\partial_t p_{00}\Big[c^I_{00}\Big]_{\partial\mathcal{B}}|\mathcal{B}^I| = 0 \quad (2.59a)$$

$$-b^{II}\partial_t\langle c^I_{1,-1}\rangle_1 - b^{II}\partial_t p_{00}\langle c^I_{1,-1}\rangle_1 -$$
$$b^{II}\partial_t p_{00}\int_\Gamma a^{II}_{ij}(\partial_j\mu)n_i(c^I_{1,-1} - c^{II}_{1,-1})dy -$$
$$-b^{II}\partial_t p_{00}\int_{\partial\mathcal{B}} a^{II}_{ij}(\partial_j\mu)n_i\Big[c^I_{1,-1}\Big]_{\partial\mathcal{B}}dy =$$
$$-\frac{1}{\varepsilon_p Pe}\int_\Gamma D^{II}_{ij}(\partial_j c^I_{00})n_i dy \equiv \frac{1}{\varepsilon_p Pe}q_D \quad (2.59b)$$

herein $|\mathcal{B}^I|$ is the area of the surface $\mathcal{B}^I$, α is a volume fraction of the whole block.

2.4.5. NON-CLOSED GENERAL AVERAGED EQUATIONS

Averaged Equations of Zero Order
Relations (2.56) and (2.59) constitute the system of macroscale transport equations. However this system is not closed, because the functions at the boundary Γ and the integrals over Y^I remain undefined. Nevertheless, the closed macroscale model may be deduced for the zero approximation.

Approximation of zero order for C^I and C^{II} has the following form according to (2.59a) and (2.54):

$$\partial_t C^I + C^I\partial_t P^{II} = (\partial_t P^{II})c^I|_\Gamma + (\alpha)^{-1}(\partial_t P^{II})\left[c^I\right]_{\partial\mathcal{B}}|\mathcal{B}^I| = 0 \quad (2.60a)$$

$$\mathbf{D}^{II}_{km}\partial^2_{xkm}C^{II} - B^{II}\partial_t C^{II} + A^{II}_{km}\partial_{xk}P^{II}\partial_{xm}C^{II} = B^I\partial_t P^{II}\Big[c^I|_\Gamma - C^{II}\Big] \quad (2.60b)$$

The functions $c^I|_\Gamma$, $[c^I]_{\partial\mathcal{B}}$ and $|\mathcal{B}^I|$ can be defined in various ways depending on the sign of the value $\partial_t P^{II}$. Closed relations for them will be obtained furthermore.

Herein the following definition for the pressure has been taken into account: P^{II}:

$$P^{II} = p_{00}(x,t) + O(\varepsilon_p)$$

Averaged Equations of First Order

Multiplying equation (2.59b) by $\varepsilon/\varepsilon_p$ and adding it to (2.59a), we obtain the following result for the next approximation with accuracy $O(\varepsilon_p + \frac{\varepsilon^2}{\varepsilon_p^2})$:

$$\begin{cases} \partial_t C^I + C^I \partial_t P^{II} = -(\partial_t P^{II}) Q_\Gamma - (\partial_t P^{II}) Q_{\partial\mathcal{B}} + \left(b^{II}\varepsilon_p Pe\right)^{-1} q_D \\ \mathbf{D}^{II}_{km}\partial^2_{\mathrm{x}km} C^{II} - B^{II}\partial_t C^{II} + A^{II}_{km}\partial_{\mathrm{x}k} P^{II}\partial_{\mathrm{x}m} C^{II} = \\ \qquad = B^I Q^2_\Gamma - \omega_b\left(\varepsilon_p Pe\right)^{-1} q_D \end{cases} \tag{2.61}$$

Where

$$Q_\Gamma \equiv \int_\Gamma a^{II}_{ij}(\partial_j\mu) n_i c^I dy = \int_\Gamma a^{II}_{ij}(\partial_j\mu) n_i c^I_{1,-1} dy + \alpha^{-1}\partial_t P^{II} c^I_{00}|_\Gamma, \tag{2.62a}$$

$$Q_{\partial\mathcal{B}} \equiv \int_{\partial\mathcal{B}} a^{II}_{ij}(\partial_j\mu) n_i [c^I]_{\partial\mathcal{B}} dy =$$
$$\int_{\partial\mathcal{B}} a^{II}_{ij}(\partial_j\mu) n_i [c^I_{1,-1}]_{\partial\mathcal{B}} dy + \alpha^{-1}(\partial_t P^{II})[c^I_{00}]_{\partial\mathcal{B}}|\mathcal{B}|; \tag{2.62b}$$

$$Q^{II}_\Gamma \equiv \int_\Gamma a^{II}(\partial_i\mu) n_i (c^I|_\Gamma - C^{II}) dy,$$

$$q_D \equiv \int_\Gamma D^{II}_{ij}(\partial_j c^I) n_i dy = \int_\Gamma D^{II}_{ij}(\partial_j c^I_{00}) n_i dy \tag{2.62c}$$

$$c^I \equiv c^I_{00}(x,y,t) + \frac{\varepsilon}{\varepsilon_p} c^I_{1,-1}(x,y,t) \tag{2.62d}$$

Corrected Averaged Equations Keeping the Diffusion Boundary Layer

In general case, the solution inside the block undergoes sharp variations near the block boundary, due to diffusion which becomes considerable there. To take into account the diffusion boundary layer, the asymptotic expansions for concentration are needed to be modified in such a manner that will be explained furthermore, in section 2.5.1.

This will affect modification in the averaged equations.

Formally, the "corrected" macroscale equations of zero order taking into account for the boundary layer may be obtained by adding the diffusion terms from the averaged model of first order (2.61) to the original averaged model of zero order (2.60). This manipulation can be strictly proved according to the results of averaging based on the "corrected" asymptotic expansions obtained furthermore in section 2.5.1.

Finally, the corrected macroscale equations of zero approximation (2.60) keeping the diffusion boundary layer get the following form which is intermediate between (2.60) and (2.61):

$$B^I \partial_t C^I + B^I(\partial_t P^{II}) q_c^{\mathcal{B}} - \frac{\omega_m}{\varepsilon_p Pe} q_D^{\mathcal{B}} = B^I(\partial_t P^{II}) q_c^I - \frac{\omega_m}{\varepsilon_p Pe} q_D \quad (2.63a)$$

$$B^{II}\partial_t C^{II} - A_{km}^{II}\partial_{\mathrm{x}k} P^{II}\partial_{\mathrm{x}m} C^{II} - \mathbf{D}_{km}^{II}\partial_{\mathrm{x}km}^2 C^{II} =$$
$$= -\Big[B^I(\partial_t P^{II}) q_c^{II} - \frac{\omega_m}{\varepsilon_p Pe} q_D\Big]; \quad (2.63b)$$

where:

$$q_c^{\mathcal{B}} = \alpha_{\mathcal{B}} [c^I]_{\partial\mathcal{B}}; \quad (2.64a)$$
$$q_D^{\mathcal{B}} = \alpha_{\mathcal{B}} [D_{ij}^{II}\partial_j c^I n_i]_{\partial\mathcal{B}}; \qquad \alpha_{\mathcal{B}} = |\mathcal{B}| \quad (2.64b)$$
$$q_c^I = c^I|_\Gamma - C^I(x,t) \quad (2.64c)$$
$$q_c^{II} = c^I|_\Gamma - C^{II}(x,t) \quad (2.64d)$$
$$q_D \equiv -\int_\Gamma D_{ij}^{II}(\partial_j c^I) n_i dy \quad (2.64e)$$

The right-hand part of (2.63) contains the terms describing the exchange by concentration. They are splitted onto two groups where $q_c = (q_c^I,\, q_c^{II})$ is the convective exchange flow rate caused by the source flow in the block, while q_D is a diffusion exchange flow rate.

It is important that the convective exchange terms for the blocks, q_c^I, and for the matrix, q_c^{II}, are non symmetric. This effect may be explained.

Let an inkwell and a glass of water be the model of block/matrix interaction. Let us begin to pour ink in water, thus performing the exchange process. The ink concentration in the glass will be varying in time, while it remains constant in the inkwell. Thus, the exchange process exerts asymmetrical effect on fluid composition in the inkwell and in the glass.

Equation (2.65a) shows that variations in averaged block concentration are caused by the convective transport of concentration discontinuity inside the block, by the diffusion flux across the surfaces when the concentration derivative undergoes jumps, as well as by the mass exchange through the block boundary.

Macroscale equations (2.63) get too general form and are not closed yet. To close it, the cell problem (2.55) must be solved.

2.5. C-S SYSTEM: PARTICULAR SOLUTIONS OF CELL PROBLEMS

The model of zero order (2.60) may be closed, if the cell problem (2.54) was solved. The solution can be easily constructed, however it depends on the direction of wave propagation inside the block, or on the sign of derivative $\partial_t P^{II}$. General solution will be examined later, in the next chapter. Here the simplest solutions will be constructed for three cases when the time derivative of pressure P^{II} is positive (1), negative (2) or changes the sign (3).

2.5.1. GENERAL PROPERTIES OF CELL PROBLEM IN THE BLOCK

Corrected Problem of Zero Order in Diffusion Boundary Layer
Examine the main problem in block (2.54) respectively to the function c^I_{00}. It describes only the convective-source transport of concentration without any diffusion. Diffusion effects enter only in the second approximation c^I_{00} determined by problem (2.55). This circumstance results from the a priori assumption about the type of the system, in which the convective terms prevail in blocks. However, diffusion in the matrix is of order one. Hence, a diffusion boundary layer may arise near the block border.

In the last case, expansion (2.46) for concentration should be corrected by introducing a new fast variable $\xi{=}(y^* - y)/(\varepsilon_p Pe)$ which is of order one near the block border:

$$c^I = c^I_{00}(x, y, \xi, t) + \frac{\varepsilon}{\varepsilon_p} c^I_{1,-1}(x, y, \xi, t) + ...$$

where y^* is the local coordinate of the block boundary.

The condition at the boundary Γ allowing solution discontinuity may be replaced by a continuity condition, if one takes into consideration the diffusion boundary layer near Γ.

Corrected problem for concentration $c^I_{00}(x, y, \xi, t)$ may be easily deduced from (2.54) and (2.55) and gets the following form:

$$b^{II}\partial_t c^I_{00} - a^{II}_{ij}(\partial_t P^{II})\partial_i \mu \partial_j c^I_{00} = \frac{1}{\varepsilon_p Pe} L^D_0 c^I_{00}, \quad y \in Y^I; \qquad (2.65a)$$

$$c^I_{00}|_\Gamma {=} C^{II}(x, t), \qquad (2.65b)$$

$$c^I_{00}|_{t=0} = c^0(x) \qquad (2.65c)$$

The boundary conditions and the initial conditions must be self-consistent:

$$C^{II}\big|_{t=0} = c^0(x) \tag{2.66}$$

Due to Eq. (2.39) we get: $1/(\varepsilon_p Pe)\sim \varepsilon/\varepsilon_p \to 0$, then equation (2.65a) has a boundary layer near the block boundary where the diffusion term becomes important.

A Priori Properties of Solution

The following features of block problem (2.65) may be detected before any solution.

1. Concentration c_{00}^I is described by a parabolic equation, hence problem (2.65) is correctly set. However, the order of the equation reduces when $\varepsilon/\varepsilon_p \to 0$. Then the problem becomes overdetermined. One of conditions (2.65b) appears superfluous and may be neglected. Therefore the concentration jumps may appear on the interface between the blocks and matrix.

2. The solution of (2.65) depends on transport direction. If the concentration wave propagates from the block boundary into the block center, the boundary value (2.65b) and (2.65c) determines the overall wave profile inside the block. Otherwise, when the fluid escapes from the block, the boundary concentration can not be putted arbitrary, as it is formed by an escaping wave. Hence, in the last case the boundary condition is not important.

The flow direction can be described by a value:

$$\gamma(x,t) \equiv \text{sign}\left(\frac{\partial P^{II}}{\partial t}\right) = \text{sign}\left(\text{grad}_y p^I\right) \tag{2.67}$$

where $\text{grad}_y p^I$ is the pressure gradient in the block written via variable y.

When $\gamma=1$ the matrix pressure is growing, therefore the pressure gradient inside the block is directed towards its center. Hence, the fluid is injected into the block.

When $\gamma=-1$ the fluid escapes the block.

3. The diffusion term in (2.65) may be always neglected except the boundary layer near the block border.

1D Formulation of the Block Problem

To construct an effective solution of problem (2.65) in the block taking into account the boundary layer, we have to approximate the block by a sphere of radius r^*, such that $(4/3)\pi(r^*)^3=\alpha$.

Problem (2.65) written via spherical coordinates is as follows:

$$\begin{cases} b^{II}\left[\dfrac{\partial c^I}{\partial t} - \left(\dfrac{1}{3}\dfrac{\partial P^{II}}{\partial t}\right) r \dfrac{\partial c^I}{\partial r}\right] = \dfrac{1}{\varepsilon_p Pe} L_0^D c^I; \quad r \in Y^I \\ c^I\big|_{r=r^*} = C^{II}, \qquad c^I\big|_{t=0} = c^0(x) \end{cases} \tag{2.68}$$

where operators L_0^D and grad are written via variable r.

This one-dimensional problem has analytical solutions.

Solution of (2.68) will be constructed for three main cases, differing by the direction of the local transport velocity

$$V_c \equiv -\frac{r}{3b^{II}}\frac{\partial P^{II}}{\partial t} \tag{2.69}$$

in convection term, or by the sign of γ in (2.67).

2.5.2. CONCENTRATION WAVE PROPAGATION IN A BLOCK. INJECTION MODE

Definition of the Injection Mode of Block Operating

Let γ=1. Solution of (2.54) is continuous, so solution of (2.68) does not contain the boundary layer. The diffusion term is small everywhere and can be neglected.

Let us introduce new variables, as:

$$\tau(x,t) = \frac{1}{3}\int_0^t \frac{\partial P^{II}}{\partial t'} dt' = \frac{1}{3}[P^{II}(x,t) - p^0] \tag{2.70a}$$

$$u(r) = c^I - c^0, \qquad \eta = \ln(R/r) \tag{2.70b}$$

Note that τ is various for various macroscopic points x and for various times t.

Then problem (2.68) gets the following form:

$$\frac{\partial u}{\partial \tau} + \frac{\partial u}{\partial \eta} = 0, \quad \tau > 0, \ \eta \geq 0; \tag{2.71a}$$

$$u\big|_{\eta=0} = u_*(\tau) \equiv C^{II}(x,\tau) - c^0, \tag{2.71b}$$

$$u\big|_{\tau=0} = 0, \tag{2.71c}$$

The macroscale variable x plays the role of a parameter of the problem.

Since the boundary and the initial conditions are self-consistent (2.66), it follows that $u_*(0)=0$. Then the solution $u \to 0$, when $\tau \to 0$, that follows from homogeneity of equation (2.71a). Condition (2.71c) is satisfied automatically. Thus, the problem is correctly posed.

Solution of Cell Problem in Block

Examine problem (2.71a), (2.71b).

For the sake of convenience, let us prolong the boundary function $u_*(\tau)$ onto the left-hand semi-axes τ in the continuous way such that $u_*(\tau)\equiv 0$ when $\tau \leq 0$.

Characteristic lines for equation (2.71a) satisfy the relationship $\frac{d\tau}{d\eta}=1$ and have the form of straight lines, as shown in Fig. 2.10.

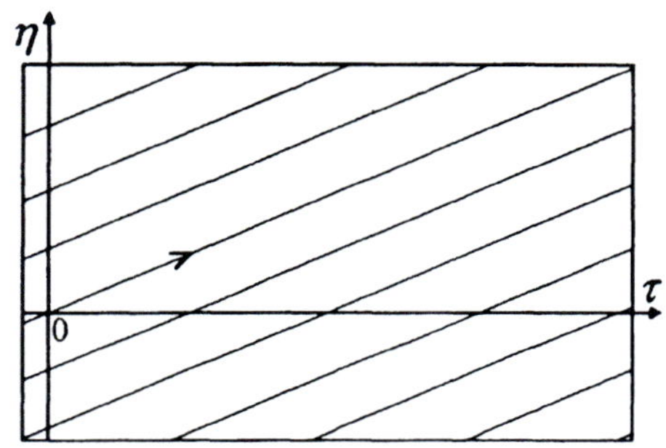

Figure 2.10. Characteristic lines for the equation of concentration wave propagation in a block operating in injection mode ($\gamma = +1$)

Solution takes the form:

$$u(\eta,\tau) = \begin{cases} u_*(\tau-\eta), & \tau>\eta \\ 0, & \tau\leq\eta \end{cases}$$

or

$$c^I(x,\eta,\tau) = \begin{cases} C^{II}(x,\tau-\eta), & \eta<\tau \\ c^0(x), & \eta\geq\tau \end{cases}$$

Returning back from η to radial coordinate r, we deduce:

$$c^I(x,r,t) = \begin{cases} C^{II}\Big((x,\tau-\ln(\frac{R}{r})\Big), & 1 \geq \frac{r}{R} > \mathrm{e}^{-\tau} \equiv \zeta_* \\ c^0, & 0 \leq \frac{r}{R} \leq \zeta_* \end{cases} \tag{2.72}$$

The boundary function C^{II} being continuous, the solution c^I is continuous too at any r, τ, but its derivative is discontinuous in the point ζ_*, as shown in Fig. 2.11.

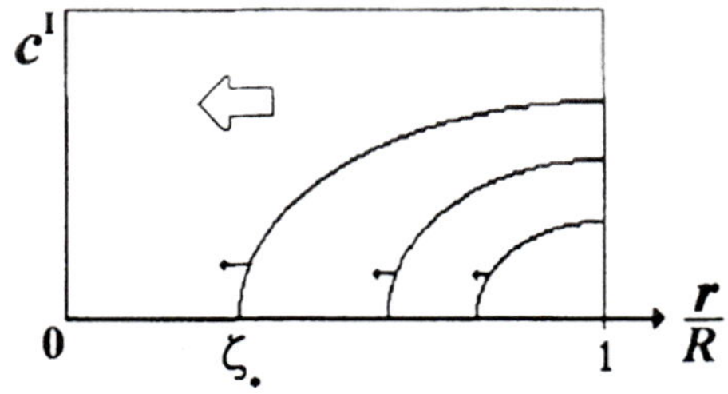

Figure 2.11. Concentration wave propagation inside a block when $\gamma=1$ (three consequent instants are shown)

This solution describes a wave traveling from the matrix into the block center. The front coordinate ζ_* is determined as

$$\zeta_*(t;x)=\exp\left[-\frac{1}{3}(P^{II}(x,t)-p^0)\right] \tag{2.73}$$

and it is various in various blocks, as it parametrically depends on x. The wave shape is determined by the boundary function C^{II}.

The following conclusions result from (2.73):

1) the wave front reaches the block center when $\tau\to\infty$, the front velocity being exponentially decreasing;

2) the front motion is determined by the function $P^{II}(x,t)$, thereby its velocity is equal to

$$\frac{d\zeta_*}{dt}=-\zeta_*\frac{dP^{II}}{dt},$$

3) when the pressure field is steady-state in the matrix, the front remains immobile, because the motion in blocks is governed by source flow which disappears in steady-state mode.

2.5.3. CONCENTRATION WAVE PROPAGATION IN BLOCK. EXTRACTION MODE

Definition of the Extraction Mode

Let $\gamma=-1$ in (2.69). Hence, the pressure gradient is directed from the block center towards the matrix; the matrix pressure is decreasing in time. This situation is appropriate to extraction of fluid from the porous reservoir.

Let us introduce the local time τ defined in another way:

$$\tau(x,t)=\frac{-1}{3}\int_0^t\frac{\partial P^{II}}{\partial t'}dt'=-\frac{1}{3}[P^{II}(x,t)-p^0]$$

Solution Away from the Boundary Layer
Problem (2.68) written via new time τ and variables (2.70b) obtains the following form, if we neglect the diffusion term:

$$\frac{\partial u}{\partial \tau}-\frac{\partial u}{\partial \eta}=0, \quad \tau>0, \ \eta\geq 0; \tag{2.74a}$$

$$u\big|_{\tau=0}=0, \tag{2.74b}$$

Condition (b) in (2.71) can not be satisfied, as the characteristic lines does not fall within the area Y^I. Thus, the boundary concentration corresponding to η=0 may be transported into the area of negative values of η, as shown in Fig. 2.12. This generates a concentration discontinuity at the block boundary.

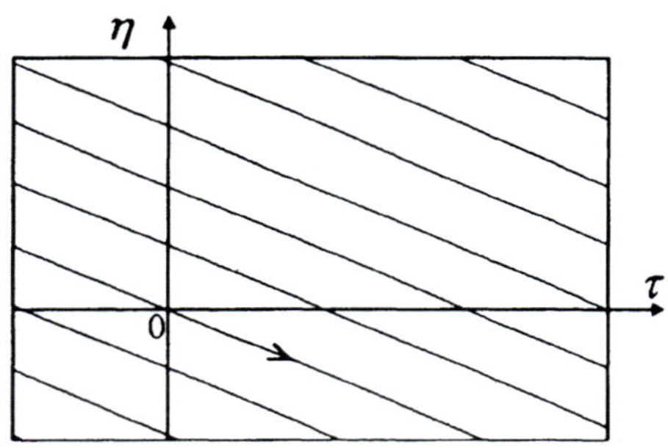

Figure 2.12. Characteristic lines for the equation of concentration wave propagation in a block operating in extraction mode ($\gamma= -1$)

Solution of (2.74) gets the form:

$$c^I(x,r,\tau)=\begin{cases} c^0, & 0\leq r/R<1 \\ C^{II}, & r/R=1 \end{cases} \tag{2.75}$$

Inside the block the solution remains unperturbed and undergoes a jump on the boundary, as shown in Fig. 2.13.

Solution within Diffusion Boundary Layer. Matching Asymptotics
To construct the solution in the boundary layer near the block border, the method of matching asymptotic expansions will be used.

Examine full equation (2.65) with small diffusion term:

$$\frac{\partial c^I}{\partial t}+f(x,t)r\frac{\partial c^I}{\partial r}=\lambda\frac{1}{r^2}\frac{\partial}{\partial r}\left(r^2\frac{\partial c^I}{\partial r}\right) \tag{2.76}$$

where

$$f(x,t)\equiv\frac{1}{3}\left|\frac{\partial P^{II}}{\partial t}\right|, \quad \lambda\equiv\left(\frac{\epsilon}{\epsilon_p}\right)\frac{D^{II}}{b^{II}Pe^*}, \quad \lambda<<1$$

The diffusion parameter is assumed to be scalar, D^{II}.

Now we shall construct the asymptotic solution when $\lambda\to 0$.

The external expansion, which is true everywhere except the boundary layer, coincides with the diffusionless solution (2.75) examined above.

To construct the boundary layer expansion, let us introduce a new stretched coordinate being of order one near the block boundary:

$$\rho=\frac{R-r}{\lambda}$$

Problem (2.76) yields:

$$\lambda\frac{\partial c^I}{\partial t}-f(x,t)(R-\lambda\rho)\frac{\partial c^I}{\partial\rho}=\frac{1}{(R-\lambda\rho)^2}\frac{\partial}{\partial\rho}\left((R-\lambda\rho)^2\frac{\partial c^I}{\partial\rho}\right) \tag{2.77}$$

From this equation we get for the first term $\tilde{c}^I$ of the boundary layer expansion when $\lambda\to 0$:

$$-Rf(x,t)\frac{\partial\tilde{c}^I}{\partial\rho}=\frac{\partial^2\tilde{c}^I}{\partial\rho^2}, \quad \rho>0; \quad \tilde{c}^I\Big|_{\rho=0}=C^{II} \tag{2.78}$$

The last problem appears steady-state, therefore the initial condition can not be satisfied. Instead of it, a condition of matching with the external expansion holds.

Solution of (2.78) gets the form:

$$\tilde{c}^I=\frac{G}{R}\left[1-e^{-Rf\rho}\right]+C^{II} \tag{2.79}$$

where G is a constant of integration. It should be defined by means of matching asymptotic expansions (2.79) and (2.75). In this case, this conditions is equivalent to a requirement that the boundary layer expansion (2.79) would tend to c^0, when $\rho\to\infty$. Then

$$c^0=\frac{G}{R}+C^{II}, \Rightarrow G=R(c^0-C^{II})$$

Matched Solution
Definitely, the solution of the full problem (2.68) is as follows:

$$c^I(x,r,t)\begin{cases} c^0, & 0\leq r\ll\lambda; \\ C^{II}+(c^0-C^{II})\exp\left\{-\dfrac{R(R-r)}{3\lambda}\right\}\dfrac{\partial P^{II}}{\partial t}, & R-\lambda\leq r\leq R \end{cases} \tag{2.80}$$

It is shown in Fig. 2.13.

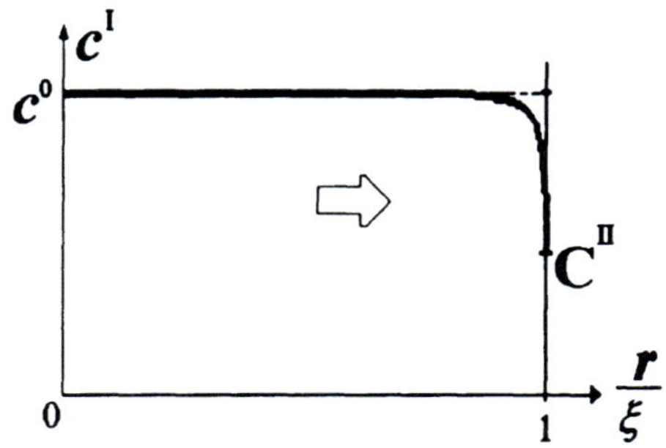

Figure 2.13. Concentration distribution inside the block when $\gamma=-1$

2.5.4. CONCENTRATION WAVES PROPAGATION INSIDE THE BLOCK. INJECTION-EXTRACTION MODE

Let the direction of process be varying in time:

$$\gamma(x,t)=\begin{cases} 1, & t\in[0,t_1) \\ -1, & t>t_1 \end{cases} \tag{2.81}$$

such, that the injection into a block is performed at the first time stage, after which the extraction from block happens.

At the second time-stage the concentration jump appears at the block boundary. During all time stages the point of derivative discontinuity moves inside the block.

Let us introduce the dimensionless time τ as:

$$\tau(x,t)=\frac{1}{3}\int_0^t\left|\frac{\partial P^{II}}{\partial t'}\right|dt'$$

which depends still on the macroscale variable x.

Problem (2.68) without small diffusion term may be written using dimensionless variables (2.70b) in the following way:

$$\gamma(\tau)\frac{\partial u}{\partial \tau}+\frac{\partial u}{\partial \eta}=0, \quad \tau>0,\ \eta\geq 0; \tag{2.82a}$$

$$u\big|_{\eta=0}=u_*(\tau)\equiv C^{II}(x,\tau)-c^0, \tag{2.82b}$$

$$u\big|_{\tau=0}=0, \tag{2.82c}$$

Until the instant t_1 the solution has been already deduced in previous sections. Concentration distribution is the following, according to (2.80):
- inside the block:

$$c^I(x,\zeta,\tau_1)=C^{II}(x,\tau_1+\ln\zeta), \qquad \zeta\equiv r/R$$

- inside the matrix: $C^{II}(x,\tau_1)$, where $\tau_1\equiv\ \tau(t_1)$.

The characteristic lines of equation (2.71a) are shown in Fig. 2.14.

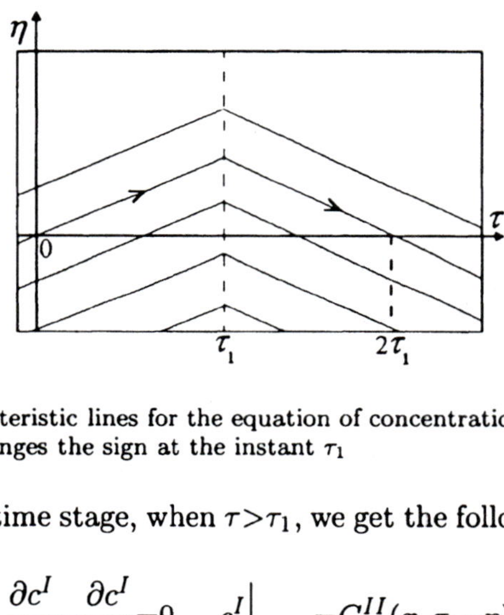

Figure 2.14. Characteristic lines for the equation of concentration wave propagation in the block when γ changes the sign at the instant τ_1

At the second time stage, when $\tau>\tau_1$, we get the following problem from (2.82):

$$\frac{\partial c^I}{\partial \tau}-\frac{\partial c^I}{\partial \eta}=0, \quad c^I\Big|_{\tau=\tau_1}=C^{II}(x,\tau_1-\eta)$$

where $\eta=\ln R/r$. Boundary condition is not satisfied again.

Solution gets the form

$$c^I(x,\eta,\tau)=C^{II}(x,2\tau_1-\tau-\eta)$$

As earlier, the coordinate of the probable front may be defined from the condition that the second argument would be converted to zero in the last relationship. Then:

$$\zeta_*(\tau)=\exp[-2(2\tau_1-\tau)]$$

At the second time stage, a concentration jump exists on the block boundary, because

$$c^I(x,\zeta,\tau)\Big|_{\zeta=1} = C^{II}(x,2\tau_1-\tau) \neq C^{II}(x,\tau)$$

Definitely

$$c^I(x,\zeta,\tau) = \begin{cases} C^{II}(x,\tau+\ln\zeta), & 0\leq\tau<\tau_1, \quad \forall\zeta\in(\zeta_*,1) \\ C^{II}(x,2\tau_1-\tau+\ln\zeta), & \zeta\in(\zeta_*,1), \quad \tau_1\leq\tau<2\tau_1 \\ C^{II}(x,\tau), & \zeta=1, \quad \tau_1\leq\tau<2\tau_1 \\ c^0(x), & \zeta\in[0,\zeta_*], \quad \forall\tau \end{cases} \tag{2.83}$$

Radial coordinate of the forward wave front r_* (or ζ_*, in dimensionless form) varies according to the law:

$$\zeta_*(\tau) \equiv \frac{r_*(\tau)}{R} = \begin{cases} e^{-\tau}, & \tau<\tau_1 \\ e^{-(2\tau_1-\tau)}, & \tau_1\leq\tau<2\tau_1 \\ 1, & \tau\geq 2\tau_1 \end{cases} \tag{2.84}$$

The multiple change of sign for the time derivative of pressure may ne described in the same way. After the second change of sign (repeat injection into the block) the solution inside the block has a discontinuity. In fact, the discontinuity has been formed at the block boundary during the second time stage (extraction from the block). Now, if the flow direction is reversed, then this boundary jump will be transported into the block center.

2.5.5. CLOSURE RELATIONS FOR HOMOGENIZED MODEL

The solutions obtained give closure relations for model (2.63).

General Relationship for Diffusion Exchange
Dependence of the concentration c^I on the coordinate r results from (2.80):

$$c^I = f\,[T + \ln(r/R)] \tag{2.85}$$

where $T(\tau)$ is some function dependent on time only.

The value q_D will be calculated from (2.64) assuming that $D^{II}_{ij} = D^{II}\delta_{ij}$, where D^{II} is a scalar diffusion coefficient.

The following relation holds:

$$q_D \equiv \int\limits_{\partial Y^I} D^{II} \frac{\partial c^I}{\partial y_i} n_i dy = 4\pi R^2 D^{II} \frac{\partial c^I}{\partial r}\Big|_{r=R}$$

Because

$$\frac{\partial c^I}{\partial r} = \frac{1}{r}\frac{df}{dz}\Big|_{z=T+\ln(r/R)}$$

then

$$q_D = 4\pi\xi D^{II} \frac{df}{dz}\Big|_{z=T} \tag{2.86}$$

Similarly, if $\partial\mathcal{B}$ is a surface of weak discontinuity, such that $\mathcal{B} = \{0 \le r \le R\zeta_*(\tau)\}$, while the concentration is constant inside $\mathcal{B}$ and is described by relation (2.85) outside of $\mathcal{B}$, then we get for $q_D^{\mathcal{B}}$ from (2.64):

$$q_D^{\mathcal{B}} = -4\pi R\zeta_* D^{II} \frac{\partial f(\tau)}{\partial\tau}\Big|_{\tau=0} \tag{2.87}$$

because the following is always true for ζ_*: $T + \ln\zeta_* = 0$.

Closure Relations for Injection Mode

Only one closed curve $\partial\mathcal{B} = \{r = \zeta_*(t)\}$ exists, the derivative becomes discontinuous on. The concentration is continuous everywhere. Closure relations are as follows:

$$\begin{aligned}
&c^I|_\Gamma = C^{II}; \quad [c^I]_{\partial\mathcal{B}} = 0\\
&q_c^{\mathcal{B}} \equiv 0;\\
&q_D^{\mathcal{B}} = -12\pi R\zeta_*(t)\mathbf{D}^{II}\Big(\partial_t C^{II}(x,t)\Big)_{t=0}\Big(\partial_t P^{II}|_{t=0}\Big)^{-1}\\
&q_c^I = C^{II}(x,t) - C^I(x,t),\\
&q_c^{II} \equiv 0\\
&q_D = 12\pi R\mathbf{D}^{II}\partial_t C^{II}(x,t)\Big(\partial_t P^{II}\Big)^{-1}
\end{aligned} \tag{2.88}$$

Closure Relations for Extraction Mode

There are no of concentration discontinuity in the block, therefore:

$$\begin{aligned}
&c^I|_\Gamma = c^0 \equiv C^I; \quad [c^I]_{\partial\mathcal{B}} = 0\\
&q_c^{\mathcal{B}} \equiv 0; \qquad q_D^{\mathcal{B}} \equiv 0;
\end{aligned} \tag{2.89}$$

$$q_c^I \equiv 0;$$
$$q_c^{II} \equiv C^I - C^{II} = c^0 - C^{II}(x,t);$$
$$q_D \equiv 0$$

Closure Relations for Injection-Extraction Mode
One surface $\partial\mathcal{B}$ with radial coordinate $r_* = R\zeta_*(t)$ exists which is gone according to law (2.84). Then:

$$\left[c^I\right]_{\partial\mathcal{B}} = 0; \quad c^I|_\Gamma = \begin{cases} C^{II}(x,\tau), & \tau < \tau_1 \\ C^{II}(x, 2\tau_1 - \tau), & \tau_1 \leq \tau < 2\tau_1 \\ c^o, & \tau \geq 2\tau_1 \end{cases} \tag{2.90}$$

Closure relationships get the following form:

$$q_c^{\mathcal{B}} \equiv 0; \tag{2.91a}$$

$$q_D^{\mathcal{B}} = -12\pi R\zeta_*(t)\mathbf{D}^{II}\left(\frac{\partial C^{II}(x,t)}{\partial t}\right)|_{t=0}\left[\frac{\partial P^{II}}{\partial t}|_{t=0}\right]^{-1} \tag{2.91b}$$

$$q_c^I = \begin{cases} C^{II}(x,\tau) - C^I(x,\tau), & \tau < \tau_1 \\ C^{II}(x, 2\tau_1 - \tau) - C^I(x,\tau), & \tau_1 \leq \tau < 2\tau_1; \\ 0, & \tau \geq 2\tau_1 \end{cases} \tag{2.91c}$$

$$q_c^{II} = \begin{cases} 0, & \tau < \tau_1; \\ C^{II}(x, 2\tau_1 - \tau) - C^{II}(x,\tau), & \tau_1 \leq \tau < 2\tau_1; \\ c^o - C^{II}(x,\tau), & \tau \geq 2\tau_1 \end{cases} \tag{2.91d}$$

$$q_D = \begin{cases} 12\pi R\mathbf{D}^{II}\partial_t C^{II}(x,t)(\partial_t P^{II})^{-1}, & \tau < \tau_1 \\ -12\pi R\mathbf{D}^{II}\partial_t C^{II}(x,t))|_{t=\varphi(2\tau_1-\tau)}\left((\partial_t P^{II})|_{t=\varphi(2\tau_1-\tau)}\right)^{-1}, & \tau_1 \leq \tau < 2\tau_1 \\ 0, & \tau \geq 2\tau_1 \end{cases} \tag{2.91e}$$

where $\zeta_*(\tau)$ is determined by relations (2.84); the function $t = \varphi(\tau)$ is inverse to $\tau = \tau(t)$; $\tau_1 = \tau(t_1)$, $C^i(x,\tau) \equiv C^i(x,t)|_{t=\varphi(\tau)}$.

2.5.6. MAIN RESULTS

i) The averaged model depends on the process history and constitutes a system of delaying partial differential equations.

ii) The exchange process is splited onto the convective component (q_c^I, q_c^{II}), caused by source flow, and the diffusion component q_D.

iii) In the examined media non-stationarity of the pressure field is the basic force of convection flow in blocks. If the pressure field becomes steady-state, then any convection flows disappear in blocks.

iv) The exchange terms for blocks and matrixes are not symmetrical. Moreover, they cannot be expressed in term of the difference between two concentrations.

2.6. C-S SYSTEM: GENERAL SOLUTIONS OF CELL PROBLEMS. GENERAL CLOSURE RELATIONS

In general case the derivative $\partial_t P^{II}$ changes sign many times. This corresponds to the nonmonotonous in time boundary-value conditions respectively to the pressure field. This reflects on the cell problem of concentration transport in such a way that the transport velocity inside each block changes sign. However, its general solution may be obtained in the analytical form, which will be done in this chapter.

2.6.1. GENERAL CELL PROBLEM WITH CHANGING SIGN OF TRANSPORT VELOCITY

Let x be a fixed macroscale point. In this macro-point, let the pressure derivative $\partial_t P^{II}$ changes sign many times during the process evolution. Then the time interval $(0,t]$ consists of $(n+1)$ segments $I_i(x)=(t_i(x),\ t_{i+1}(x)]$, $i=0,...,n$, where the sign of the time derivative $\partial_t P^{II}$ remains invariable. For the sake of simplicity, let $\partial_t P^{II}>0$ in the zero interval $(0,t_1]$. Let $\gamma_i(x)$ be the value of $\gamma(t;x)$ in the segment $I_i(x)$, then

$$\gamma_i = (-1)^i \tag{2.92}$$

Hence, using the nomenclature of the previous chapter, we can say that during each time period I_i the fluid is injected into the porous reservoir when i is even, whereas the fluid is extracted when i is odd.

The case when $\gamma_i=0$ is trivial and is then omitted.

For an arbitrary segment $I_i(x)$ the general cell problem takes the form analogous to (2.71), or (2.82):

$$\frac{\partial u_i}{\partial \tau}+\gamma_i\frac{\partial u_i}{\partial \eta}=0, \quad t_i<\tau\leq t_{i+1},\ \eta\geq 0; \tag{2.93a}$$

$$u_i\big|_{\eta=0}=u_*(\tau)\equiv C^{II}(x,\tau)-c^0, \tag{2.93b}$$

$$u_i\big|_{\tau=\tau_i}=u_{i-1}(\eta,\tau) \tag{2.93c}$$

where

$$u_i(\eta,\tau)=c^I(\eta,\tau)-c^0, \quad \text{when } \tau_i<\tau\leq\tau_{i+1};$$

$$\eta=\ln\frac{R}{r}; \qquad \tau(t;x)=3\int_0^t\left|\frac{\partial P^{II}}{\partial t'}\right|dt' \tag{2.94}$$

As the variable x plays the role of a parameter, it will be omitted in order to simplify the notations.

2.6.2. RECURRENT FORM FOR GENERAL SOLUTION OF CELL PROBLEM

Two types of solutions exist, depending on the value γ_i.

When $\gamma_i = -1$ boundary-value condition (2.93b) can not be satisfied, and we get the simple initial problem (2.93a), (2.93c). The solution is:

$$u_i(\eta,\tau)=u_{i-1}\left(\eta-\int_{\tau_i}^{t}\gamma_i(s)ds,\ \tau_i\right),\quad \text{when } i \text{ is odd} \tag{2.95}$$

When $\gamma_i=1$, both conditions in (2.93) are important. In general case they are not consistent, such that the solution has a discontinuity at the boundary $\eta=0$ at the instant $t=t_i$. This jump will be kept during consequent evolution. Solution gets the form (i is even):

$$u_i(\eta,\tau)=\begin{cases} u_*\left(\tilde{T}_i(\eta,\tau)\right), & \eta_i^{*i}<\eta; \\ u_{i-1}\left(\eta-\int_{\tau_i}^{\tau}\gamma_i(s)ds,\ \tau_i\right), & \eta>\eta_i^{*i} \end{cases} \tag{2.96a}$$

$$\eta_j^{*i}(\tau)=\eta_{j-1}^{*i}(\tau_j)+(-1)^j(\tau-\tau_j) \tag{2.96b}$$

The value $\eta_j^{*i}(\tau)$ is the i-th *front* within the j-th time segment, which is equal to the coordinate of the surface transporting the concentration discontinuity inside the block.

The front has the upper number i, if it has arisen in the time segment I_i.

The function $\tilde{T}_i(\eta,\tau)$ is defined as:

$$\int_{\tilde{T}_i}^{t}\gamma(s)ds=\eta \tag{2.97}$$

The recurrent sequence of formulas (2.95) and (2.96) will be closed, if the definition of the zero order term is added:

$$u_0(\eta,\tau)=u_*\left(\tilde{T}_0(\eta,\tau)\right)=u_*(\tau-\eta) \tag{2.98}$$

This recurrent form is not constructive, but is a tool to deduce the explicit form of the solution.

2.6.3. EXPLICIT FORM OF SOLUTION FOR CELL PROBLEM

The explicit form of solution may be deduced from (2.95)-(2.98) using the induction method:

$$u_{2k+1}(\eta,\tau)=\begin{cases} u_*\left(T_{2k}^{2k+1}(\eta,\tau)\right), & \eta\leq\eta_{2k+1}^{*2\kappa} \\ u_*\left(T_{2k-2}^{2k+1}(\eta,\tau)\right), & \eta_{2k+1}^{*2k}<\eta\leq\eta_{2k+1}^{*2k-2} \\ u_*\left(T_{2k-2m}^{2k+1}(\eta,\tau)\right), & \eta_{2k+1}^{*2k-2m+2}<\eta\leq\eta_{2k+1}^{*2k-2m} \\ \dots & \\ u_*\left(T_{0}^{2k+1}(\eta,\tau)\right), & \eta>\eta_{2k+1}^{*2} \end{cases} \tag{2.99a}$$

$$u_{2k}(\eta,\tau)=\begin{cases} u_*\left(T_{2k}^{2k}(\eta,\tau)\right), & \eta\leq\eta_{2k}^{*2k} \\ u_*\left(T_{2k-2}^{2k}(\eta,\tau)\right), & \eta_{2k}^{*2k}<\eta\leq\eta_{2k}^{*2k-2} \\ u_*\left(T_{2k-2m}^{2k}(\eta,\tau)\right), & \eta_{2k}^{*2k-2m+2}<\eta\leq\eta_{2k}^{*2k-2m} \\ \dots & \\ u_*\left(T_{0}^{2k}(\eta,\tau)\right), & \eta>\eta_{2k}^{*2} \end{cases} \tag{2.99b}$$

(2.99c)

where $m=0,\dots,k$; $k=0,\dots,[n/2]$, $\eta^{*0}=\infty$.

The coordinate of k-th front inside the block within the time segment $(2k+m)$ is estimated as:

$$\eta_j^{*i}(\tau)=\begin{cases} \sum\limits_{q=1}^{j-i}(-1)^{j-q}\left(\tau_{j+1-q}-\tau_{j-q}\right)+(-1)^j\left(\tau-\tau_j\right), & i=2,4,6,\dots;\ j\geq i \\ \infty, & i=0,1,3,5,7,\dots,\ \forall j \end{cases} \tag{2.100}$$

when $\tau_j<\tau\leq\tau_{j+1}$.

The function $T_{2k+\alpha}^{2k+\beta}(\eta,\tau)$ $(\beta\geq\alpha)$, is defined by the relation:

$$\int\limits_{T_{2k+\alpha}^{2k+\beta}}^{\tau}\gamma(s)ds=\eta;\qquad \tau\in I_{2k+\beta},\quad T_{2k+\alpha}^{2k+\beta}\in I_{2k+\alpha} \tag{2.101}$$

The last relation may be rewritten in the explicit form, if Eq. (2.92) is used:

$$T^{2k+\beta}_{2k+2\alpha}(\eta,\tau)=2\sum_{j=2k+2\alpha+1}^{2k+\beta}(-1)^{j-1}\tau_j-\eta+(-1)^{\beta}\tau \tag{2.102}$$

The following property has been used to deduce Eqs. (2.99):

$$\tilde{T}_{2k}\left(\eta-\int_{\tau_{2k+1}}^{\tau}\gamma_{2k+1}ds,\ \tau_{2k+1}\right)=T^{2k+1}_{2k}(\eta,\tau)$$

In fact,

$$\int_{\tilde{T}_{2k}}^{\tau_{2k+1}}\gamma_{2k}(s)ds=\eta-\int_{\tau_{2k+1}}^{\tau}\gamma_{2k+1}ds;\ \Rightarrow\ \int_{\tilde{T}_{2k}}^{\tau}\gamma(s)ds=\eta$$

that defines the function T^{2k+1}_{2k} according to (2.101).

The solution is completely defined.

2.6.4. CONCENTRATION AT THE BLOCK BOUNDARY

The boundary function $\tilde{u}(\tau)\equiv u(\eta,\tau)|_{\eta=0}$ is defined in various way for even and odd time segments.

Within an even segment I_{2k} the boundary function is defined as:

$$\tilde{u}(\tau)=u_*\left(T^{2k}_{2k}(0,\tau)\right)=u_*(\tau) \tag{2.103}$$

Within each odd time-segment all the fronts move from the block center towards the boundary. Then the amount of fronts is decreased. Then, in the segment I_{2k+1}, the boundary function is equal to $u_*\left(T^{2k+1}_{2k}(0,\tau)\right)$, if the last front η^{*2k}_{2k+1} exists. Condition of existence for this front is $\eta^{*2k}_{2k+1}>0$. If this front already does not exist, it is necessary to check for the existence of the next front, η^{*2k-2}_{2k+1}, which consists in the following pair of inequalities: $\eta^{*2k}_{2k+1}<0$, $\eta^{*2k-2}_{2k+1}>0$, e.t.c.

Then we can write:

$$\tilde{u}(\tau)=u_*\left(T^{2k+1}_{2k-2m}(0,\tau)\right),\ \text{when}\ \eta^{*2k-2m+2}_{2k+1}<0,\ \eta^{*2k-2m}_{2k+1}>0$$

Using (2.100), we can deduce that the system of conditions $\eta_{2k+1}^{*2k-2m+2}<0$ and $\eta_{2k+1}^{*2k-2m}>0$ is equivalent to the following relationship:

$$\begin{cases} \sum_{q=1}^{2m-1}(-1)^{q+1}\left(\tau_{2k+2-q}-\tau_{2k+1-q}\right)+\tau_{2k+1}<\tau\leq \\ \qquad \sum_{q=1}^{2m+1}(-1)^{q+1}\left(\tau_{2k+2-q}-\tau_{2k+1-q}\right)+\tau_{2k+1}, \quad \text{if } m=0,1,\ldots,k-1; \\ \sum_{q=1}^{2m-1}(-1)^{q+1}\left(\tau_{2k+2-q}-\tau_{2k+1-q}\right)+\tau_{2k+1}<\tau\leq\tau_{2k+1}, \quad \text{if } m=k \\ \qquad k=0,1,2,3,\ldots,[n/2] \end{cases} \tag{2.104}$$

Then

$$c^I\big|_\Gamma=\begin{cases} C^{II}\left(x,\,T_{2k-2m}^{2k+1}(0,\tau)\right), & \text{when } \tau\in I_{2k+1} \text{ and (2.104) is satisfied;} \\ C^{II}(x,\,\tau)), & \text{when } \tau\in I_{2k} \end{cases} \tag{2.105}$$

2.6.5. CONCENTRATION JUMPS AT MOVING FRONTS

New concentration discontinuity arises only within the even time-segments when the discontinuity at the block boundary begins to propagate into the block. During the next odd time-segment new fronts are not formed, while all previously formed fronts reverse the direction of their motion.

The number of fronts transporting concentration jumps inside a block is equal to k within the time-segment I_{2k} or I_{2k+1}.

For the front η_{2k}^{*2k-2m} the concentration jump is equal to

$$[u]_{2k}^{*2k-2m}=u\left(\eta_{2k}^{*2k-2m}-0,\tau\right)-u\left(\eta_{2k}^{*2k-2m}+0,\tau\right)= \\ u_*\left(T_{2k-2m}^{2k}(\eta_{2k}^{*2k-2m},\,\tau)\right)-u_*\left(T_{2k-2m-2}^{2k}(\eta_{2k}^{*2k-2m},\,\tau)\right)$$

and by analogy:

$$[u]_{2k+1}^{*2k-2m}=u_*\left(T_{2k-2m}^{2k+1}(\eta_{2k+1}^{*2k-2m},\,\tau)\right)-u_*\left(T_{2k-2m-2}^{2k+1}(\eta_{2k+1}^{*2k-2m},\,\tau)\right)$$

Using definition (2.100) and (2.102) we obtain the following simple relationships defining concentration jumps associated to the moving fronts:

$$[u]_{2k+1}^{*2k-2m}=[u]_{2k}^{*2k-2m}=u_*\left(\tau_{2k-2m}\right)-u_*\left(2\tau_{2k-2m-1}-\tau_{2k-2m}\right) \\ m=0,\ldots,k-1 \tag{2.106}$$

2.6.6. GENERAL CLOSURE RELATIONS FOR MACROSCALE MODEL

Based on Eqs. (2.105) and (2.106), the following relations result for the boundary concentration and the flow rates (2.64) in the averaged model (2.63) when $\tau \in I_{ij}$:

$$c^I\big|_\Gamma = C^{II}\left(x,\ \Phi_{ij}(x) + (-1)^i \tau\right), \tag{2.107a}$$

$$\Phi_{ij} = \begin{cases} 0, & \text{when } i \text{ is even,} \\ 2 \sum\limits_{q=i-j}^{i} (-1)^{q-1} \tau_q; & \text{when } i \text{ is odd} \end{cases} \tag{2.107b}$$

$$i=0, 1, 2, ..., n; \quad j=0, 1, 2, ..., i$$

$$q_c^{\mathcal{B}} = \sum_{q=0}^{\rho(i)/2-1} \alpha_{\mathcal{B}_q} \left[C^{II}(x, \tau_{\rho(i)-2q}) - C^{II}(x, 2\tau_{\rho(i)-2q-1} - \tau_{\rho(i)-2m}))\right],$$

$$\rho(i) = \begin{cases} i, & \text{if } i \text{ is even,} \\ i-1, & \text{if } i \text{ is odd} \end{cases} \tag{2.108a}$$

$$q_c^I = C^{II}\left(x, \Phi_{ij} + (-1)^i \tau]\right) - C^I(x, \tau), \tag{2.108b}$$

$$q_c^{II} = C^{II}\left(x, \Phi_{ij} + (-1)^i \tau]\right) - C^{II}(x, \tau) \tag{2.108c}$$

where the definition for the time-segment I_{ij} results from (2.104):

$$I_{ij} = \begin{cases} \sum\limits_{q=1}^{j-1} (-1)^{q+1} (\tau_{i+1-q} - \tau_{i-q}) + \tau_i < \tau \leq \\ \qquad \sum\limits_{q=1}^{j+1} (-1)^{q+1} (\tau_{i+1-q} - \tau_{i-q}) + \tau_i, \\ \qquad\qquad \text{if } i=1, 3, 5, ..., 2\left[\frac{n}{2}\right]+1, \quad j=0, 2, 4, ..., i-3; \\ \sum\limits_{q=1}^{j-1} (-1)^{q+1} (\tau_{i+1-q} - \tau_{i-q}) + \tau_i < \tau \leq \tau_{i+1}, \\ \qquad\qquad \text{if } i=1, 3, 5, ..., 2\left[\frac{n}{2}\right]+1, \quad j=i-1 \\ \tau_i < \tau \leq \tau_{i+1}, \qquad \text{if } i=0, 2, 4, ..., 2\left[\frac{n}{2}\right], \ \forall j \end{cases} \tag{2.109}$$

Thus, the interval I_{ij} is defined for all $i=0, 1, 2, ..., n$ and for even j, such that $j=0, 2, 4, ..., 2\left[\frac{i}{2}\right]-1$.

The value n is equal to the number of cycles. The symbol $[n]$ means the operation of selection of the integer part of the number n.

2.7. C-S SYSTEM: MACROSCALE MODEL

Let us call a process *n-cycle* when the sign of the pressure derivative $\partial_t P^{II}$ changes n times though for one point of the medium. It is necessary to note that the number of cycles may be various for various points x of the domain.

The general macroscale model for the process with any number of cycles may be obtained using previous results and neglecting the diffusion boundary layers for the moment.

For particular case of 1-cycle model the results obtained allow to construct a closed macroscale model with diffusion effects inside the blocks.

2.7.1. GENERAL MACROSCALE MODEL FOR ANY NUMBER OF CYCLES

Collecting all results obtained in previous section, we get the following macroscale model for C-S mode of concentration transport in highly heterogeneous medium.

For $\tau(t;x)\in I_{ij}(x)$, where I_{ij} is defined according to (2.109):

$$B^I\frac{\partial C^I}{\partial t}+B^I\frac{\partial P^{II}}{\partial t}\sum_{q=0}^{\frac{\rho(i)}{2}-1}\alpha_{B_q}\Big\{C^{II}(x,\tau_{\rho(i)-2q})-C^{II}(x,2\tau_{\rho(i)-2q-1}-\tau_{\rho(i)-2q})\Big\}=$$

$$B^I\frac{\partial P^{II}}{\partial t}\Big\{C^{II}\Big(x,\,\Phi_{ij}(x)+(-1)^i\tau\Big)-C^I(x,\tau)\Big\}; \tag{2.110a}$$

$$B^{II}\frac{\partial C^{II}}{\partial t}-A^{II}_{km}\partial_{\mathrm{x}k}P^{II}\partial_{\mathrm{x}m}C^{II}-\mathbf{D}^{II}_{km}\partial^2_{\mathrm{x}km}C^{II}=$$

$$-B^I\frac{\partial P^{II}}{\partial t}\Big\{C^{II}\Big(x,\Phi_{ij}+(-1)^i\tau]\Big)-C^{II}(x,\tau)\Big\} \tag{2.110b}$$

Where:

$$\Phi_{ij}=\begin{cases}0, & \text{when } i \text{ is even,}\\ 2\sum\limits_{q=i-j}^{i}(-1)^{q-1}\tau_q; & \text{when } i \text{ is odd}\end{cases}$$

$$\rho(i)=\begin{cases}i, & \text{if } i \text{ is even,}\\ i-1, & \text{if } i \text{ is odd}\end{cases} \tag{2.111}$$

2.7.2. 1-CYCLE MACROSCALE MODEL WITH DIFFUSION EFFECTS

When the sign of the pressure derivative $\partial_t P^{II}$ changes only once, the macroscale model gets more simple form, as the concentration fronts inside the blocks arise only after the second change of sign. Moreover, absence of fronts allows to describe diffusion effects inside the blocks, which are then associated only with a single surface of weak discontinuity.

The main part of the macroscale model, without diffusion layers inside the blocks, may be deduced directly from general model (2.110) with n=1 (n is the number of cycles). Then we get from (2.109): i=0, 1; j=0, e.t.c.

Influence of diffusion layers inside the blocks was described in section 2.5.

Collecting results (2.63) and those obtained in section 2.5.2 and 2.5.4, we get the following macroscale model for 1-cycle case.

For $\tau(t;x)\in I_{i0}(x)$, i=0, 1, where

$$I_{00} = \{\tau_0<\tau\leq\tau_1\}, \quad I_{10} = \{\tau_1<\tau\leq\tau_2\}$$

$$B^I\frac{\partial C^I}{\partial t}+\frac{(-1)^i 12\pi R D^{II}\omega_m}{\varepsilon_p Pe}\exp\left\{-\left(\Phi_{i0}+(-1)^i\tau\right)\right\}\times \frac{\partial C^{II}(x,\tau)}{\partial \tau}\bigg|_{\tau=0} = B^I\frac{\partial P^{II}}{\partial t}\left\{C^{II}\left(x,\, \Phi_{i0}(x)+(-1)^i\tau\right)-C^I(x,\tau)\right\}- \frac{(-1)^i 12\pi R D^{II}\omega_m}{\varepsilon_p Pe}\frac{\partial C^{II}(x,\theta)}{\partial\theta}\bigg|_{\theta=\Phi_{i0}+(-1)^i\tau}, \tag{2.112a}$$

$$B^{II}\frac{\partial C^{II}}{\partial t}-A^{II}_{km}\frac{\partial P^{II}}{\partial x_k}\frac{\partial C^{II}}{\partial x_m}-\mathbf{D}^{II}_{km}\frac{\partial^2 C^{II}}{\partial x_k\partial x_m}= -B^I\frac{\partial P^{II}}{\partial t}\left\{C^{II}\left(x,\Phi_{i0}+(-1)^i\tau\right)-C^{II}(x,\tau)\right\}+ \frac{(-1)^i 12\pi R D^{II}\omega_m}{\varepsilon_p Pe}\frac{\partial C^{II}(x,\theta)}{\partial\theta}\bigg|_{\theta=\Phi_{i0}+(-1)^i\tau} \tag{2.112b}$$

where:

$$\Phi_{00}=0, \quad \Phi_{10}=2\tau_1$$

$$\frac{\partial C(x,\tau)}{\partial t}=\frac{\partial C(x,\tau)}{\partial\tau}\frac{\partial\tau}{\partial t}$$

$$C^{II}(x,\tau)=C^I(x,\tau)=c^0, \quad \frac{\partial C^{II}}{\partial t}=0, \text{ if } \tau\leq 0$$

2.7.3. RESULTS FOR C-S SYSTEMS

1. Due to domination of convection in blocks the macroscale dispersion tensor is equivalent to the effective dispersion tensor of the matrix only.

2. Source geometry of streamlines in blocks, been convective by physical origin, influences only the structure of the effective permeability tensor which is equivalent to the effective permeability of the matrix only.

3. Source type of flow inside the blocks leads, as always, to arising of delay phenomena related to the mass exchange between the blocks and the matrix. These phenomena are described by delay differential equations. In contrast, in the next section it will be shown that domination of diffusion transport with source type geometry of flow leads to integro-differential equations.

4. Arising of concentration discontinuity inside the block gives contribution to the macroscale model for block concentration C^I, as the time variation of C^I depends on the velocity of discontinuity propagation inside this block.

5. Near the surfaces in block when the space derivative of the local concentration is discontinuous the diffusion boundary layers arise, where diffusion transport becomes comparable to the convection. This areas give some contribution to the macroscale equations. As far as the block/matrix interface is always associated to some discontinuities of concentration field, the diffusion boundary layer arises near the block boundary. It gives contribution to the macroscale equations in the form of diffusion exchange terms.

6. The form of delay operators describing the convection exchange depends on the local direction of concentration wave propagation inside the block, or on sign of the time-derivative of pressure $\gamma=\text{sign}\{\partial_t P^{II}\}$.

7. The macroscale model for the n-cycle process describes a system oscillating in time with non-averaged time-oscillations. However, these oscillations are caused by external perturbations, but not by the medium heterogeneity. Due to this fact, the scale of these oscillations is not related to the heterogeneity scale and is not small.

Example of application of such a model to the practical problems will be shown in section 2.11.

2.8. D-S SYSTEM

When diffusion is predominating in blocks, while the geometry of flow corresponds to the source type, the effective model is similar to the results obtained for the source flow of a single-phase homogeneous fluid, in chapter 1. Basic result which is expect to arise is associated to the macroscale phenomena nonlocal in time.

In some papers as [36], [55] and [56] the similar case has been studied in approximate way, when the blocks were operating as sources, while the concentration transport in blocks had a purely diffusion origin. In this section the macroscale equations will be deduced using homogenization method.

According to the diagram of flow described in Fig. 2.8, the D-S system can be observed, if:

$$\begin{aligned} &\tfrac{\varepsilon}{\varepsilon_p} < Pe < \tfrac{1}{\varepsilon_p}, \qquad \text{when } \varepsilon_p > \varepsilon \\ &\tfrac{\varepsilon}{\varepsilon_p} < Pe < \tfrac{1}{\varepsilon}, \qquad \text{when } \varepsilon_p \leq \varepsilon \end{aligned} \tag{2.113}$$

2.8.1. A PRIORI FEATURES OF D-S SYSTEMS

Results for Pressure Field

According to chapter 1, the pressure field can be defined in various ways, depending on the parameter ε_p. As the existence of a D-S system is conditioned only by the value of Peclet number, the parameter ε_p is not limited. Thus, all the models deduced in chapter 1 are valid for the case of D-S system.

For example, assuming $\varepsilon_p=1$, we have to describe the pressure field by the model explained in section 1.5.1. However, Eq. (2.113) yields the following limitations for Peclet number: $\varepsilon<Pe<1$.

In this section we assume however that the block permeability is not very small, i.e., the parameter ε_p is not very large, namely: $\varepsilon_p\leq 1$. The opposite case requires special attention due to the boundary layers arising at the block boundary. It will be studied especially in the next section.

Sub-Modes of Concentration Transport

As to the concentration transport, it is natural to expect that several subclasses exist in frameworks of D-S systems, depending on Peclet number.

In the concentration transport equation the parameter ε_c plays the same role as the parameter ε_p in the equation for pressure. In chapter 1 it has been shown that the developed source flow is characterized by a point $\varepsilon_p \sim 1$, while the point $\varepsilon_p \sim \varepsilon$ determines the source-translation flow. By the formal analogy, the similar results are expected to be valid for the concentration transport equation. In particular, the bottom curve in diagram 2.8, with $\varepsilon_c \sim \varepsilon$, is expected to be correspondent to the diffusion source-translation flow. The upper curve, where $\varepsilon_c \sim 1$, should be correspondent to the most developed form of diffusion-source flow.

According to diagram 2.8, the bottom curve $Pe = \varepsilon / \varepsilon_p$, or $\varepsilon_c = \varepsilon$, really marks an equilibrium between the source and translation geometry of flow, with keeping diffusion as the basic mechanism of transport (D-ST system).

However, the upper curve $Pe = 1/\varepsilon_p$, or $\varepsilon_c \equiv \varepsilon_p Pe = 1$, corresponds to the equilibrium between diffusion and convection, with keeping the source geometry of flow. Thus, the most developed form of diffusion-source flow is always superposed with the convective-source flow of the comparable intensity (DC-S system).

The pure form of D-S flow may be observed only in the central part of corresponding area in Fig. 2.8, away from the upper and lower interfaces, when

$$\frac{\varepsilon}{\varepsilon_p} << Pe << \frac{1}{\varepsilon_p} \tag{2.114}$$

In this case the parameter $veps_c$ is always small, such that $\varepsilon << \varepsilon_c << 1$. Thus the pure D-S flow can be reached only in the rather weak form.

The more developed form of D-S flow, the higher the role of convective form of transport.

In this section only the pure D-S flow is examined.

2.8.2. AVERAGED MODEL FOR D-S FLOW

Omitting all intermediate deductions, we explain the final result concerning the averaged transport equations.

$$C^I(x,t) = C^{II}(x,t) - \tau^{DS} \frac{\partial C^{II}}{\partial t}, \tag{2.115a}$$

$$B \frac{\partial C^{II}}{\partial t} - A_{km} \frac{\partial P^{II}}{\partial x_k} \frac{\partial C^{II}}{\partial x_m} - \mathbf{D}^{II}_{km} \frac{\partial^2 C^{II}}{\partial x_k \partial x_m} = \tau_c^{DS} \frac{\partial^2 C^{II}}{\partial t^2} \tag{2.115b}$$

The second equation describes concentration transport through the matrix, while the concentrations C^I and C^{II} are linked using kinetic relationship with characteristic relaxation time for diffusion exchange flow $\tau_c^{DS} \sim \varepsilon_c$.

Note, the second transport relation is elliptic, although it is nonstationary evolving equation.

2.8.3. EFFECTIVE PARAMETERS

Effective Dispersion Tensor

The macroscale dispersion tensor is equivalent to the effective dispersion in the matrix only and is defined as:

$$\mathbf{D}_{ik}^{II} \equiv \left\langle D_{ij}^{II} \left(\frac{\partial \psi_{Dk}^{II}}{\partial y_j} + \delta_{jk} \right) \right\rangle_2 \tag{2.116}$$

The functions $\psi_{Dk}^{II}(y)$, $k=1,2,3$ are solutions of the cell problem:

$$\begin{cases} \partial_i (D_{ij} \partial_j \psi_{Dk}^{II}) = -\partial_i D_{ik}, & i \in Y^{II}; \\ \langle \psi_{Dk}^{II} \rangle_2 = 0, & \psi_{Dk}^{II} \text{ is y-periodic} \end{cases} \tag{2.117}$$

These functions are similar to the functions $\psi_k^{II}(y)$ introduced in section 1.6.1, Eqs. (1.71a), for instance. All results, obtained for these functions and for the permeability tensor A_{ik}^{II} of the source systems in section 1.6.1 remain valid for the functions $\psi_{Dk}^{II}(y)$ and for the dispersion tensor $\mathbf{D}_{ik}^{II}$, if the parameter $a(y)$ will be replaced by D_{ij} with corresponding renumbering of indexes.

Relaxation Time of Diffusion Exchange

Two new times τ^{DS} and τ_c^{DS} appear in model (2.115). The time τ^{DS} describes the relaxation of the block concentration C^I, while the time τ_c^{DS} characterizes the relaxation of the rate of the block/matrix diffusion exchange. Both these values pertain to the same process of the source flow attenuation. Although they are of the same order, it is useful to note they are not completely equivalent.

They are determined as

$$\tau^{DS} \equiv \frac{\varepsilon_p Pe}{\alpha} \langle \varphi(y) \rangle_1, \quad \tau_c^{DS} \equiv \varepsilon_p Pe \langle b^I(y) \varphi(y) \rangle_1 \tag{2.118a}$$

where $\varphi(y)$ is the solution of the cell problem:

$$\begin{cases} \partial_i (D_{ij} \partial_j \varphi_D^I) = -b_*(y), & i \in Y^I; \\ \varphi_D^I|_\Gamma = 0 \end{cases} \tag{2.118b}$$

If b_* is constant, than τ^{DS} and τ_c^{DS} are related by a linear dependence:

$$\tau_c^{DS} = \frac{\omega_m}{\alpha} b_* \tau^{DS} \tag{2.118c}$$

All results obtained for the eigen times τ_* and τ_*^S in section 1.7.3 remain enable for the times τ^{DS} and τ_c^{DS} with replacing the parameter $a(y)$ by the diffusion parameters D_{km}.

Stability Condition

As in section 1.7.3, equation (2.115b) contains the second temporal derivative and is elliptic, as $\tau_c^{DS}>0$ and the matrix D_{ij}^{II} is positively defined. The classical initial problem for it is unstable, then the stability condition is needed to be added.

Stability condition consists in eliminating the exponentially increasing solutions:

$$|C^{II}|<M \exp(k\tau), \ \forall\tau\geq 0, \ \forall k>0 \tag{2.119}$$

where $M>0$ is a constant.

2.9. DC-S SYSTEM (WITH CONVECTIVE BOUNDARY LAYER)

At the interface $Pe=1/\varepsilon_p$ between two areas in Fig. 2.8 diffusion and convection are of the same order, with the source type of streamline geometry. The structure of exchange flow formed under the interaction between diffusion and convection is not clear and calls for investigation.

This kind of interaction will be examined for a system located near the separating curve $Pe=1/\varepsilon_p$ inside the area of D-S flow. Thus, the diffusion-source flow predominates, but the convection-source transport is rather intensive, such that it should be taken in consideration too.

The permeability of blocks is assumed to be very small, such that the pressure field inside the block remains almost unperturbed except a thin layer near the block boundary. In the diagram of flow shown in Fig. 2.12 this mode corresponds to the area located to the right of the line $\varepsilon_p=1$.

These systems possess some new properties.

On the one hand, diffusion is dominating mechanism of concentration transport inside the blocks for a D-S system. Therefore the diffusion block/matrix exchange happens under normal condition, without formation of any boundary layers and independently of the pressure field. On the other hand, the pressure gradient becomes huge in narrow boundary areas, that produces rather intensive convective transport from the block to the matrix across the block/matrix boundary. This kind of narrow areas will be called as *convective boundary layers*.

In convective boundary layer convection transport may be comparable with diffusion transport, if one takes in consideration, that, in addition, the diffusion parameter in blocks is small respectively to that in the matrix, although diffusion is dominant in blocks.

Existence of two types of exchange across the block/matrix boundary allows to study the interaction between them, that has been formulated as the objective of this section.

It will be shown that this interaction is nonlinear, but may be effectively described on the macroscale.

2.9.1. ASYMPTOTIC EXPANSIONS

The basic class of media can be obtained for the following ratio of parameters:

$$\varepsilon_p>1, \quad \frac{\varepsilon}{\varepsilon_p}<Pe<\frac{1}{\varepsilon_p}$$

According to the above analysis, it is assumed that the boundary layer arises in the block where the total perturbation of the pressure field is concentrated. In contrast, the boundary layer of concentration field can arise only in higher approximations.

Asymptotic expansions have the three-scale form:

$$p^I = p_0^I(x,y,\xi,t) + \varepsilon p^{I,I}(x,y,\xi,t) + \dots; \quad \xi \equiv (r - r^*)/\sqrt{Pe}$$

$$c^I = c_0^I(x,\xi,t) + Pe\sqrt{Pe}c_{01}^I(x,y,\xi,t) + O(\varepsilon + Pe^2)$$

General relations (2.20)–(2.23) are kept in the same form but with a new order between parameters. The following problems hold for the zero order terms:

$$\frac{b^{II}}{a^{II}}\frac{\partial p_0^I}{\partial t} = \frac{\partial^2 p_0^I}{\partial \xi^2}, \qquad p_0^I|_{t=0} = p^0, \qquad p_0^I|_{\xi=0} = p_{00}^{II} \tag{2.120a}$$

$$\varepsilon_p Pe{\cdot}b^{II}\frac{\partial c_0^I}{\partial t} = D^{II}\frac{1}{r^2}\frac{\partial}{\partial r}\left(r^2\frac{\partial c_0^I}{\partial r}\right),$$

$$D^{II}\frac{\partial^2 c_{01}^I}{\partial \xi^2} = -a^{II}\frac{\partial p_0^I}{\partial \xi}\frac{\partial c_0^I}{\partial r} \tag{2.120b}$$

$$c_0^I|_{t=0} = c^0, \quad c_0^I|_{\xi=0} = c_{00}^{II}, \quad c_{01}^I|_{\xi\to\infty} = 0, \quad c_{01}^I|_{\xi=0} = 0$$

Other deductions are omitted.

The following solutions of (2.120) result:

$$p_0^I(x,r,t) = p_0^{II}(x,t) + (p^0 - p_0^{II})\Phi\left(\frac{1}{2}(r^* - r)\sqrt{\frac{\lambda}{Pe{\cdot}t}}\right)$$

$$c_{01}^I = \gamma\frac{\partial c_0^I}{\partial r}(p^0 - p_{00}^{II})\left[\xi\left(1 - \Phi(-\frac{\xi}{A})\right) - \frac{A(t)}{\sqrt{\pi}}\left(1 - \mathrm{e}^{-\xi^2/A^2}\right)\right] + O\left(\mathrm{e}^{-\eta^2\mu^2/A}\right)$$

$$\gamma \equiv (a^{II}/D^2); \quad \xi = (r - r^*)/\sqrt{Pe}; \quad A(t) = 2\sqrt{t/\lambda};$$

$$\lambda = (b^{II}/a^{II}); \quad \Phi(u) = \frac{2}{\sqrt{\pi}}\int_0^u \mathrm{e}^{-t^2}dt$$

$$c_0^I = \sum_{k\geq 1} R_k(r)\phi_k(x,t), \quad R_k = \frac{\sin(r\sqrt{\mu_k})}{r},$$

$$\phi_k(x,t) = \beta_k\int_0^t \frac{\partial c_0^I}{\partial \Theta}\mathrm{e}^{-\mu_k(t-\Theta)/\lambda_c}d\Theta, \quad \beta_k \equiv \frac{\langle R_k\rangle_1}{\langle R_k^2\rangle_1} = (-1)^{1+k}\frac{2r^*}{\pi k}$$

2.9.2. HOMOGENIZED EQUATIONS

The homogenized equation in blocks is as follows:

$$C^I \equiv C^{II} - \int_0^t \frac{\partial C^{II}}{\partial \Theta} G(t-\Theta)d\Theta + \gamma\sqrt{Pe}(p^0 - P^{II}) \int_0^t \frac{C^{II}}{\partial \Theta} K_{II}(t-\Theta)d\Theta + \dots \tag{2.121}$$

For the matrix:

$$B^{II}\frac{\partial C^{II}}{\partial t} - A_{ij}\frac{\partial P^{II}}{\partial x_i}\frac{\partial C^{II}}{\partial x_j} - \mathrm{D}_{ij}^{II}\frac{\partial^2 C^{II}}{\partial x_i \partial x_j} = -B^I \int_0^t \frac{\partial C^{II}}{\partial \Theta} K_{II}(t-\Theta)d\Theta -$$

$$-B^I \gamma\sqrt{Pe}\frac{\partial}{\partial t}\left(\sqrt{t}(p^0 - P^{II}) \int_0^t \frac{\partial C^{II}}{\partial \Theta} K_{II}(t-\Theta)d\Theta\right) \tag{2.122}$$

$$\tag{2.123}$$

$$K_I(t) = \frac{6}{\pi^2}\sum_{k\geq 1}\frac{\mathrm{e}^{-\pi^2 k^2 t/\lambda_c r^*}}{k^2}, \quad K_{II}(t) \equiv -\frac{dK_I}{dt} = \frac{6}{\lambda_c r^*}\sum_{k\geq 1}\mathrm{e}^{-\pi^2 k^2 t/\lambda_c r^*}$$

$$\gamma \equiv \frac{2\varepsilon_p Pe a^{II}\sqrt{a^{II} b^{II}}}{\sqrt{\pi}(D^{II})^2}, \quad \lambda_c \equiv \frac{\varepsilon_p Pe b^{II}}{D^{II}}, \quad r^* = \left(\frac{3\alpha}{4\pi}\right)^{1/3}$$

2.9.3. ANALYSIS

The first exchange term in (refeq6.4) is associated to the diffusion exchange which is dominating.

The second exchange term is induced by convective flow of the whole fluid in the boundary layer near the block border. However, this convective term includes a component dependent on the diffusion parameter. Thus, convection exchange can not happen in pure form and is nonlinearly superposed with diffusion exchange. In other words, the convective and diffusive exchange are nonlinearly interacted and can not be splitted onto two independent addends.

2.10. C-T SYSTEM

2.10.1. NON-SPLITTED CHARACTER OF C-T FLOW

Chemical or heat transport in highly heterogeneous media seams to be the most complicated when the system belongs to the Convection-Translation class.

However, the diagram in Fig. 2.8 shows that C-T mode of flow is appropriated to media with limitation for heterogeneity degree: $\varepsilon^2 \leq \varepsilon_p \leq \varepsilon$, or $\varepsilon \leq \omega_K \leq 1$. Such media may be considered rather as moderately heterogeneous. Then, it is natural to wait that homogenization does not lead to splitting the concentration transport equation onto two equations for blocks and matrix.

Fig. 2.15 shows the results of simulations of concentration transport through a medium with $\omega_K = 0.1$.

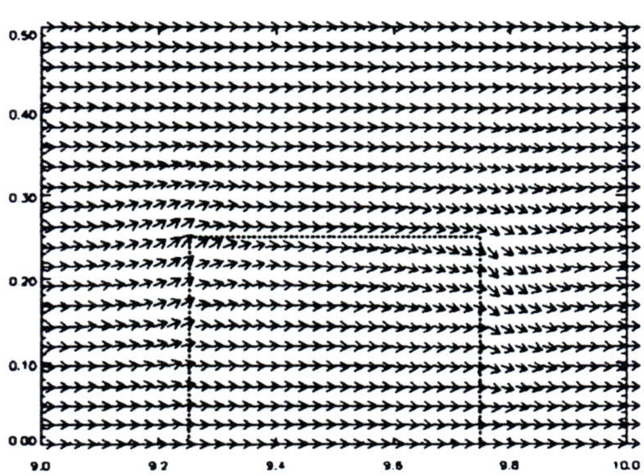

Figure 2.15. Flow streamlines in highly heterogeneous C-T medium

The upper half of the cell is shown. The flow is directed to the right from the left. The central block is low permeable.

It can be seen that the streamline pattern in rather homogeneous in the overall medium and does not undergo sharp distortions at the block/matrix interface. Then, to study C-T system, it is possible to assume that the transport velocity field and, consequently, the concentration field is moderately heterogeneous, such that the averaged equations are expected to be uniform for the whole medium.

According to the ideas mentioned in section 1.3, splitting onto two macroscale continuous media is ensured by the limit transition $\omega \to 0$. When splitting does not happen the process regularly depends on the parameter ω. Moreover, the parameter ω_K is small but rather moderately. Thus, for the examined C-T system the solution may be developed at fixed parameter ω_K, assuming $\omega_K \simeq 1$ when $\varepsilon \to 0$ in all asymptotic expansions. The results obtained may be expanded into series over the parameter ω_K. As to the Peclet number Pe, it will be accepted as

$$Pe \geq \frac{1}{\varepsilon} \tag{2.124}$$

according to the diagram in Fig. 2.8.

Usually the convection transport is examined when the transport velocity field is assumed to be periodic, or ergodic in random case. Periodicity of velocity allows to keep similarity of flow in all the cells. Abandonment of periodicity leads to another results. In this book, C-T transport will be considered for an arbitrary velocity field.

Summarizing, the following equation will be subjected to the procedure of averaging:

$$b\left(\frac{x}{\varepsilon}\right)\frac{\partial c}{\partial t}+v_i\left(x,\frac{x}{\varepsilon},t;\omega_K\right)\frac{\partial c}{\partial x_i}=\frac{1}{Pe}\frac{\partial}{\partial x_i}\left(D_{ij}\left(\frac{x}{\varepsilon};\omega_K\right)\frac{\partial c}{\partial x_j}\right) \tag{2.125}$$

2.10.2. DEFINITION OF TRANSPORT VELOCITY

We shall notice:

$$v_i\left(x,\frac{x}{\varepsilon},t;\omega_K\right)\equiv -a\left(\frac{x}{\varepsilon}\right)\frac{\partial p}{\partial x_i}, \quad i=1,2,3 \tag{2.126}$$

where v_i is Darcy's velocity of the mixture, which undergoes the macroscale variations in space produced by fast oscillations of medium heterogeneity.

Definition for $\vec{v}$ is done by continuity equation:

$$\frac{\partial v_i}{\partial x_i}=F\left(x,t,\frac{x}{\varepsilon}\right) \tag{2.127}$$

with summation over i. When the fluid is weakly compressible the right-hand part in (2.127) is defined as

$$F\equiv -\beta b(x)\partial p/\partial t$$

whereas for an incompressible fluid:

$$F\equiv 0$$

As the pressure field is always independent, the function F will be considered as known and getting the rather general form.

When ε is small the following expansion is valid for $\vec{v}$:

$$v_i = v_{i0}(x, y, t) + \varepsilon v_{i1}(x, y, t) + \varepsilon^2 ..., \quad y = x/\varepsilon \tag{2.128a}$$

The coefficients v_{ik} can be determined using (2.127):

$$\frac{\partial v_{i0}}{\partial y_i} = 0, \quad \frac{\partial v_{i1}}{\partial y_i} = -\frac{\partial v_{i0}}{\partial x_i} + F_0 \quad \text{e.t.c.} \tag{2.128b}$$

where $F = F_0(x, y, t) + \varepsilon F_1(x, y, t) + \varepsilon^2 ...$

The velocity at the domain boundary is assumed to be known, then the field $\vec{v}$ and its expansion are entirely defined.

Velocity $\vec{v}$ is not periodic, but may be represented in the form with splitted fast and slow variables:

$$v_{ik} = \sum_{m=1}^{N(i,k)} u_{ik}^m(x, t) \zeta_{ik}^m(y) \tag{2.129}$$

Herein the numbers $N(i, k)$, as well as the functions u_{ik}^m and $\zeta_{ik}^m(y)$ are given. For example, in case (2.126):

$$v_{i0} = -a(y)\left(\frac{\partial \psi_k}{\partial y_i} + \delta_{ik}\right)\frac{\partial p_0}{\partial x_k},$$

$$v_{i1} = -a(y)\left(\frac{\partial \psi_k}{\partial y_i} + \delta_{ik}\right)\frac{\partial p_{10}}{\partial x_k} - a\left(\frac{\partial \psi_{km}}{\partial y_i} + \psi_k \delta_{mi}\right)\frac{\partial^2 p_0}{\partial x_k \partial x_m}$$

2.10.3. STRUCTURE OF EXPANSIONS

Let a new parameter be introduced:

$$\sigma = \frac{1}{\varepsilon Pe} << 1 \tag{2.130}$$

Expansion (2.125), when $\varepsilon, \sigma \rightarrow 0$, gets the form:

$$\begin{aligned} c(x, y, t) &= c_0(x, t; \sigma) + \varepsilon c_1(x, t; \sigma) + \varepsilon^2 ... = \\ &[c_{00}(x, t) + \sigma c_{01}(x, t) + \sigma^2 ...] + \varepsilon[c_{10}(x, y, t) + \\ &\sigma c_{11}(x, y, t) + \sigma^2 ...] + \varepsilon^2 ... \end{aligned} \tag{2.131}$$

Two-parametrical asymptotic expansion at $\varepsilon, \sigma \to 0$ may be developed in two stages: a) first, the expansion is constructed with $\varepsilon \to 0$, while σ is fixed ($\sigma \simeq 1$); b) the results obtained will be expanded into asymptotic series at $\sigma \to 0$.

After introducing the fast variables $y = x/\varepsilon$ Eq. (2.125) gets the form:

$$
\begin{aligned}
-\partial_i(D_{ij}\partial_j c) &= -\sigma^{-1} v_i \partial_i c + \varepsilon \left[\partial_i(D_{ij}\partial_{x_j} c) + \partial_{x_i}(D_{ij}\partial_j c) - \right. \\
&\left. \sigma^{-1}(\partial_t c + v_i \partial_{x_i} c)\right] + \varepsilon^2 \partial_{x_i}(D_{ij}\partial_{x_j} c);
\end{aligned} \tag{2.132}
$$

Using (2.131) we get from (2.132):

$$
\begin{aligned}
&\varepsilon\Big[\partial_i(D_{ij}\partial_j c_1) + \partial_i(D_{ij}\partial_{x_j} c_0) - \sigma^{-1}\left(b\partial_t c_0 + v_i(\partial_i c_1 + \partial_{x_i} c_0)\right)\Big] + \\
&\sum_{k\geq 2} \varepsilon^k \Big[\partial_i(D_{ij}\partial_j c_k) + \partial_i(D_{ij}\partial_{x_j} c_{k-1}) + \partial_{x_i}(D_{ij}\partial_j c_{k-1}) + \\
&\partial_{x_i}(D_{ij}\partial_{x_j} c_{k-2}) - \sigma^{-1}\Big(b\partial_t c_{k-1} + v_i(\partial_i c_k + \partial_{x_i} c_{k-1})\Big)\Big] = 0;
\end{aligned} \tag{2.133}
$$

$$
[c_k]_\Gamma = 0, \ k \geq 1; \quad D_{ij} \sum_{k\geq 1} \varepsilon^k (\partial_i c_k + \partial_{x_i} c_k) n_j|_\Gamma = 0;
$$

$$
c_0(x,0) = c^0(x), \quad c_k(x,y,0) = 0, \ k \geq 1
$$

The necessary condition of existence for the y-periodic solution of Eq. (2.132) requires that the averaged right-hand part is equal to zero, that yields:

$$
\begin{aligned}
&-\varepsilon\sigma^{-1}\Big[\langle b\rangle \partial_t c_0 + \langle v_i s_i^0\rangle\Big] + \textstyle\sum_{k\geq 2} \varepsilon^k \Big\langle \partial_{x_i}(D_{ij}\partial_j c_{k-1}) + \\
&\partial_{x_i}(D_{ij}\partial_{x_j} c_{k-2}) - \sigma^{-1}\Big(b\partial_t c_{k-1} + v_i s_i^{k-1}\Big)\Big\rangle = 0
\end{aligned} \tag{2.134}
$$

where it is designated: $s_i^k \equiv \partial_i c_{k+1} + \partial_{x_i} c_k$.

Another relationship will be used instead of (2.134), which may be obtained in the following way. Dividing Eq. (2.133) by b and deducting Eq. (2.134) divided by $\langle b\rangle$, we exclude the derivative $\partial_t c_0$:

$$
\begin{aligned}
0 = \varepsilon &\left[\partial_i(D_{ij}\partial_j c_1) + \partial_i(D_{ij}\partial_{x_j} c_0) - \sigma^{-1}\left(v_{i0} s_i^0 - \frac{b}{\langle b\rangle}\langle v_{i0} s_i^0\rangle\right)\right] + \\
&\sum_{k\geq 1} \varepsilon^k \Big\{\partial_i(D_{ij}\partial_j c_k) + \partial_i(D_{ij}\partial_{x_j} c_{k-1}) + \partial_{x_i}(D_{ij}\partial_j c_{k-1}) + \partial_{x_i}(D_{ij}\partial_{x_j} c_{k-2}) -
\end{aligned}
$$

$$\frac{b}{\langle b\rangle}\Big\langle\partial_{\mathrm{x}i}(D_{ij}\partial_j c_{k-1})+\partial_{\mathrm{x}i}(D_{ij}\partial_{\mathrm{x}j}c_{k-2})\Big\rangle-\sigma^{-1}\Big[b\partial_t c_{k-1}-\frac{b}{\langle b\rangle}\Big\langle b\partial_t c_{k-1}\Big\rangle\Big]+$$

$$\sum_{m=0}^{k-1}\Big(s_i^{k-m-1}v_{im}-\frac{b}{\langle b\rangle}\langle s_i^{k-m-1}v_{im}\rangle\Big)\Big\};\qquad(2.135\text{a})$$

$$-\varepsilon\sigma^{-1}\Big[\langle b\rangle\partial_t c_0+\langle v_{i0}s_i^0\rangle\Big]+\sum_{k\geq 1}\varepsilon^k\Big\langle\partial_{\mathrm{x}i}(D_{ij}\partial_j c_{k-1})+$$

$$\partial_{\mathrm{x}i}(D_{ij}\partial_{\mathrm{x}j}c_{k-2})-\sigma^{-1}\Big[b\partial_t c_{k-1}+\sum_{m=0}^{k-1}v_{im}s^{k-m-1}\Big]\Big\rangle=0\qquad(2.135\text{b})$$

This yields for the first approximations:

$$\sim\varepsilon:\begin{cases}\partial_i(D_{ij}\partial_j c_1)+\partial_i(D_{ij}\partial_{\mathrm{x}j}c_0)-\sigma^{-1}\Big(v_{i0}s_i^0-b\langle b\rangle\langle v_{i0}s_i^0\rangle\Big)=0,\\ \langle b\rangle\partial_t c_0+\langle v_{i0}s_i^0\rangle=0\end{cases}\qquad(2.136\text{a})$$

$$\sim\varepsilon^2:\begin{cases}\partial_i(D_{ij}\partial_j c_2)+\partial_i(D_{ij}\partial_{\mathrm{x}j}c_1)+\partial_{\mathrm{x}i}(D_{ij}\partial_j c_1)+\partial_{\mathrm{x}i}(D_{ij}\partial_{\mathrm{x}j}c_0)-\\ \quad\frac{b}{\langle b\rangle}\Big\langle\partial_{\mathrm{x}i}(D_{ij}\partial_j c_1)+\partial_{\mathrm{x}i}(D_{ij}\partial_{\mathrm{x}j}c_2)\Big\rangle-\sigma^{-1}\Big(b\partial_t c_1+v_{i0}s_i^1+\\ \quad v_{i1}s_i^0-\frac{b}{\langle b\rangle}\Big\langle b\partial_t c_1+v_{i0}s_i^1+v_{i1}s_i^0\Big\rangle\Big)=0;\\ \Big\langle\partial_{\mathrm{x}i}(D_{ij}\partial_j c_1)+\partial_{\mathrm{x}i}(D_{ij}\partial_{\mathrm{x}j}c_0)\Big\rangle-\sigma^{-1}\Big\langle b\partial_t c_1+v_{i0}s_i^1+v_{i1}s_i^0\Big\rangle=0\end{cases}$$

(2.136b)

2.10.4. CELL PROBLEMS. VIOLATION OF PROCESS SIMILARITY

The following representation is valid for the coefficients c_k:

$$c_1=\varphi_k(x,y,t)\partial_{\mathrm{x}k}c_0+c_{10}(x,t),\qquad(2.137)$$

$$c_2=\varphi_{km}(x,y,t)\partial^2_{xk,xm}c_0+\varphi_k(x,y,t)\partial_{\mathrm{x}k}c_{10}+\zeta_k(x,y,t)\partial_{\mathrm{x}k}c_0+c_{20}(x,t)$$

where the functions φ_k, φ_{km} and ζ_k already depend on slow variables which can not be splitted with fast variables.

The consequent techniques is traditional. The intermediate deductions are omitted.

The functions φ_k are defined as solutions of the following problems:

$$\begin{cases}-\partial_i(D_{ij}\partial_j\varphi_k)=\partial_i D_{ik}-\frac{1}{\sigma}\Big[v_{i0}\Big(\partial_i\varphi_k+\delta_{ki}\Big)-\frac{b}{\langle b\rangle}V_{k0}\Big],\\ \langle\varphi_k\rangle=0,\quad(k=1,2,3)\end{cases}\qquad(2.138\text{a})$$

$$V_{k0}\equiv\langle v_{i0}(\partial_i\varphi_k+\delta_{ki})\rangle=\langle v_{k0}\rangle\qquad(2.138\text{b})$$

The last relation results from the first relation in (2.128b) and periodicity conditions. In fact,

$$\langle v_{i0}(\partial_i\varphi_k+\delta_{ki})\rangle=\langle\partial_i(v_{i0}\varphi_k)-\varphi_k\partial_i v_{i0}+v_{k0}\rangle=\langle v_{k0}\rangle$$

Expressions for φ_{km} and ζ_k are too cumbersome, therefore they are omitted, taking into account that only the function φ_k is sufficient for the averaged equation.

In problem (2.138a) the function v_{i0} depends on slow variables, hence the function φ_k also depends on x, t, therefore the function φ_k appears various in various cells. It means that the similarity of transport in the cells is violated.

Splitting of slow and fast variables of the function φ_k is found impossible, even using the splitted form (2.129) for the functions v_i.

This situation is observed only for -T systems.

2.10.5. AVERAGED MODEL FOR $PE \sim \varepsilon^{-1}$

The following system may be obtained from (2.136a) and (2.136b):

$$B\frac{\partial c_0}{\partial t}+V_{k0}\frac{\partial c_0}{\partial x_k}=0, \qquad B\equiv\langle b\rangle, \tag{$*$}$$

$$B\frac{\partial c_{10}}{\partial t}+V_{k0}\frac{\partial c_{10}}{\partial x_k}+\frac{1}{\varepsilon}\Delta V_k\frac{\partial c_0}{\partial x_k}=\sigma\frac{\partial}{\partial x_k}\left(\mathbf{D}_{km}\frac{\partial c_0}{\partial x_m}\right),$$

Notations are given furthermore, after Eq. (2.139)

Let C be the averaged concentration:

$$C\equiv\langle c\rangle=c_0+\varepsilon c_{10}+\varepsilon^2...,$$

Multiplying the second equation (*) by ε and summing the result with the first equation (*), we get for C:

$$B\frac{\partial C}{\partial t}+V_k(x,t)\frac{\partial C}{\partial x_k}=\frac{1}{Pe}\frac{\partial}{\partial x_k}\left[\mathbf{D}_{km}(x,t)\frac{\partial C}{\partial x_k}\right]+O(\varepsilon^2), \tag{2.139}$$

where

$$V_k(x,t)=V_{k0}(x,t)+\Delta V_k,$$

$$\Delta V_k(x,t)\equiv\varepsilon\left\langle b\frac{\partial\varphi_k}{\partial t}+\frac{b}{B}V_{m0}\frac{\partial\varphi_k}{\partial x_m}+v_{k1}\right\rangle,$$

$$\mathbf{D}_{km}(x,t)=\left\langle D_{mj}\left(\frac{\partial\varphi_k}{\partial y_i}+\delta_{ki}\right)-\varepsilon Pe\left[\langle v_{m0}\varphi_k\rangle-\frac{\langle b\varphi_m\rangle}{B}V_{k0}\right]\right\rangle$$

where the function V_{k0} is defined according to Eq. (2.138b).

The components of the effective dispersion tensor are not constant and depend on the variable field of transport velocity. It is important to note that the effective transport velocity V is inconsistent with the averaged convective transport velocity. The difference between them is equal to a value ΔV which depends on diffusion, in particular. The velocity shift ΔV is of the same order as the dispersion term. Moreover, the value ΔV is dominating when $Pe>>1/\varepsilon$.

Effect of diffusion on the averaged convective transport velocity was not observed in $C-S$ and $D-S$ systems.

2.10.6. CASE OF PERIODIC TRANSPORT VELOCITY

If $v_k=v_k(y)$ is the periodic steady-state field, then the results can be simplified:

1) $V_{k0}=$ const;

2) $\Delta V_k\equiv 0, \forall k$;

3) Problem (2.138a) becomes local and identical for all cells; the functions $\varphi_k=\varphi_k(y)$, $\forall k$, i.e., depend on fast variable y only.

4) Each component of transport velocity $v_i=v_i(y)$, moreover, $v_i=v_{i0}(y)$, $\forall i$, i.e., does not depend on slow variables x, t;

5) Each component of the effective dispersion tensor, $\mathbf{D}_{km}$, is constant, due to previous four items.

2.10.7. AVERAGED MODEL FOR $PE>>\varepsilon^{-1}$

Model (2.139) has been deduced with assuming $Pe\sim\varepsilon^{-1}$, or $\sigma\sim 1$. When $Pe>>\varepsilon^{-1}$, or $\sigma\ll 1$, the results obtained may be expanded into the series over the small parameter σ.

Let $\sigma\to 0$. Then the functions $\varphi_k(y)$ have the following structure:

$$\varphi_k(y)=\varphi_k^{(0)}(y)+\sigma\varphi_k^{(1)}(y)+\sigma^2...$$

which may be developed using (2.138a):

$$v_{i0}\frac{\partial\varphi_k^{(0)}}{\partial y_i}=\frac{b}{\langle b\rangle}V_{k0}-v_{k0};\quad \langle\varphi_k^{(0)}\rangle=0, \tag{2.140a}$$

$$v_{i0}\frac{\partial\varphi_k^{(1)}}{\partial y_i}=\frac{\partial}{\partial y_i}\left(D_{ik}\frac{\partial\varphi_k^{(0)}}{\partial y_k}\right)+\frac{\partial D_{ik}}{\partial y_i},\quad \langle\varphi_k^{(1)}\rangle=0; \tag{2.140b}$$

where the functions $\varphi_k^{(0)}$ and $\varphi_k^{(1)}$ are y-periodic.

It is clear that the necessary condition of existence for the periodic solution is satisfied.

The effective dispersion tensor has the following structure:

$$\mathbf{D}_{km}=\frac{1}{\sigma}\mathbf{D}_{km}^{(0)}+\mathbf{D}_{km}^{(1)}+\sigma... \tag{2.141a}$$

$$\begin{cases} \mathbf{D}_{km}^{(0)}=\dfrac{\langle b\varphi_k^{(0)}\rangle}{\langle b\rangle}V_{m0}-\langle v_{m0}\varphi_k^{(0)}\rangle, \\ \\ \mathbf{D}_{km}^{(1)}=\left\langle D_{mi}\left(\dfrac{\partial\varphi_k^{(0)}}{\partial y_i}+\delta_{ik}\right)\right\rangle+\dfrac{\langle b\varphi_k^{(0)}\rangle}{\langle b\rangle}V_{m0} \end{cases} \tag{2.141b}$$

The main part $\mathbf{D}_{km}^{(0)}$ does not depend on the microscale diffusion D_{ij}. This is the dispersion tensor in pure form. It will be calculated for lattice media in chapter 3.

2.10.8. CORRECTNESS OF CELL PROBLEMS

Problems (2.138a) with $\sigma\sim 1$ and (2.140b) are similar to classical cell problems resulting from homogenization of moderately heterogeneous parabolic equations, which have been proved as correctly set in [10], for instance.

Problem (2.140) is qualitatively distinct, been formulated for the equation of first order. Resolvability of such a problem is not obvious and is studied on a 1D variant of problem (2.140a) with periodic transport velocity:

$$\begin{cases} v\dfrac{\partial\varphi^{(0)}}{\partial y}=bV-v; \quad y\in(-1/2,1/2)=Y \\ \\ \langle\varphi^0\rangle=0, \quad \varphi^{(0)} - \text{y-periodic} \end{cases} \tag{**}$$

where $V\equiv\langle v\rangle$, $B\equiv\langle b\rangle$.

As the continuity equation yields $\frac{\partial v}{\partial y}=0 \Rightarrow v=\text{const}$, and $V=v$, then:

$$\varphi^{(0)}=M+\frac{\Phi(y)}{B}-y; \quad \Phi(y)\equiv\int_{-1/2}^{y} b(\overline{y})d\overline{y};$$

It follows from the last condition in (**): $M=-\frac{\langle\Phi\rangle}{B}$.

It remains to check out the periodicity condition:

$$\varphi^{(0)}(-1/2)=M+1/2; \quad \varphi^{(0)}(1/2)=M-1/2+\Phi(1/2)/B,$$

but $\Phi(1/2)=\langle b\rangle=B$, therefore

$$\varphi^{(0)}(1/2)=M-1/2+1=M+1/2=\varphi^{(0)}(-1/2)$$

Thus, the solution exists and is unique.

In 2D case, the situation is more complicated, in particular the problem may not have as smooth as continuous solution. This situation is analyzed in chapter 3 for lattice media.

2.10.9. "SUPER-DISPERSION" IN HIGHLY HETEROGENEOUS MEDIA

According to macroscale equation (2.139), the "complete" dispersion coefficient may be defined as

$$\mathcal{D} \equiv \frac{1}{Pe}\mathbf{D} = \frac{\varepsilon Pe}{Pe}\mathbf{D}^{(0)} = \varepsilon \mathbf{D}^{(0)} \tag{2.142}$$

As $Pe \leq \varepsilon^{-1}$, the complete dispersion parameter seems to be small. But this is only true when the medium is moderately heterogeneous, i.e., when $\omega_K \sim 1$.

When $\omega_K \ll 1$ the dispersion parameter $\mathcal{D}$ is of order one. It may be easy proved basing on cell problem (2.140).

Examine the first equation in (2.140) inside a low permeable block, Y^I:

$$v_{i0}^I \frac{\partial \varphi_k^{(0)}}{\partial y_i} = \frac{b^I}{\langle b \rangle} V_{k0} - v_{k0}^I, \qquad y \in Y^I \tag{2.143}$$

The local velocity v is proportional to permeability, then $v_{i0}^I \sim \omega_K$. However, the averaged velocity, V, is of order one. In fact, the parameter beforehand the derivative in the left-hand part of (2.143) is of order ω_K, while the right-hand part is of order one. As the order of both the parts must be equivalent, the function φ_k^0 must be large, such that:

$$\varphi_k^0 \sim \frac{1}{\omega_K}, \qquad y \in Y^I$$

Then it follows from (2.141a) that:

$$\mathbf{D}_{ij}^(0) \sim \varphi_k^0 \sim \frac{1}{\omega_K}, \qquad y \in Y^I$$

Then:

$$\mathcal{D} \sim \frac{\varepsilon}{\omega_K}, \qquad y \in Y^I$$

For C-T media, $\varepsilon \leq \omega_K \leq 1$. Then, if $\omega_K \sim \varepsilon$, we get:

$$\mathcal{D} \sim 1, \qquad y \in Y^I$$

This proves that dispersion tensor becomes considerable in highly heterogeneous media.

It is necessary to note, that the dispersion term arising after averaging the first-order differential equation of convective transport is related to spatial smoothing of functions within each cell and, consequently, with increasing of the order of their differentiability. On the other hand, the same phenomena can be explained in term of non-equilibrium thermodynamics. Averaging leads to an irreversible loss of information within each cell, which is related, in its tour, to an irreversible loss of the system energy within each cell. As the result, the reversible Hamiltonian system is transformed into an irreversible dissipative one, where the dissipation is displayed in term of dispersion. In classical case, the irreversibility, the dissipation, or the dispersivity are of small order equal to the heterogeneity scale. In highly heterogeneous medium, although the heterogeneity scale is kept small, all these phenomena become of order one. Such an effect is unusual and can be called as "super-dispersion".

2.10.10. RESULTS FOR C-T SYSTEMS

1. In case of C-T systems it is natural to expect that all the medium is described by a unique macroscale concentration averaged over the overall cell. Splitting onto block and matrix concentrations does not happen.

2. Interaction between the heterogeneous diffusion field and permeability field generates a phenomenon of the macroscale transport velocity shift, in addition to the well known macroscale dispersion effect.

3. The macroscale velocity shift does not exist when the local transport velocity field is periodic and steady-state.

4. The similarity of transport in various cells is violated, when the transport velocity is non-periodic. The cell problem depends on the macroscale variable, so that the macroscale system becomes spatially nonlocal.

5. The macroscale dispersion tensor is shown to be large (of order one), when the medium is highly heterogeneous.

2.11. MEMORY EFFECTS IN DELAY MODEL OF TRACER TRANSPORT

Various classes of macroscale behavior presented in previous sections are characterized by multiple nonlocal phenomena. Behavior of the system with predomination of diffusion is similar to that examined in chapter 1, however the macroscale behavior of the system with domination of convection is not studied yet in details.

An example of such behavior for a C-S system is described below. The macroscale model of convection transport is the delay differential equation of first order. The examined initial problem may be interpreted as tracer injection in oil reservoir. Analytical solution of a 1D problem of tracer injection is obtained.

To understand the influence of nonlocality the process with inversion of direction will be considered, because various memory effects may be more clearly detected when the direction of the process is varying in time.

The analyzed problem has practical significance. Its solution gives possible to develop some tool to determine the heterogeneity parameters using special inverse method of tracer injection.

In section 1.8 the principle of uncertainty has been proved which says that the simultaneous determination of the block/matrix volume ratio and the block/volume permeability ratio is impossible, if only the information concerning the pressure field is available. However these two parameters can be determined when the supplement information about the concentration field will be used. This information may be obtained by means of a special in-situ experiment consisting in tracer injection in porous reservoir with flow rate variable in time. This experiment is named as the "inverse method".

Theory of the method is developed basing on the model of C-S flow.

2.11.1. MODEL PROBLEM OF TRACER INJECTION WITH INVERSION OF PROCESS DIRECTION

Formulation of the Problem Using C-S Model

Let the highly heterogeneous, double porosity, medium is saturated by a single phase fluid containing one chemical component. The fluid phase will be named as "liquid", while the chemical component is "oil".

The same phase as liquid but with different composition is injected in the medium. It contains two chemical components, oil and tracer ("the agent"). Tracer concentration is C.

The 1D problem is studied when the liquid flows from the left to the right, the tracer is injected through the left-hand inlet section of the medium.

For the sake of simplicity, the macroscale matrix pressure is assumed to be linear along the coordinate x. It is linearly growing in time until the instant t_1, nd next is linearly decreasing:

$$\frac{\partial P^{II}}{\partial x}=-\beta_1,\ \forall t; \qquad \frac{\partial P^{II}}{\partial t}=\begin{cases}\beta_2, & 0\leq t\leq t_1\\ -\beta_2, & t>t_1\end{cases}$$

$$\beta_1,\beta_2=\text{const} \tag{2.144}$$

Thus, the tracer injection rate exceeds the rate of tracer extraction until the instant t_1, and vice versa after t_1.

According to (2.70), (2.81) and (2.67), the times τ linearly depends on t:

$$\tau=\frac{1}{3}\beta_2 t,\ \forall t, \qquad \text{then} \quad \tau_1=\frac{1}{3}\beta_2 t_1$$

The mode of flow is assumed to be of C-S type. In frameworks of this model diffusion can be neglected. Then the 1D-problem for (2.112) gets the form:

$$\frac{\partial C^{II}}{\partial \tau}+\frac{\partial C^{II}}{\partial \xi}=0,\ 0<\tau\leq\tau_1 \tag{2.145a}$$

$$\frac{\partial C^{II}}{\partial \tau}+\frac{\partial C^{II}}{\partial \xi}=-\alpha_0\left[C^{II}(\xi,\tau)-C^{II}(\xi,2\tau_1-\tau)\right],\ \tau>\tau_1 \tag{2.145b}$$

$$C^{II}(\xi,0)=1-H(\xi) \tag{2.145c}$$

$$\xi=x/L, \quad L\equiv 3\beta_1 A^{II}/B^{II}\beta_2, \quad \alpha_0\equiv 3\beta_1/B^{II}$$

$$C^{II}(\xi,\tau)\equiv C^{II}(x,t)\Big|_{\substack{x=\xi L,\\ t=3\tau/\beta_2}}$$

Herein A^{II} is the macroscale permeability of the matrix.

2.11.2. SOLUTION OF PROBLEM

First Time Stage

Solution at the first time stage, i.e. $t\leq t_1$, is as follows:

$$C^{II}=1-H(\xi-\tau), \quad \tau\leq\tau_1 \tag{2.146}$$

where $H(x)=1$ when $x\geq 0$, $H(x)=0$ when $x<0$.

Second Time Stage

Solution for the second time stage may be also obtained in the analytical form.

Note that in (2.145b) the following holds: $C^{II}(\xi,\theta)=1-H(\xi-\theta)$, where $\theta=2\tau_1-\tau$ if $\tau>\tau_1$, because $\theta<\tau_1$. Then (2.145b) and (2.145c) get the form:

$$\begin{cases} \dfrac{\partial C^{II}}{\partial \tau}+\dfrac{\partial C^{II}}{\partial \xi}=-\alpha_0 C^{II}(\xi,\tau)+\alpha_0[1-H(\xi-2\tau_1+\tau)], \quad \tau>\tau_1 \\ C^{II}(\xi,\tau_1)=1-H(\xi-\tau_1) \end{cases} \tag{2.147}$$

Problem (2.147) is solved using the method of characteristics:

$$C^{II}(\xi,\tau)=[1-H(\xi-\tau_1)]\exp\{-\alpha_0(\tau-\tau_1)\}+ \alpha_0\int_{\tau_1}^{\tau}[1-H(\xi-2\tau_1+2\tau'-\tau)]\exp\{-\alpha_0(\tau-\tau')\}\,d\tau'$$

or:

$$C^{II}(\eta,\theta)=[1-H(\eta)]e^{-\alpha_0\theta}+ \underbrace{\frac{\alpha_0}{2}\int_{\eta-\theta}^{\eta+\theta}[1-H(z)]\exp\left[-\frac{\alpha_0}{2}(\theta+\eta-z)\right]dz}_{I} \tag{2.148}$$

where the new time θ and the new coordinate η are introduced in such a way, that the axis origin is superposed with the front location at the instant τ_1: $\theta=\tau-\tau_1$, $\eta=\xi-\tau_1$.

Due to properties of the function $H(z)$, the integral I in (2.148) gets three distinct values depending on the value of η.

If $\eta\leq-\theta$, then

$$I=\int_{\eta-\theta}^{\eta+\theta}\exp\left[-\frac{\alpha_0}{2}(\theta+\eta-z)\right]dz=\frac{2}{\alpha_0}\left(1-e^{-\alpha_0\theta}\right)$$

If $-\theta<\eta\leq\theta$, then:

$$I=\int_{\eta-\theta}^{0}\exp\left[-\frac{\alpha_0}{2}(\theta+\eta-z)\right]dz=\frac{2}{\alpha_0}\left(e^{-\alpha_0\frac{\eta+\theta}{2}}-e^{-\alpha_0\theta}\right)$$

If $\eta>\theta$, then $I\equiv0$.

Definitely, the solution takes the following form at $\tau>\tau_1$:

$$C^{II}(\xi,\tau)=\begin{cases} 1, \quad 0\leq\xi\leq\xi_*(\tau), \\ \exp\left[-\frac{\alpha_0}{2}(\xi+\tau-2\tau_1)\right], \quad \xi_*(\tau)<\xi\leq\tau_1, \\ \exp\left[-\frac{\alpha_0}{2}(\xi+\tau-2\tau_1)\right]-\exp\left[-\alpha_0(\tau-\tau_1)\right], \quad \tau_1<\xi\leq\xi^*(\tau); \\ 0, \quad \xi>\xi^*(\tau) \end{cases} \tag{2.149}$$

Where $\xi_*(\tau)\equiv2\tau_1-\tau, \quad \xi^*(\tau)\equiv\tau.$

2.11.3. PHYSICAL PATTERN OF PROCESS

Formation of the Forward and the Reverse Concentration Wave
Concentration profile $^{II}(\xi)$ for the first time stage is shown in Fig. 2.16, which corresponds to solution (2.146). The profile for the second stage is shown in Fig. 2.17.

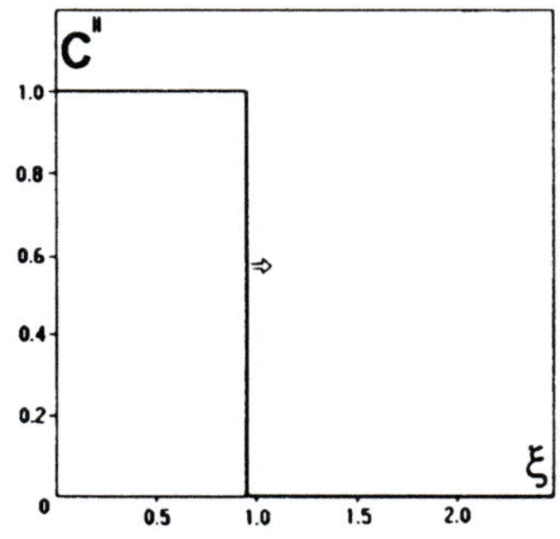

Figure 2.16. Injection-extraction of tracer in oil reservoir. C-S mode of flow. Concentration wave propagation at the first stage

After the process is reversed at the instant τ_1, the discontinuity $\xi=\tau_1$ is desintegrated onto two fronts. One of them, ξ^*, runs forwards with zero concentration ahead, while the second front, ξ_*, runs back with the concentration ahead equal to one. These two profiles are named as the forward and the reverse concentration wave. Both the fronts are weak discontinuities. A strong discontinuity which is disappearing in time exists in the point $\xi=\tau_1$. The amplitude of the discontinuity, $f=\exp\{-\alpha_0(\tau-\tau_1)\}$, is varying in time, being monotonously decreasing.

The concentration of the reverse wave monotonously tends to zero, whereas the concentration grows in all points of the forward wave until

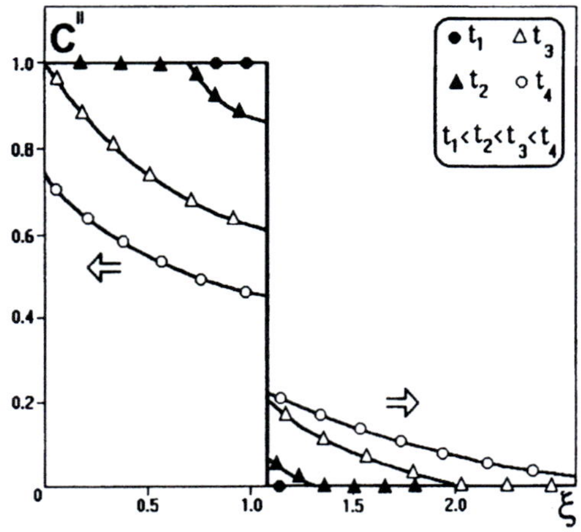

Figure 2.17. Injection-extraction of tracer in oil reservoir. C-S mode of flow. Concentration wave propagation at the second stage

the instant $\tau_{max}(\xi)$. Since this instant the values of concentration in the forward wave begin to decrease up to zero.

Then an extreme mode arises. The point $\tau_{max}(\xi)$ determining the extremum is defined as:

$$\tau_{max}(\xi)=\xi+\frac{2}{\alpha_0}\ln 2 \tag{2.150}$$

Thus, two time stages may be distinguished.

At the first time stage the concentration wave, which has the form of a step-function, advances along the matrix. The exchange flow is directed from the matrix toward the blocks, that changes the fluid composition inside the blocks, but does not influence the fluid composition inside the matrix.

The second time stage starts just after the instant τ_1. The direction of exchange flow is reversed: the flow current goes out of blocks and changes the mixture composition inside the matrix in such a manner, that the concentration C^{II} decreases.

After the process is reversed the concentration propagates within the matrix in the form of two waves which run up in opposite sides from the point $\xi=\tau_1$. That is the point of the forward front at the instant of process inversion. After inversion, all the field is divided onto two areas separated by an interface $\xi=\tau_1$ which remains immobile in time. To the left of this point the reverse wave moves towards the injection well. To the right, the forward wave is propagating towards the extraction well.

It is useful to note that solution (2.149) is unable to examine the asymptotics when $t\to\infty$, because the assumption about linear variation of pressure in time becomes incorrect at large values of time.

Structure of Reverse Wave

Formation of the reverse wave which goes from the front towards the injection well after the instant of inversion may be easily explained.

After inversion, the tracer injected into the block goes out of the block with the same concentration as in the matrix. Therefore it does not influence the fluid composition in the matrix. The larger is the volume of tracer injected at the first time stage in blocks, the longer is the duration of the stage when the fluid composition in the matrix remains invariable. At the first time stage, the concentration wave (2.145) deeply penetrates in those blocks which are located near the injection well. Therefore near the injection well the matrix does not feel any influence of blocks over a long period. In contrast, the matrix composition near the forward front changes just after inversion.

After the tracer abandons the block, the liquid flowing out of the block into the matrix carries zero concentration. Mixing of this flow with the flow in matrix leads to the exponential decreasing of the matrix concentration.

Variation of concentration in any point ξ near the injection well is shown in Fig. 2.18. Initially the matrix concentration is equal to zero. The tracer front reaches this point at the instant $\tau_0 \equiv \xi$. At the instant of inversion, τ_1, injection has been replaced by extraction, the reverse wave starts to move towards the injection well. At the instant $\tau_2 \equiv 2\tau_1 - \xi$ the reverse wave attains the point ξ. Since this instant the exponential attenuation of the local concentration is observed. Note, that during the period $\tau_1 - \tau_0$ tracer penetrated into the blocks in the examined point ξ.

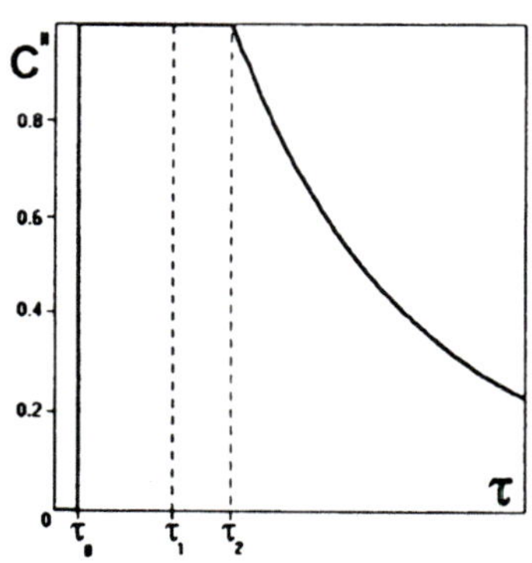

Figure 2.18. Time variation of matrix concentration near the injection well

Variation of concentration away from the injection well, but just near the forward front is shown in Fig. 2.19. Initially the matrix concentration is equal to zero. It may be seen that the period $\tau_1 - \tau_0$ of tracer penetration into the blocks is missing in this point.

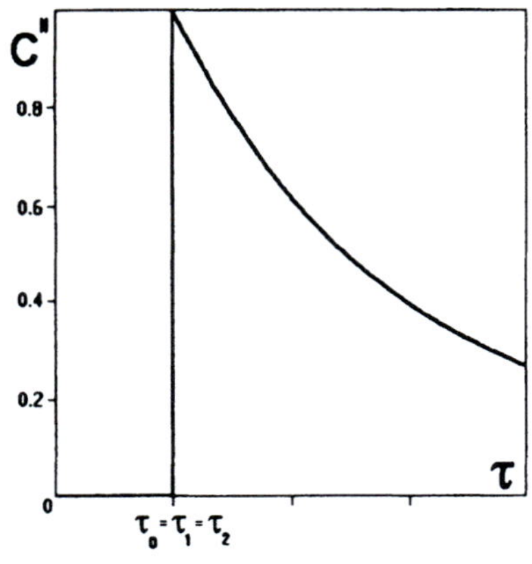

Figure 2.19. Time variation of matrix concentration near the forward front

The velocity $\frac{d\xi_*}{d\tau}$ of the front of the reverse wave is constant. The associated discontinuity is weak.

Structure of Forward Wave

Velocity of the forward front is not changed after inversion. However the shape of the wave is considerably modified. Dissipation of the wave happens due to attenuation of the matrix concentration explained above.

Concentration evolution in three spatial points to the right of the forward front is shown in Fig. 2.20.

The front approaches to the point ξ at the instant $\tau_0=\xi$.

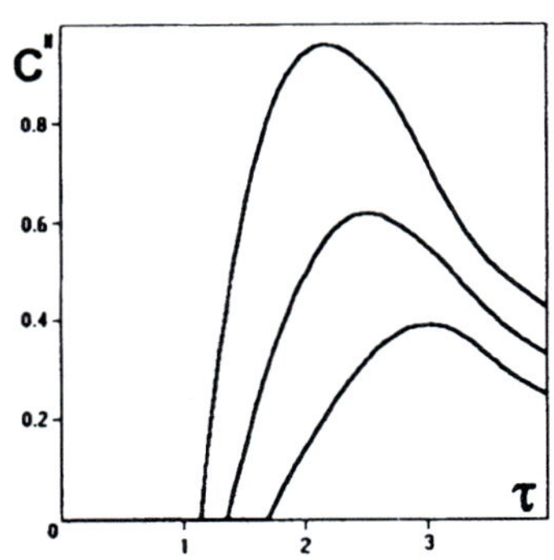

Figure 2.20. Concentration evolution in forward wave

The matrix concentration C^{II} is formed by two factors. First, the tracer inflow (flow-I) from the injection well tends to increase the value of C^{II}. Secondly, the flow current from the blocks (flow-II) tends to decrease it.

After inversion, the intensity of flow-I decreases, while that of the flow-II increases in time. Thus, an extreme instant τ_{max} exists, which corresponds to the maximum of concentration.

2.11.4. INVERSE METHOD TO DETERMINE THE PARAMETERS OF HETEROGENEITY

Theory of Method

Let the process of inverse tracer injection described above be performed in a groundwater or oil reservoir. Let the tracer concentration C^{II} be measured at the extraction well. Let the dimensionless coordinate of this well be ξ_c.

It follows from (2.149) for the point of extraction:

$$C^{II}(\xi_c,\tau)=\begin{cases} 0, & \tau\leq\xi_c \\ \exp\left[-\frac{\alpha_0}{2}(\xi_c+\tau-2\tau_1)\right]-\exp\left[-\alpha_0(\tau-\tau_1)\right], & \tau>\xi_c \end{cases} \tag{2.151}$$

Examine the asymptotics when $\tau\to\xi_c+0$. Let $z\equiv\tau-\xi_c$, then:

$$C^{II}=-\exp\{-\alpha_0(\xi_c-\tau_1)\}\left(\exp\left\{-\frac{\alpha_0 z}{2}\right\}-\exp\{-\alpha_0 z\}\right)\sim \frac{\alpha_0}{2}(\tau-\xi_c)\exp\{-\alpha_0(\xi_c-\tau_1)\}$$

Returning back to the variable with dimension, we get:

$$C^{II}(x_c,t)\sim a_1(t-t_0) \tag{2.152}$$

where

$$a_1(\lambda)\equiv\frac{\lambda\beta_2}{2}\exp\left[\lambda\beta_2(t_1-t_0)\right],\quad t_0\equiv\frac{X_cB^{II}}{A^{II}\beta_1},\quad \lambda\equiv B^I/B^{II} \tag{2.153}$$

where t_0 is the instant when the front reaches the extraction well.

The parameter λ, which is equal to the ratio of block/matrix water (oil) reserves should be determined as the result of the experiment.

The parameter λ may be calculated by nonlinear regression methods, if the instant t_0, the derivative $\beta_2=|\frac{\partial P^{II}}{\partial t}|$ at the extraction well, and the real plot $C^{II}(t)$ are known.

Method Performance

To ensure the necessary modes of flow in the reservoir, the injection rate Q_1 of the injection well must exceed the extraction rate Q_2 of the extraction well during the first time stage. The ratio $Q_1/Q_2>1$ is kept invariable up to the instant t_1. This ensures increasing of pressure in time and penetration of tracer into the blocks.

After the instant t_1, the following ratio should be established: $Q_1/Q_2<1$. Then decreasing of pressure is ensured.

At the extraction well, it is necessary to fix the instant t_0 when the tracer front reaches the well, to measure the dependence $C^{II}(t)$ and $P^{II}(t)$. Then the coefficient a_1 in (2.152) may be calculated as the tangent of the function $C^{II}(t)$, the parameter $\beta_2=|\partial P^{II}/\partial t|$ is directly estimated, and λ is obtained as solution of the nonlinear equation using formula (2.153) for $a_1(\lambda)$.

The simplest way to estimate λ is when $t_1=t_0$, as in Fig. 2.19. However this mode can not be exactly reached in practice, because the change of the flow rates requires some time, that distorts the total picture. Therefore more suitable modes correspond to the cases when $t_1 \gg t_0$, however the time t_1 should not be too small. Otherwise the exponential functions in (2.152) become very small, and the method occurs insensitive to the reservoir parameters. Thus, $(t_1 \neq 0)$, therefore the inversion is obligatory.

Use of Extreme Modes

To define the ratio of water (oil) reserves λ, the non monotonous dependence $C^{II}(t)$ may be used. The time instant t_{max} fixing the maximum of concentration at the extraction well may be defined using (2.150):

$$t_{max}(X_c)=t_0+\frac{2\ln z}{\beta_2\lambda}$$

then:

$$\lambda=\frac{12\ln 2}{\beta_2(t_{max}-t_0)} \tag{2.154}$$

Fixing the instants t_0 and t_{max} and measuring the derivative $\beta_2=|\frac{\partial P^{II}}{\partial t}|$ averaged over the interval $t\in[t_0,t_1]$, we can obtain the ratio λ.

References

1. Allaire, G. (1992) Homogenization and two-scale convergence, *SIAM J. Math. Anal.*, **23**, no.6, pp. 1482–1518.
2. Amaziane, B., Bourgeat A., Koebbe J.V. (1990) *Proc. 2nd European Conf. Mathematics of Oil Recovery, Arles, France, Sept.11–14 , 1990*, Ed.Technip, Paris, pp. 75–81.
3. Amirat, Y., Hamdache, K. and Ziani, A. Homogénéisation d'équations hyperboliques du premier ordre: Application aux milieux poreux, *Rapport I.N.R.I.A.*, no. 803.
4. Amirat, Y., Hamdache, K. and Ziani, A. (1989) Homogénéisation d'équations hyperboliques du premier ordre et application aux écoulements miscibles en milieu poreux, *Ann. Inst. H. Poincaré. Analyse non linéaire.*, **6**, no. 5, pp. 397–417.
5. Amirat, Y., Hamdache, K. and Ziani, A. (1990) Comportemet limite de modèles d'équations de convection-diffusion dégénérées, *C. R. Acad. sci. Paris, Ser. A.*, **310**, pp. 76–78.
6. Amirat, Y., Hamdache, K. and Ziani, A. (1990) Etude d'une équation de transport à mémoire, *C. R. Acad. sci. Paris, Ser. A.*, **311**.
7. Amirat, Y., Hamdache, K. and Ziani, A. (1991) Kinetic formulation for a transport equation with memory, *Commun. Partial Differ. Equations*, **16**, no. 8–9, pp. 1287–1311.
8. Arbogast T., Douglas J., Hornung U. *Derivation of The Double Porosity Model of Single Phase Flow via Homogenization Theory* - SIAM J. Math. Anal., 1990, **21**, pp. 823-836.
9. Arbogast T., Douglas J., Hornung U. *Modelling of Naturally Fractured Reservoirs by Formal Homogenization Thechniques* - In: "Frontiers in Pure and Applied Mathematics", Ed. R.Dautray, Elsevier, Amsterdam, 1991, pp. 1-19.
10. Bakhvalov N., Panasenko G. *Homogeneisation: Averaging Processes in Periodic Media* - Kluwer, Dordrecht, 1989.
11. Barenblatt G.I., Entov V.M., Ryjik V.M. *Theory of Transient Liquide and Gas Flow Through Porous Media* - Moscow, Nedra, 1972, 288 p. (translated in Kluwer, Dordrecht).
12. Barenblatt G.I., Zheltov Iu.P., Kochina I.N. *Basic Concepts in The Theory of Seepage of Homogeneous Liquids in Fissured Rocks* - PMM, **24**, 1960, pp. 852-864.
13. Barker J.A. *Block-Geometry Functions Characterizing Transport in Densely Fissured Media* - J. Hydrol., 1985, **77**, pp. 263-279.
14. Barker J.A. *Modelling the Effects of Matrix Diffusion on Transport in Densely Fissured Media* - Mem. of 18th Congress of the Inter.

Assoc. of Hydrogeol., Cambridge, 1985, pp. 250-269.

15. Bedrikovetski P.G., Istomin G.V., Kniazeva M.B. *Miscible Displacement From The Fractured Porous Media* - Izvstiya Academii Nauk SSSR, Ser. MJG, 1986, N 6, pp. 100-110 (translated in: Fluid Mechanics, 1987).
16. Bohm M., Showalter R.E. *Diffusion in Fissured Media* - SIAM J. Math. Anal., 1985, **16**, N 3, pp. 500-509.
17. Bourgeat A. *Homogenization Method Applied to the Behaviour of a Naturally Fissured Reservoir* - Mathematical Method in Energy Research, Ed. SIAM (K.J.Kross), 1984, pp. 181-193.
18. Bourgeat A. *Homogenized Behaviour of Diphasic Flow in a Naturally Fissured Reservoir with Uniform Fractures* - Comput. Methods in Applied Mechanics and Engineering, 1984, N 47, pp. 205-217.
19. Businov S.N., Umrihin I.D. *Gydrodynamics Methods of Well and Field Investigations* - Moscow, Nedra, 1973, 243 p. (in Russian).
20. Charlaix E., Hulin J.P., Plona T.J. *Experimental Study of Tracer Dispersion in Sintered Glass Porous Materials of Variable Compaction* - Phys. Fluids, 1987, **30**, pp. 1690-1698.
21. Coats K.H., Smith B.D. *Dead-End Pore Volume and Dispersion in Porous Media* - Trans. Soc. Pet. Eng., 1964, **231**, N 3, pp. 73-84.
22. Deans H.A. *A Mathematical Model For Dispersion in the Direction of Flow in Porous Media* - Trans. Soc. Pet. Eng., 1963, **v.288**, pp. 49-52.
23. De Swaan A.O. *Analytic Solutions for Determining Naturally Fractured Reservoir Properties by Well Testing* - SPE J., 1976, June, pp. 117-122.
24. Douglas J., Arbogast T. *Dual Porosity Models for Flow in Naturally Fractured Reservoirs* - In: "Dynamics of Fluids in Hierarchical Porous Media", Ed. J.H.Cushman, Academic Press, 1990, pp. 177-220.
25. Duguid I.O., Lee P.C., *Water Resour. Res.*, **13**, N3, (1977), pp. 558–566.
26. Egorov A.G., Pudovkin M.A., in :*Applied Problems of Math. Phys.*, Riga, Ed. Latvien State University, (1983), pp. 98–107 (in Russian).
27. Egorov A.G., Salamatin A.N., *Teplophysika Vysokikh Temperatur*, **22**, N5, (1984), pp. 919–923 (in Russian).
28. Evans R.D., *SPE J.*, **22**, N5, (1982), pp. 669–680.
29. Hornung U., in :*Nonlinear Partial Differential Equations* (P. File, R. Bates, Eds.), Springer-Verlag, Berlin/New York, (1988).
30. Hornung U., Jager W., in :*Complex Chemical Reactions, Modelling and Simulation* (P. Denflhard, w. Jager, Eds.), Springer - Verlag, Berlin/New York, (1987).

31. Hornung U., Showalter R.E., *J. Math. Analysis and Applications*, **147**, N1, (1990), pp. 69–80.
32. Hornung U., Showalter R.E., Walkington N.J., *J. Math. Analysis and Applications*, **155**, N1, (1991), pp. 1–20.
33. Kazemi H., Seth, Thomas G.W., *SPE J.*, (1969), Dec., pp. 463–472.
34. Khruslov E.Ya., *Averaged Model of Non Stationnary Flow through Fractured Porous Media*, USSR, Ed. Acad. Sci. Ukraine, Preprint N50, (1988), 34p. (in Russian).
35. Khruslov E.Ya., *Uspekhi Math. Nauk*, **45**, N1(277), (1990), pp. 197–198 (in Russian).
36. Kutliarov V.S., *J. PMTF*, (1967), N1, pp. 84–88 (in Russian).
37. Levy T., *European J. of Mechanics. Ser. B : "Fluids"*, **9**, N4, (1990), pp. 309–327.
38. Nguetseng G., *SIAM J. Math. Anal.*, **20**, N3, (1989), pp. 608–628.
39. Odeh A.S., *SPE J.*, (1965), Janv., pp. 60–66.
40. Palatnik B.., *Proc. 7th All Russian Congress on Theoretics and Applied Mechanics*, Moscow, Aug. 15–21, (1991), p. 275 (in Russian).
41. Panasenko G.P. *Averaging Fields in Composite Materials with High-Modulus Reinforcement* - Vestnik MGU, Ser. 15, N 2, 1983, pp. 20-27 (in Russian).
42. Panasenko G.P. *Averaging of Processes in Highly Heterogeneous Structures* - Dokladi AN SSSR, **298**, 1988, pp. 76-79 (translated in: Soviet Phys. Dokl., 1988.)
43. Panasenko G.P., *J. Vychislitelnoi Mathematiki i Mathematicheskoi Physiki*, **30**, N2, (1990), pp. 243–253 (translated on English)
44. Panfilov M., *Proc. 6th All Russian Congress on Theoretical and Applied Mechanics, Tashkent, Sept. 24–30*, (1986), pp. 503–504 (in Russian).
45. Panfilov M., *Proc. 3rd Ukraine Conf. of Ukraine Acad. Sci. "Integral Equations in Applied Modelling", Odessa, Nov. 14–16*, (1989), part.2, pp. 102–103 (in Russian).
46. Panfilov M., *Soviet Physics Dokl.*, **35 (3)**, (1990), pp. 225–227.
47. Panfilov M.B., *Proc. 2nd European Conference on the Mathematics of Oil Recovery, Arles, France, Sept. 11–14, 1990*, Ed. Technip, (1990), pp. 347–350.
48. Panfilov M., *Izvestiya Academii Nauk (Russia)*, Mekhanika Zhidkosti i Gasa, N6, (1992), pp. 103–116; (translated in Fluid Dynamics, N3, (1993), pp. 112–120).
49. Panfilov M., *C.R. Acad. Sci. Paris*, **318**, Serie II, (1994), pp. 1437–1443.
50. Panfilov M., *Homogenised Models of Convection-Diffusion transport in Periodic Network Media*, Moscow, Ed. Oil & Gas Research Insti-

tute, Russian Acad.Sci., preprint N26, (1995), 72p.

51. Panfilov M., *Averaged Models of Convection-Diffusion Transfer Through Highly Heterogeneous Porous Media, in : Math. Modelling of Flow Through Porous Media*, Proc. Int. Conf., World Scientific Pub., Singapore, pp. 276–300.
52. Panfilov, M. and Panfilova, I. (1996) *Averaged models of flows with heterogeneous internal structure*. Ed. Nauka, Moscow (in Russian).
53. Prues K., Narasimhan T.N., *SPE J.*, (1985), Feb., pp. 14–26.
54. Showalter R.E., Walkington N.J., *J. of Mathematical Analysis and Applications*, **155**, N1, (1991), pp. 1–20.
55. Volkov I.A., in :*Problems of Appl. Math. Geometr. Modelling*, Leningrad, Ed. Ingenerno-Stroitelnogo Instituta, (1967), pp. 33–35 (in Russian).
56. Volkov I.A., in :*Theor. and Experim. Research of Hydrocarbon Migration Laws*, Leningrad, Ed. VNIIGRI, (1980), pp. 104–117 (in Russian).
57. Warren J.E., Root J., *SPE J.*, **3**, Sept., (1963), pp. 245–255.
58. Wu Y.-S., Prues K., *SPE Res. Eng.*, **3**, N1, (1988), pp. 327–336.
59. Teodorovich E.V., *Fluid Dynamics*, no. 4, (1994).

DISPERSION TENSOR IN ANISOTROPIC NETWORK MEDIA. STREAM CONFIGURATION METHOD

3.1. DESCRIPTION OF THE LATTICE MODEL OF POROUS MEDIUM

Calculation of the effective dispersion tensor is one of the basic problems in the theory of chemical or heat transport through heterogeneous media. A number of paper was devoted to this problem using various mathematical approaches. A review of the results obtained may be found in the book of P.Adler, 1991 (see list of references at the end of chapter 2).

Almost all approaches are based on numerical simulations of the dispersion tensor. Analytical solutions are practically not available.

In this book, we present a method allowing to obtain analytical relationships for the dispersion tensor for media of network structure. Unique restriction concerns the thickness of the channels where the transport happens. Their thickness should be small with respect to the scale of the heterogeneity cell. Then the cell problem defining the structure of dispersion tensor can be solved in the analytical way. However, solution is not trivial, because the cell problem becomes incorrect when the transport channels are thin.

In fact, the flow in each narrow channel tends to be one-dimensional, therefore the flow structure becomes undefined in the nodes of the channel intersection.

In this section, the regularization method is described, named as method of stream configurations. It offers some additive conditions in the nodes, which describe the probable structure of flow streams there. Definition of stream structures is related to taking in consideration all geometrical configurations (graphs) formed by the flow streams entering into and going out of each node. For a periodic lattice, the number of all stream configurations is shown to be small enough, such that all the configurations can be renumbered.

The method is explained for 2D case, however it may be applied to 3D case in the analogous way.

Only this part of dispersion tensor is studed, which does not depend on the microscale diffusion, i.e., Peclet number is assumed to be large enough.

3.1.1. PHYSICAL MODEL OF THE PERIODIC LATTICE

The main part $\mathbf{D}_{km}^{(0)}$ of the tensor $\mathbf{D}_{km}$, that is a convective dispersion tensor, can be defined analytically in media with thin channels (we shall use the different names for them: throats, channels, fractures), where the principle of averaged operator decomposition (Bakhvalov, Panasenko, 1984) may be applied. The analogous approach was applied in (Egorov, 1983, 1984), but without taking into account stream interactions inside the nodes.

Geometrical Description of the Lattice

Consider the plate medium Ω constituting a square lattice of channels, as shown in Fig. 3.1. The fluid flows inside the channels only, while the remaining part of the medium is impermeable. The lattice is turned for the angle γ around the axis y_1. From the periodicity of the system it follows that the angle may get only two magnitudes, either 0, or $\pi/4$.

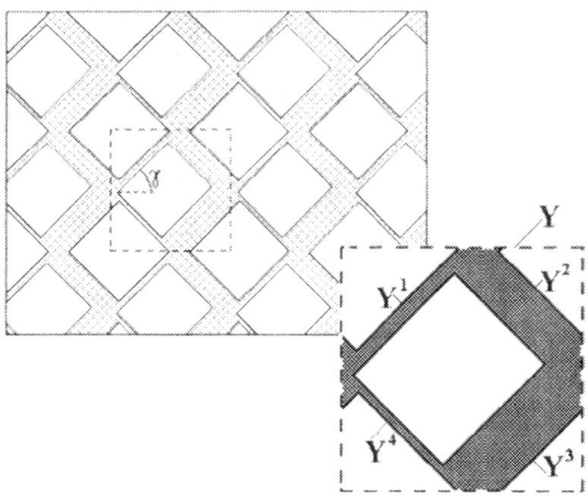

Figure 3.1. Structure of the lattice model

Each channel is assumed to be formed by a sequence of "the segments" of identical length equal to $2l=1/(\sin\gamma+\cos\gamma)$, and of various transverse thickness h (the values h and l are dimensionless and correspond to the scale of the cell Y). Only four different magnitudes of thickness are allowed, $h^{(i)}$, $i=1-4$. For the sake of determinacy, each segment will be accepted as a 3D cylinder or a 2D band.

Based on these premises, a unique cell Y has the interior structure which is shown in Fig. 3.1. It constitutes a square with a site equal to 1. The channel structure inside a cell is formed by four segments Y^i, i=1−4.

The domain of superposition of two neighbor segments, Y^i and Y^j, forms a "node" U^{ij}. The lattice structure is such, that each node touches the cell boundary. The coordinates of nodes in the system (η_1, η_2) are the following:

$$U^{12}=(l,l), \quad U^{23}=(l,-l), \quad U^{34}=(-l,-l), \quad U^{41}=(-l,l). \tag{3.1}$$

where the semi-length of each segment l is defined as:

$$l=\sqrt{2}/4, \text{ when } \gamma=\pi/4;$$
$$l=1/2, \text{ when } \gamma=0;$$

Along with coordinates (y_1, y_2), another coordinate system (η_1, η_2) turned on the angle γ counter-clockwise will be used:

$$\eta_1=\lambda_1 y_1+\lambda_2 y_2, \quad \eta_2=\lambda_2 y_1+\lambda_1 y_2, \quad \lambda_1\equiv\cos\gamma, \quad \lambda_2\equiv\sin\gamma.$$

where $\lambda\equiv\lambda_1=\lambda_2=\sqrt{2}/2$, as shown in Fig. 3.2.

Interior Structure of the Channels

Each channel may be filled by porous medium or be empty. This fact has not practically any importance to study the problem of chemical or temperature transport. Only the transport velocity will be defined in different ways, i.e., either using Darcy's law, or using Poiseuille's law.

In any case, the interior porosity of each channel, $b(y)$, as well as the interior permeability $K(y)$ should be defined. The porosity outside of the channels is $b(y)\equiv 0$. When the channels are empty, the porosity inside the channels b is equal to 1. The effective porosity of the medium is $B=2l\sum_{i=1}^{4} b^{(i)}h^{(i)}$.

3.1.2. DEFINITION OF THE VELOCITY FIELD

In any point of the domain Y the velocity vector v will be defined by its y-components (v_1, v_2), or (u_1, u_2) in coordinates η. The following relations may be written:

$$\begin{cases} u_1=\lambda_1 v_1+\lambda_2 v_2 \\ u_2=\lambda_1 v_2-\lambda_2 v_1 \end{cases} \quad \text{or} \quad \begin{cases} v_1=\lambda_1 u_1-\lambda_2 u_2 \\ v_2=\lambda_1 u_2+\lambda_2 u_1 \end{cases}$$

Herein and furthermore, the lower index i means projection on the appropriate cell axis, whereas the upper indexes mean the number of the cell segment.

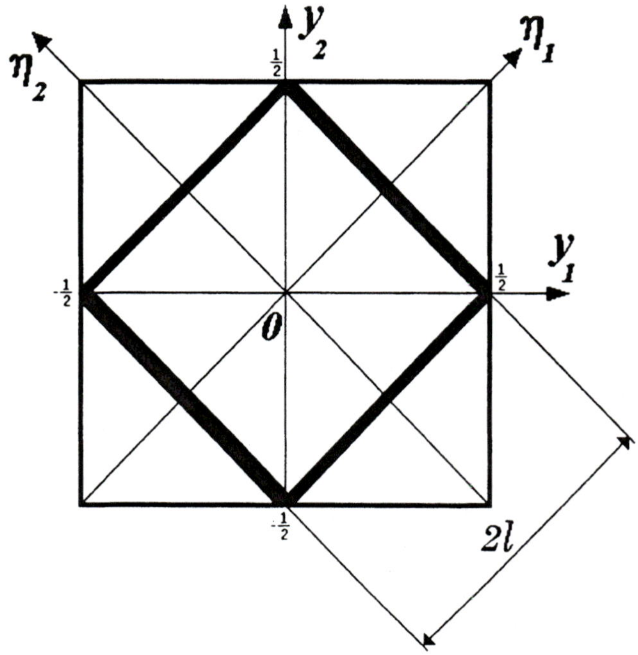

Figure 3.2. Two coordinate systems on a cell

The operator $v_i\partial/\partial y_i$ (with summation over repeated indexes) is invariable respectively to the axis turning:

$$v_i\frac{\partial}{\partial y_i}=u_i\frac{\partial}{\partial \eta_i} \tag{3.2}$$

herein the repeated index i means summation from 1 to 2.

Let's determine the velocity field in the cell. The flow in each segment Y^i is one-dimensional along the segment axis, then the velocity is constant within the segment because of the flow continuity. The velocity in the segment Y^i is denoted as $v^{(i)}$, i=1, ..., 4, where the magnitude of $v^{(i)}$ is positive, if the velocity vector direction in i-th segment corresponds to the scheme of Fig. 3.3.

Thus, it is sufficient to determine four values $v^{(i)}$. To define the velocity field, the boundary-value conditions should be given. In a periodic medium described above and getting rectangular form, it is sufficient to set two flow velocities (flow rates) at two neighbor sites of the domain, or the averaged velocity vector.

The averaged velocity vector $\mathbf{W}\equiv\langle v\rangle$ is assumed to be known and be defined by means of its components W_1, W_2 along the axes (y_1, y_2). On the

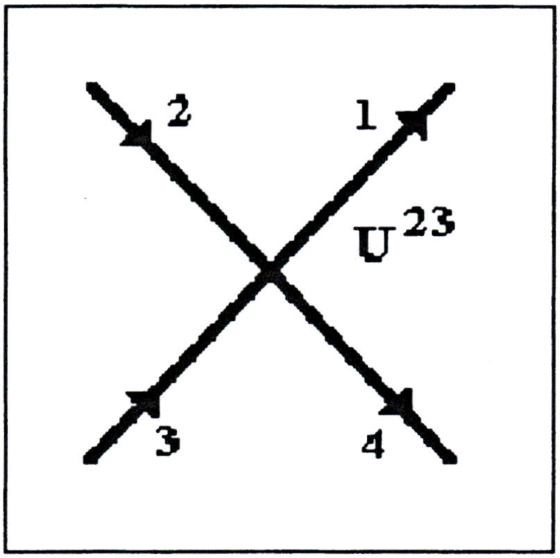

Figure 3.3. Positive direction of flow in the segments

other hand, the vector **W** may be defined by its modules W and its angle γ_w with respect to the axis y_1, where $W_1=W\cos\gamma_w$, $W_2=W\sin\gamma_w$.

Along with the vector **W**, two boundary values of the flow rate will be used, namely the specific flow rate q of the fluid injection into the medium through the left-hand side, and the rate q^* of the fluid flowing out of the medium through the top side. These two values are not independent and will be defined via the averaged vector **W** later.

The velocity field is defined by the following equations

$$q^{(1)}+q^{(4)}=q^{(2)}+q^{(3)}, \tag{3.3a}$$

$$q^{(1)}+q^{(4)}=q, \tag{3.3b}$$

$$q^{(1)}-q^{(2)}=q^{(3)}-q^{(4)}=q^*, \tag{3.3c}$$

$$[(q^{(1)}+q^{(3)})\lambda_1+(q^{(2)}+q^{(4)})\lambda_2]2l=W_1, \tag{3.3d}$$

$$[(q^{(1)}+q^{(3)})\lambda_1-(q^{(2)}+q^{(4)})\lambda_1]2l=W_2, \tag{3.3e}$$

$$q^{(i)}=v^{(i)}h^{(i)}b^{(i)}, \qquad i=1,...,4. \tag{3.3f}$$

Here the first three equations are Kirchhoff's laws for any node which is located inside the domain Ω (a), for the nodes located at the left boundary of the domain (b), for the node located at the top limiting site (c). Relations (d) and (e) are the averaged velocity definition, while equation (f) relates the flow rate and the flow velocity.

System (3.3) is not closed.

Case of Poiseuille's Flow

Let's use the motion equations in the form of Poiseuille's law, for example for a segment Y^1:

$$v^{(1)}=(h^{(1)})^2\delta(P^{41}-P^{12}) \tag{3.4}$$

where P^{mn} is a pressure in the node U^{mn}, δ is a constant value. Then after excluding the pressures we obtain one closing equation:

$$\frac{q^{(1)}}{(h^{(1)})^3}+\frac{q^{(2)}}{(h^{(2)})^3}=\frac{q^{(4)}}{(h^{(4)})^3}+\frac{q^{(3)}}{(h^{(3)})^3} \tag{3.5}$$

Other relations resulting from other paths of nodes circuits are not independent.

Then the following results from (3.3)-(3.5) for any γ:

$$v^{(i)}=q^{(i)}/(h^{(i)}b^{(i)}), \qquad q^{(i)}=WZ^{(i)}, \qquad i=1,...,4 \tag{3.6}$$

where

$$Z^{(1)} \equiv \frac{1}{A}\left\{\cos\gamma_w\left[\frac{1}{(h^{(3)})^3}+\frac{1}{(h^{(4)})^3}\right]+\sin\gamma_w\left[\frac{1}{(h^{(2)})^3}+\frac{1}{(h^{(3)})^3}\right]\right\} \tag{3.7a}$$

$$Z^{(2)} \equiv \frac{1}{A}\left\{\cos\gamma_w\left[\frac{1}{(h^{(3)})^3}+\frac{1}{(h^{(4)})^3}\right]-\sin\gamma_w\left[\frac{1}{(h^{(1)})^3}+\frac{1}{(h^{(4)})^3}\right]\right\} \tag{3.7b}$$

$$Z^{(3)} \equiv \frac{1}{A}\left\{\cos\gamma_w\left[\frac{1}{(h^{(1)})^3}+\frac{1}{(h^{(2)})^3}\right]+\sin\gamma_w\left[\frac{1}{(h^{(1)})^3}+\frac{1}{(h^{(4)})^3}\right]\right\} \tag{3.7c}$$

$$Z^{(4)} \equiv \frac{1}{A}\left\{\cos\gamma_w\left[\frac{1}{(h^{(1)})^3}+\frac{1}{(h^{(2)})^3}\right]-\sin\gamma_w\left[\frac{1}{(h^{(2)})^3}+\frac{1}{(h^{(3)})^3}\right]\right\} \tag{3.7d}$$

$$A\equiv\sum_{i=1}^{4}\frac{1}{(h^{(i)})^3}$$

The boundary flow rates q and q^* are defined through W_1 and W_2 as:

$$\begin{aligned} &W_1=q, \qquad W_2=q^*, \qquad \text{when } \gamma=\pi/4; \\ &W_1=q+q^*, \quad W_2=q^*-q, \qquad \text{when } \gamma=0 \end{aligned} \tag{3.8}$$

It is necessary to note, that:

$$Z^{(1)}+Z^{(4)}=Z^{(2)}+Z^{(3)}=\cos\gamma_w, \qquad Z^{(1)}-Z^{(2)}=Z^{(3)}-Z^{(4)}=\sin\gamma_w \tag{3.9}$$

Certainly, the flow properties must be invariable respectively to the choice of the coordinate system, but they can depend on the direction of the vector **W**. Therefore, the lattice location will be fixed below as $\gamma=\pi/4$, but the angle γ_w will be examined as variable.

The inverse approach may be used too, when the orientation of the vector **W** would be fixed (for example, $\gamma_w=0$), but the lattice location might be varied. However, in this case, the intermediate deductions appear more cumbrous.

Case of Darcy's Flow

When the channels are filled by porous medium, the transport velocity is defined from Darcy's law, which may be written for all segments Y^i as:

$$v^{(1)}=K^{(1)}\tilde{\delta}(P^{41}-P^{12}), \quad v^{(2)}=K^{(2)}\tilde{\delta}(P^{12}-P^{23}),$$
$$v^{(3)}=K^{(3)}\tilde{\delta}(P^{34}-P^{23}), \quad v^{(4)}=K^{(4)}\tilde{\delta}(P^{41}-P^{34})$$

where $K^{(i)}$ is the permeability of the i-segment; P^{mn} is the pressure in the node U^{mn}, $\tilde{\delta}$ is a constant value (inverse to viscosity and to the segment length l).

Summation of four above equations allows to exclude the pressures, that yields one closing relation for (3.3):

$$\frac{q^{(1)}}{h^{(1)}b^{(1)}K^{(1)}}+\frac{q^{(2)}}{h^{(2)}b^{(2)}K^{(2)}}=\frac{q^{(3)}}{h^{(3)}b^{(3)}K^{(3)}}+\frac{q^{(4)}}{h^{(4)}b^{(4)}K^{(4)}} \tag{3.10}$$

Then the following results from (3.3) and (3.10) for any γ:

$$v^{(i)}=q^{(i)}/(h^{(i)}b^{(i)}), \qquad q^{(i)}=WZ^{(i)}, \qquad i=1,...,4 \tag{3.11}$$

where

$$Z^{(1)} \equiv \frac{1}{A}\left\{\cos\gamma_w\left[\frac{1}{h^{(3)}b^{(3)}K^{(3)}}+\frac{1}{h^{(4)}b^{(4)}K^{(4)}}\right]+\right.$$
$$\left.\sin\gamma_w\left[\frac{1}{h^{(2)}b^{(2)}K^{(2)}}+\frac{1}{h^{(3)}b^{(3)}K^{(3)}}\right]\right\}, \tag{3.12a}$$

$$Z^{(2)} \equiv \frac{1}{A}\left\{\cos\gamma_w\left[\frac{1}{h^{(3)}b^{(3)}K^{(3)}}+\frac{1}{h^{(4)}b^{(4)}K^{(4)}}\right]-\right.$$
$$\left.\sin\gamma_w\left[\frac{1}{h^{(1)}b^{(1)}K^{(1)}}+\frac{1}{h^{(4)}b^{(4)}K^{(4)}}\right]\right\}, \tag{3.12b}$$

$$Z^{(3)} \equiv \frac{1}{A}\left\{\cos\gamma_w\left[\frac{1}{h^{(1)}b^{(1)}K^{(1)}}+\frac{1}{h^{(2)}b^{(2)}K^{(2)}}\right]+\right.$$
$$\left.\sin\gamma_w\left[\frac{1}{h^{(1)}b^{(1)}K^{(1)}}+\frac{1}{h^{(4)}b^{(4)}K^{(4)}}\right]\right\}, \tag{3.12c}$$

$$Z^{(4)} \equiv \frac{1}{A}\left\{\cos\gamma_w\left[\frac{1}{h^{(1)}b^{(1)}K^{(1)}}+\frac{1}{h^{(2)}b^{(2)}K^{(2)}}\right]-\right.$$

$$\sin\gamma_w \cdot \left[\frac{1}{h^{(2)}b^{(2)}K^{(2)}} + \frac{1}{h^{(3)}b^{(3)}K^{(3)}}\right]\Bigg\}, \quad (3.12d)$$

$$A \equiv \sum_{i=1}^{4} \frac{1}{h^{(i)}b^{(i)}K^{(i)}}$$

Relations (3.8) between q, q^* and $\mathbf{W}_1$, $\mathbf{W}_2$ are kept, as well as the property (3.9).

3.2. CELL PROBLEM OF CHEMICAL TRANSPORT THROUGH THE LATTICE MEDIUM

3.2.1. STATEMENT OF THE PROBLEM. DECOMPOSITION PRINCIPLE

In 2D case, two cell functions $\varphi_k(y)$, $k=1,2$ define the macroscale parameters via cell problems (2.140a).

Let us consider the first function $\varphi_1^{(0)}(y_1,y_2)$ in (2.140a). To simplify the notations, in this section this function will be designated simply as $\varphi(y_1,y_2)$. It may be written via variables (η_1,η_2) in the following form:

$$u_1\frac{\partial\varphi}{\partial\eta_1}+u_2\frac{\partial\varphi}{\partial\eta_2}=C_0-\lambda_1u_1+\lambda_2u_2,\quad \langle\varphi\rangle=0,\quad C_0\equiv W_1/B \tag{3.13}$$

where $\lambda\equiv\lambda_1=\lambda_2=\sqrt{2}/2$.

Herein, relations (3.2)-(3.3) have been taken into account.

The areas of segment intersections may be neglected because they have a disappearing measure, if the segment thickness is small. Equation (3.13) is assumed to be defined along a closed broken line being the axes of the segment Y^i. We notice that the jump changes of velocity vector can occur in the nodes. Then the problem will get the form:

$$u_1\frac{\partial\varphi^{(i)}}{\partial\eta_1}+u_2\frac{\partial\varphi^{(i)}}{\partial\eta_2}=C_0-\lambda_1u_1+\lambda_2u_2,\ \eta\in Y^i,\ i=1,...,4 \tag{3.14a}$$

$$\langle\varphi\rangle\equiv\sum_{i=1}^{4}h^{(i)}\int_{Y^i}\varphi d\eta=0 \tag{3.14b}$$

where it's implied that Y^i is an open interval with the length equal to $2l$, which is located on the axis of a corresponding segment. Here

$$u_1=\begin{cases} v^{(1)}, & \eta\in Y^1\\ 0, & \eta\in Y^2,\, Y^4\\ v^{(3)}, & \eta\in Y^3\end{cases}\qquad u_2=\begin{cases} 0, & \eta\in Y^1,\, Y^3\\ -v^{(2)}, & \eta\in Y^2\\ -v^{(4)}, & \eta\in Y^4\end{cases} \tag{3.15}$$

Eq. (3.13) may be written in the divergent form:

$$\frac{\partial(u_i\varphi)}{\partial\eta_i}=C_0-\lambda_1u_1+\lambda_2u_2 \tag{3.16}$$

because the vector u satisfies the conservation law $\operatorname{div}u=0$.

The problem (3.14) must be supplemented by a relation between the functions in the nodes.

3.2.2. INCORRECTNESS OF THE CELL PROBLEM

For each function $\varphi^{(i)}$ problem (3.14) becomes one-dimensional, so linear representation of the solution is valid:

$$\begin{aligned}
&\varphi^{(1)}=\beta^{(1)}\eta_1+\alpha^{(1)}, \quad \varphi^{(2)}=\beta^{(2)}\eta_2+\alpha^{(2)},\\
&\varphi^{(3)}=\beta^{(3)}\eta_1+\alpha^{(3)}, \quad \varphi^{(4)}=\beta^{(4)}\eta_2+\alpha^{(4)}
\end{aligned} \tag{3.17}$$

Coefficients β_i can be found immediately from (3.14a):

$$\begin{aligned}
\beta^{(1)}&=\left(\frac{C_0}{v^{(1)}}-\lambda_1\right)=\frac{h^{(1)}\cos\gamma_w}{Z^{(1)}B}-\lambda_1,\\
\beta^{(2)}&=-\left(\frac{C_0}{v^{(2)}}-\lambda_2\right)=-\frac{h^{(2)}\cos\gamma_w}{Z^{(2)}B}+\lambda_2\\
\beta^{(3)}&=-\left(\frac{C_0}{v^{(3)}}-\lambda_1\right)=-\frac{h^{(3)}\cos\gamma_w}{Z^{(3)}B}+\lambda_1,\\
\beta^{(4)}&=-\left(\frac{C_0}{v^{(4)}}-\lambda_2\right)=-\frac{h^{(4)}\cos\gamma_w}{Z^{(4)}B}+\lambda_2
\end{aligned} \tag{3.18}$$

Four coefficients $\alpha^{(i)}$ in (3.17) remain indefinite. For them we have only one condition (3.14b). Additional relation will be found using conservation laws in the nodes, as well as conditions determining the flow structure in the nodes. Below it will be shown that continuity of solution in the nodes is usually broken down.

3.2.3. CONSERVATION CONDITION IN THE NODE

We shall obtain the relations in the nodes ω using the integral analog of (3.13):

$$\int_{\partial\omega} u_n\varphi d\eta=\iint_{\omega}(C_0-\lambda_1 u_1+\lambda_2 u_2)d\eta$$

where $\partial\omega$ is the boundary of the area ω that is shaded in Fig. 3.4, n is the exterior normal vector. The divergent form (3.16) is used.

Further calculations don't require any explanations:

$$-u_1^{(1)}h^{(1)}\varphi^{(3)}+u_1^{(3)}h^{(3)}\varphi^{(3)}-u_2^{(2)}h^{(2)}\varphi^{(2)}+u_2^{(4)}h^{(4)}\varphi^{(4)}=(C_0-\lambda_1\overline{u}_1+\lambda_2\overline{u}_2)|\omega|$$

where $\overline{u}_i$ are the averaged values over the element ω, $|\omega|$ is the node volume, the upper index i means belonging to the appropriate channel segment Y^i.

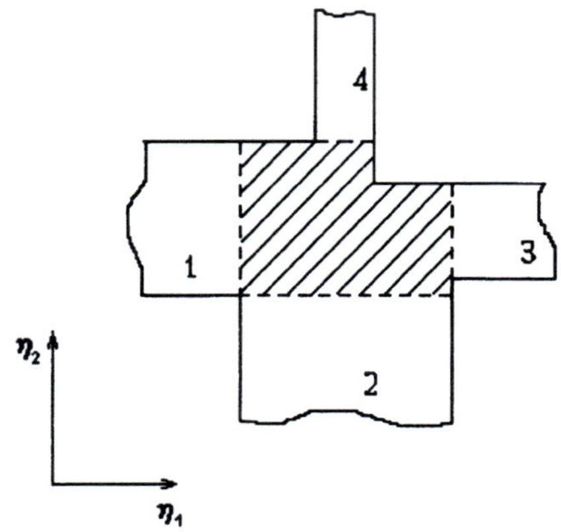

Figure 3.4. Scheme of channel intersection

For narrow channels the thickness $h^{(i)}$ tends to zero. Taking into account that $|\omega|\sim(h^{(i)})^2$, the following limiting equation results:

$$q^{(1)}\varphi^{(1)}+q^{(4)}\varphi^{(4)}=q^{(2)}\varphi^{(2)}+q^{(3)}\varphi^{(3)} \tag{3.19}$$

$$q^{(1)}=u_1^{(1)}h^{(1)},\quad q^{(2)}=-u_2^{(2)}h^{(2)},\quad q^{(3)}=u_1^{(3)}h^{(3)},\quad q^{(4)}=-u_2^{(4)}h^{(4)}$$

The last relation describes the component mass conservation in a node. In fact, the function $c\equiv\varphi+y_1$ is the concentration of chemical component inside a cell, which is normalized in such a way that its mean value is equal to zero. In a node with fixed coordinate y_1 (let us designate it as y_1^*), relation (3.19) takes the form:

$$q^{(1)}c^{(1)}+q^{(4)}c^{(4)}-q^{(2)}c^{(2)}-q^{(3)}c^{(3)}=y_1^*(q^{(1)}+q^{(4)}-q^{(2)}-q^{(3)})=0$$

due to relation (3.3a).

Eq. (3.19) can be written in another form via the flow rates q and q^*:

$$q^{(1)}(\varphi^{(1)}+\varphi^{(3)}-\varphi^{(2)}-\varphi^{(4)})=q(\varphi^{(3)}-\varphi^{(4)})+q^*(\varphi^{(3)}-\varphi^{(2)}) \tag{3.20}$$

For the internal nodes U^{12} and U^{23} conservation equation (3.19) takes the following form:

$$q^{(1)}\varphi^{(1)}(l,\, l)+q^{(4)}\varphi^{(4)}(-l,\, -l)=q^{(3)}\varphi^{(3)}(-l,\, -l)+q^{(2)}\varphi^{(2)}(l,\, l) \tag{3.21a}$$

$$q^{(1)}\varphi^{(1)}(-l,\, l)+q^{(4)}\varphi^{(4)}(-l,\, l)=q^{(3)}\varphi^{(3)}(l,\, -l)+q^{(2)}\varphi^{(2)}(l,\, -l) \tag{3.21b}$$

Two other relations for the internal nodes U^{34} and U^{41} coincide with Eq. (3.21) due to medium periodicity.

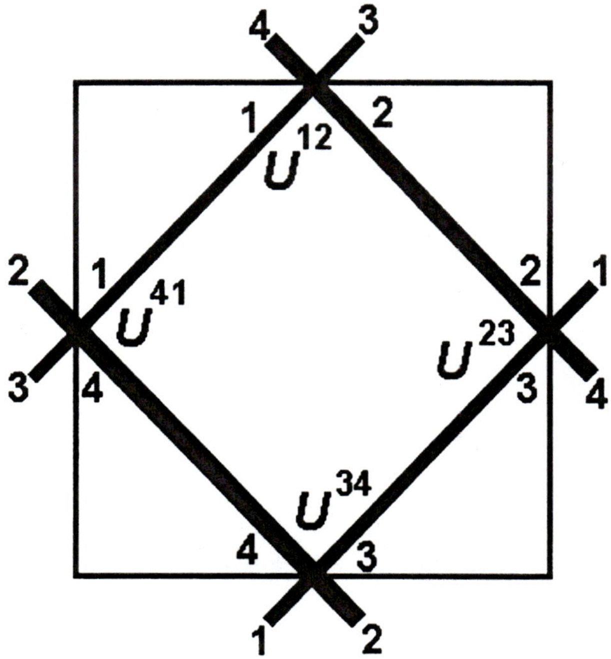

Figure 3.5. The segments and the nodes in a cell

In fact, in the periodic medium described above the nodes $U^{(41)}$ and $U^{(23)}$ are identical, as well as the nodes $U^{(12)}$ and $U^{(34)}$, as seen from Fig. 3.5.

However the interior structure of flow streams in the nodes U^{12} and U^{23} is distinct. This is seen in Fig. 3.6, which shows the whole lattice.

It can be shown that conditions (3.21a) and (3.21b) for equation (3.14) are not independent too. Substituting representation (3.17) - (3.18) in (3.21), we will obtain:

$$\beta_*^{(1)}l+\alpha_*^{(1)}-\beta_*^{(4)}l+\alpha_*^{(4)}=-\beta_*^{(3)}l+\alpha_*^{(3)}+\beta_*^{(2)}l+\alpha_*^{(2)}$$

$$\beta_*^{(1)}l+\alpha_*^{(1)}+\beta_*^{(4)}l+\alpha_*^{(4)}=\beta_*^{(3)}l+\alpha_*^{(3)}-\beta_*^{(2)}l+\alpha_*^{(2)};\ \alpha_*^{(i)}\equiv\alpha^{(i)}q^{(i)},\ \beta_*^{(i)}\equiv\beta^{(i)}q^{(i)}$$

The expressions for $\alpha^{(i)}$ follow from this:

$$\alpha_*^{(1)}+\alpha_*^{(4)}=\alpha_*^{(3)}+\alpha_*^{(2)},\quad \text{or}\quad \alpha^{(1)}q^{(1)}+\alpha^{(4)}q^{(4)}=\alpha^{(2)}q^{(2)}+\alpha^{(3)}q^{(3)} \qquad (3.22)$$

and a relation for already determined coefficients $\beta_*^{(i)}$:

$$Q\equiv-\beta_*^{(1)}+\beta_*^{(4)}-\beta_*^{(3)}+\beta_*^{(2)}=0$$

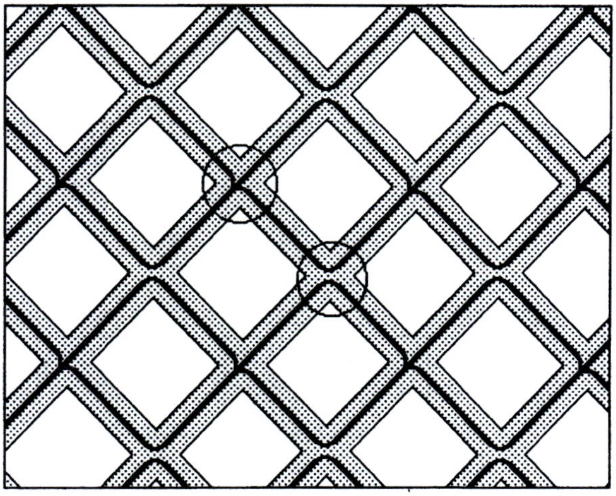

Figure 3.6. Independent nodes of the lattice

It is easy to prove that this relation is an identity. In fact, substituting Eq.(3.18) in it, we will obtain:

$$Q \equiv -C_0(h^{(1)}+h^{(2)}+h^{(3)}+h^{(4)})+\lambda_1(q^{(1)}+q^{(3)})+\lambda_2(q^{(2)}+q^{(4)})$$

that leads to the following result, using definition (3.13) and (3.3b) for C_0, W_1 and B:

$$Q \equiv -C_0B/2l+W_1/2l=-W_1/2l+W_1/2l=0$$

so the assumption is proved.

Thus, one of two conditions (3.21a) or (3.21b) is unnecessary. The mass conservation conditions in the nodes give only one additional relationship: either (3.21a), or (3.21b).

3.3. METHOD OF STREAM CONFIGURATIONS

3.3.1. PHYSICAL RAISONS OF PROBLEM INCORRECTNESS

The problem (3.14), (3.21) is not yet completely determined, as only two conditions (3.14b) and (3.21b) hold for 4 unknown values $\alpha^{(i)}$, i=1−4 in representation (3.17), (3.18).

This phenomenon is caused by the features of the lattice structure. It is clear that introduction of diffusion does not solve the closure problem, because arising of the second derivatives in transport equation requires additive boundary conditions.

Let two streams, bearing different magnitudes of concentration, enter in a node. Let diffusion be absent in the system. What concentrations will be in two other streams at the outlet of this node? The answer depends on the flow structure inside the node. If the streams are not intersected within the node, then the concentrations bearing by these streams are not changed at the outlet. If the input streams are intermixed within the node, then the concentrations at the outlet become averaged.

Physical nonuniqueness of the probable flow structures in the nodes causes mathematical nonuniqueness of the cell problem. Therefore to close the cell problem it is necessary to fixe some scheme of stream interaction inside each node, that will define some additional conditions relating the inlet and the outlet concentrations for each node. Those additional conditions will be named as "structural conditions" in nodes.

Multiple types of flow structures are caused by the geometrical form of the node, which is not regular and may be considerably varying. Some types of stream structure in the nodes will be considered in the next section. In general case, investigation of all probable situations is impossible. However, for the lattices with narrow channels the stream structure within each node is universal, that allows to derive the universal closing relations for the cell problem.

3.3.2. FLOW STRUCTURES IN NODES. DISCONTINUITIES

Before deriving closing relations for narrow channels it is usual to show what types of stream structure inside the nodes and resulting closing relations might be observed in case when the thickness of channels is finite.

First, it is necessary to note that the intersection of the streams within a node often leads to concentration discontinuity. An example in Fig. 3.7b proves it. In fact, the mean concentration in segment 3 which is formed by two streams going from the segments 1 and 4, each of them is transporting its own concentration, is not equivalent to the concentration in segment 4. Thus, the physical structure of streams in each node is characterized by its own system of concentration discontinuities. The formal analysis of allowed discontinuities for solution of Eqs. (3.14) and conditions (3.19) leads to define some classes of flow structures in the nodes.

Expression (3.19), being written for a fixed node, assumes three types of solutions: 1) continuous solutions where all four functions $\varphi^{(i)}$ in the node are identical; 2) partially continuous solutions, when only some pairs of $\varphi^{(i)}$ coincide in the node; 3) discontinuous solutions, when all functions $\varphi^{(i)}$ are distinct in the node. Various types of solutions correspond to different stream structures in the node.

Fig. 3.7 shows the physical meaning of discontinuity.

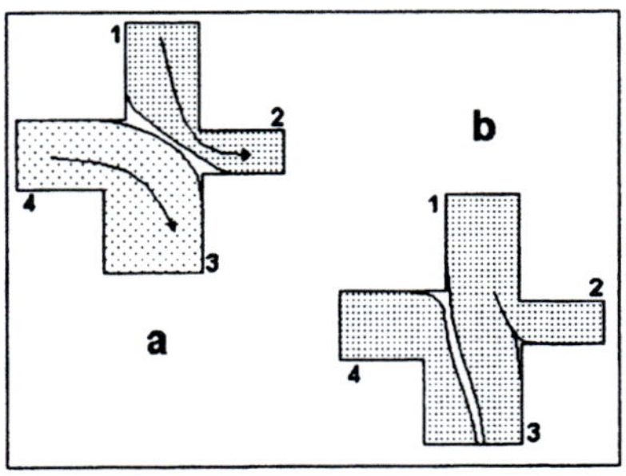

Figure 3.7. Probable stream structure in a node

When the flow stream is not branching (3.7a), the flow continuity is kept along the paths 1-2 and 4-3, so the following is true in this node:

$$\varphi^{(1)}=\varphi^{(2)}, \quad \varphi^{(3)}=\varphi^{(4)}$$

however $\varphi^{(1)}\neq\varphi^{(3)}$, because the concentrations $\varphi^{(1)}$ and $\varphi^{(3)}$ are transported into the node by two independent streams. Thus, the solution becomes partially continuous (it has only two jumps).

If the streams in the node are branched (Fig. 3.7b), then all four concentrations are distinct and the solution has four jumps.

It is interesting to consider two following limiting cases.

A: the node is formed by an intersection of two channels ($q^{(3)}=q^{(4)}=0$, $q^{(1)}=q^{(2)}$). In that node the streams are never branched, the solution in the node is always continuous: $\varphi^{(1)}=\varphi^{(2)}$.

B: the node is formed by three channels ($q^{(4)}=0$, $q^{(1)}=q^{(2)}+q^{(3)}$). In that node even one stream is branched and the discontinuities of concentration are inevitable. It follows from (3.19): $(q^{(2)}+q^{(3)})\varphi^{(1)}=q^{(3)}\varphi^{(3)}+q^{(2)}\varphi^{(2)}$. It results from this, that if any two concentrations are equal, then the third function coincides with those. Therefore, in this node partially continuous solutions are impossible.

Thus, assuming some allowed types of stream structures in nodes, it is easy to associate with them some mathematical relations for concentrations.

3.3.3. THE METHOD IDEA

Now begin to derive closing relations for the case of narrow channels.

Let the channels be narrow. The thinner channel, the smaller the node volume. In a small node the independence of streams becomes low probable. Thus, only one flow structure inside the nodes is allowed for the lattices with thin channels. That is a full mixing of streams. Therefore the process is defined in a unique manner.

After all the input streams are mixed inside the node, the magnitudes of all concentrations at the exit must be equivalent. This fact constitutes the additional "structural condition" for the cell problem.

Various ways are possible to a stream would enter in a node and would go out of the node, therefore each node of an irregular lattice is characterized by an individual stream configuration. However, the number of stream configurations for a periodic lattice becomes finite and their complete classification may be constructed.

3.3.4. STREAM STRUCTURE AND STREAM CONFIGURATION IN THE NODE

"Stream structure" in a node means the type of stream interaction: whether they are mixed or they are moving independently one of other. Thus, a mixed and an independent stream structure may exist in nodes. Some intermediate types can arise too. The typical examples of independent streams in the nodes are shown in Figs. 3.4, 3.8b.

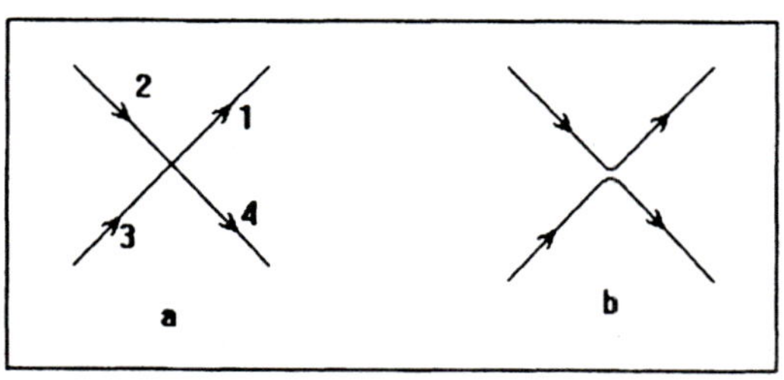

Figure 3.8. Mixed (a) and independent (b) stream structures in the node

"Stream configuration" in a node means a graph, which describes what stream is input or output for this node. An example of such a graph is shown in Fig. 3.8a.

System of relations between input and output concentrations for all the nodes, reflecting stream structure inside the node with any probable stream configuration, constitutes *"structural conditions"*, which are necessary to close the cell problem.

3.3.5. PRINCIPLE OF CONSTRUCTION OF THE STRUCTURAL CONDITIONS

Consider an example illustrating the technique to derive the structural conditions. Let two streams enter the arbitrary node along the segments 2 and 3, afterwards they are completely mixing in this node, and at last they leave the node in the form of two new streams 1 and 4, as shown in Fig. 3.8a.

As the $\varphi^{(i)}$-concentration transport is described by hyperbolic equation (3.14a) without dissipation, the following considerations hold.

The concentration beforehand the node in an input stream is independent of the neighboring input stream, as well as it is independent on any process happening behind this node. It is defined only by the own transport rate in this stream and by the concentration magnitude at the beginning of this stream. Hence, whatever the structure would be in the node, it does not influence the concentrations beforehand the node in the input streams.

The concentrations in the output streams are entirely defined by the fact that these streams go out of the node with mixed flow structure, so $\varphi^{(1)}=\varphi^{(4)}$.

Now the physical principle may be formulated that imposes complementary constraints on the flow in the lattice:

$$\left\{ \begin{array}{c} \textit{The flow structure in the node of channel intersection} \\ \textit{determines the concentration values } \varphi^{(i)} \\ \textit{at the exit out of the node, in the output streams} \end{array} \right\} \qquad (3.23)$$

For a mixed structure this principle yields:

$$\Phi^{(i)}=\Phi^{(j)} \quad \text{in the node,} \quad \Phi^{(i)}\equiv\varphi^{(i)}+y_1 \qquad (3.24)$$

where i, j are the segment numbers which correspond to the output streams, $\Phi^{(i)}$ is the pseudo-concentration (normalized to the zero mean value).

This rather general condition sufficiently depends on the stream configuration in the node. When the node has only one output stream, the structural conditions are unnecessary. If the node possesses three output streams, then Eq. (3.24) gives two independent conditions. Existence of more of three output streams for one node is impossible.

The principle (3.23) formulated above allows to write the structural conditions for other stream structures in the nodes. For example, if the stream structure in the node is independent, the structural conditions can be written in the same form as (3.24), but now i, j are the numbers of channels where the streams are continuous.

3.3.6. DOUBLE STREAM CONFIGURATIONS ON THE LATTICE

Conditions like (3.14) must be written for each node of the lattice, so the configuration graph must be defined for each node. For a periodic lattice the collection of possible stream configurations is finite.

Earlier it was noted, the plate square lattice has only two types of independent nodes, U^{12} and U^{23}, different stream configurations are allowed in.

One pair of stream configurations for two independent nodes will be named *"double stream configuration"* on the lattice. An example of the double configuration is shown in Fig. 3.9a. The general flow picture on the lattice, which corresponds to the given double configuration, is imaged on the same figure (b).

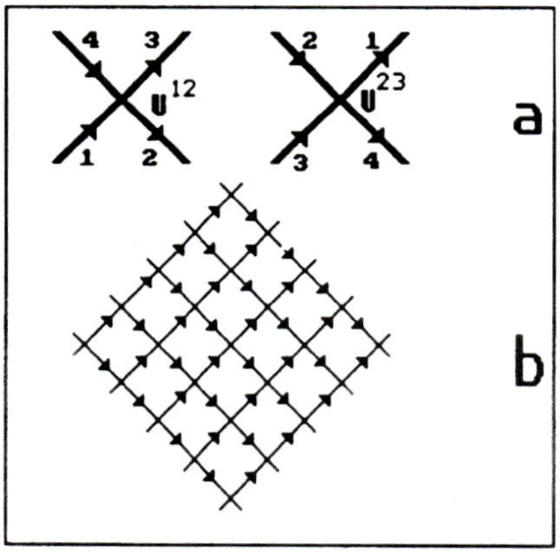

Figure 3.9. Double stream configuration (a) and corresponding total flow picture (b)

For the lattice described above, a finite number of independent double configurations exists. All they can be determined from expressions (3.6) and (3.7), or (3.11) and (3.12).

The form of a double configuration depends on the orientation of the averaged flow velocity vector $\mathbf{W}$, or on the parameter γ_w. The number of possible configurations is very large, when γ_w varies from 0 to 2π, but it is not necessary to examine all cases. The range of variations of the angle γ_w can be reduced to the so-called "main sector" due to the system symmetry.

3.3.7. MAIN SECTOR OF THE AVERAGED VELOCITY ORIENTATION. SYMMETRICAL TRANSFORMATIONS OF THE LATTICE

The range of possible orientations for the vector **W**, which corresponds to $\gamma_w \in [0, \pi/4]$ (shaded in Fig. 3.10), will be called the "*main sector*".

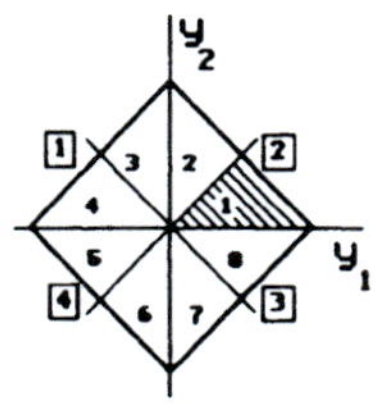

Figure 3.10. Main sector of the averaged velocity orientation (shaded)

It is sufficient to construct the solution for the cell problem in the main sector. The solution correspondent to other vector **W** orientation may be obtained from the solution in the main sector by linear transformations of the coordinates or the lattice. Lattice transformations consist in turning of the whole lattice on the plane, or in turning of each cell with respect to one of the coordinate axis.

The cell solution in the main sector will be denoted by the upper star. To determine the solution when the averaged velocity orientation angle γ_w falls outside the limits of the main sector, the following formal transformations should be made:

When $\gamma_w \in (\pi/4, \pi/2]: \quad y_1^* \to y_2, \quad y_2^* \to y_1, \quad \gamma_w^* \to \pi/2 - \gamma_w,$

$$\boxed{1}^* \to \boxed{3}, \quad \boxed{3}^* \to \boxed{1}; \tag{3.25a}$$

When $\gamma_w \in (\pi/2, 3\pi/4]: \quad y_1^* \to y_2, \quad y_2^* \to -y_1, \quad \gamma_w^* \to \gamma_w - \pi/2,$

$$\boxed{1}^* \to \boxed{4}, \quad \boxed{2}^* \to \boxed{1}, \quad \boxed{3}^* \to \boxed{2}, \quad \boxed{4}^* \to \boxed{3}; \tag{3.25b}$$

When $\gamma_w \in (3\pi/4, \pi]: \quad y_1^* \to -y_1, \quad \gamma_w^* \to \pi - \gamma_w,$

$$\boxed{1}^* \to \boxed{2}, \quad \boxed{2}^* \to \boxed{1}, \quad \boxed{3}^* \to \boxed{4}, \quad \boxed{4}^* \to \boxed{3}; \tag{3.25c}$$

When $\gamma_w \in (\pi, 5\pi/4]: \quad y_1^* \to -y_1, \quad y_2^* \to -y_2, \quad \gamma_w^* \to \gamma_w - \pi,$

$$\boxed{1}^* \to \boxed{3}, \quad \boxed{2}^* \to \boxed{4}, \quad \boxed{3}^* \to \boxed{1}, \quad \boxed{4}^* \to \boxed{2}; \tag{3.25d}$$

When $\gamma_w \in (5\pi/4,\ 3\pi/2]$: $y_1^* \to -y_2,\ \ y_2^* \to -y_1,\ \ \gamma_w^* \to 3\pi/2 - \gamma_w,$

$$\boxed{2}^* \to \boxed{4},\quad \boxed{4}^* \to \boxed{2}; \tag{3.25e}$$

When $\gamma_w \in (3\pi/2,\ 7\pi/4]$: $y_1^* \to -y_2,\ \ y_2^* \to y_1,\ \ \gamma_w^* \to \gamma_w - 3\pi/2,$

$$\boxed{1}^* \to \boxed{2},\quad \boxed{2}^* \to \boxed{3},\quad \boxed{3}^* \to \boxed{4},\quad \boxed{4}^* \to \boxed{1}; \tag{3.25f}$$

When $\gamma_w \in (7\pi/4,\ 2\pi]$: $y_2^* \to -y_1,\ \ \gamma_w^* \to 2\pi - \gamma_w,$

$$\boxed{1}^* \to \boxed{4},\quad \boxed{2}^* \to \boxed{3},\quad \boxed{3}^* \to \boxed{2},\quad \boxed{4}^* \to \boxed{1} \tag{3.25g}$$

The record $\boxed{i}^* \to \boxed{j}$ means that in the solution obtained for the main sector, it is necessary to replace all indexes corresponding to the segment Y^i with indexes concerning to the segment Y^j.

In the main sector, the dominate flow is always oriented along the axis y_1, hence $W_1 > W_2$.

3.3.8. CLASSIFICATION OF DOUBLE CONFIGURATIONS ON A SQUARE PLANE LATTICE

Because $W_1 > 0$ and $W_2 > 0$ in the main sector, then

$$\cos\gamma_w > 0,\quad \sin\gamma_w > 0 \quad \text{in the main sector}$$

Then if follows from (3.7) or from (3.12):

$$q^{(1)} > 0,\quad q^{(3)} > 0 \tag{3.26}$$

while the remaining flows $q^{(2)}$ and $q^{(4)}$ can change the sign. Therefore four various double stream configurations are possible, which are shown in Figs. 3.11 and 3.12 (graphs "a"), with the graph "b" imaging the total flow picture in the lattice.

The basic criterion convenient to construct the classification of double configurations is the number of the reflexive streams within the cell. According to the note at the end of the previous section, the averaged flow tends from the left to the right, then the reflexive flow tends from the right to the left.

Three independent double configurations can be distinguished:

I: a non reflexive configuration (reflexive flows are missing),
II: a reflexive configuration (reflexive flow is observed in one segment),
III: a twice reflexive configuration (reflexive flow happens in two segments).

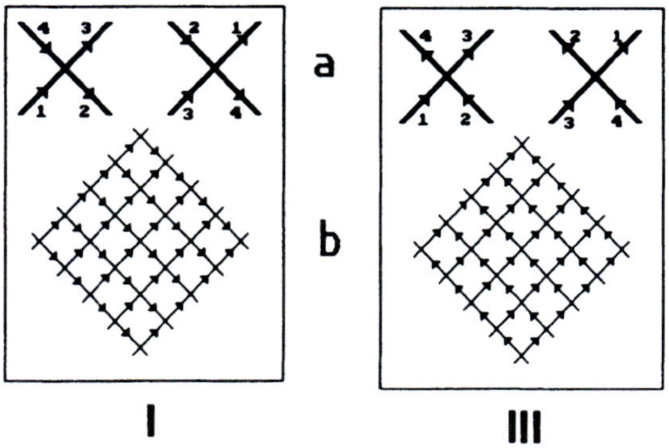

Figure 3.11. Non reflexive double configuration I and twice reflexive double configuration III

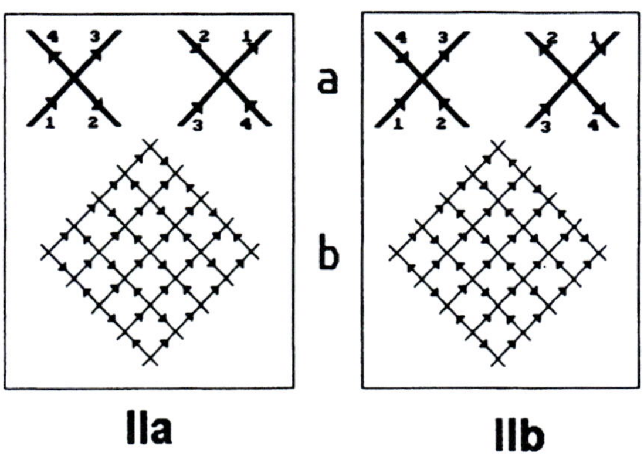

Figure 3.12. Reflexive double configurations IIa and IIb

Configuration II-b differs from II-a only by the reverse order of channel numeration, but the total flow pictures are equivalent.

It will be shown later, that configuration III is impossible to be achieved.

Note that the stream configuration within Poiseuille flow can be modified by variation of the channel thickness.

Conditions of configuration existence result from (3.6) and (3.7), or from

(3.11) and (3.12):

$$\begin{array}{lll} \text{Configuration} & I: & 0 \le \operatorname{tg}\gamma_w < \pi_1 \\ & & \\ \text{Configuration} & II: & \pi_1 \le \operatorname{tg}\gamma_w < \pi \\ & II-a: & \text{when} \quad \pi_1{=}H,\ \pi_2{=}G \\ & II-b: & \text{when} \quad \pi_1{=}G,\ \pi_2{=}H \\ & & \\ \text{Configuration} & III: & \pi_2 \le \operatorname{tg}\gamma_w < 1 \end{array} \tag{3.27}$$

The following notations are accepted:

$$\pi_1 \equiv \min(G,\, H), \quad \pi_2 \equiv \max(G,\, H)$$

$$G = \frac{F_3^{-1} + F_4^{-1}}{F_1^{-1} + F_4^{-1}}, \quad H = \frac{F_4^{-1} + F_2^{-1}}{F_2^{-1} + F_3^{-1}}$$

where

$$F_i \equiv \begin{cases} (h^{(i)})^3, & \text{for the case of Poiseuille's flow;} \\ h^{(i)} K^{(i)} b^{(i)}, & \text{for the case of Darcy's flow} \end{cases}$$

Conditions (3.27) are corollary of signs of the flow rates $q^{(i)}$, $i{=}1{-}4$. For instance, the following is valid for the configuration IIa: $q^{(2)}{>}0$, $q^{(4)}{<}0$ etc.

3.3.9. REDUCTION THE CLASSES OF ALLOWED DOUBLE CONFIGURATIONS

The configuration III in the main sector is impossible. Let us prove it.

1. Let $H{<}G$, then $\pi_1{=}H$, $\pi_2{=}G$. Then the configuration III exists if

$$H < G < 1 \tag{3.28}$$

that follows from (3.27).

We will designate $h^{(3)}/h^{(1)}{=}\lambda$, $\lambda{\ge}0$, then condition (3.28) may be written in the following form:

$$\underbrace{\frac{1+F_{23}F_{31}}{1+F_{23}}}_{H} < \underbrace{\frac{1+F_{43}}{1+F_{43}F_{31}}}_{G} < 1, \qquad F_{ij} \equiv \left(\frac{F_i}{F_j}\right)$$

The system of the last inequalities is always incompatible. In fact,

- if $0{<}F_{31}{<}1$, then $G{>}1$;

- if $F_{31}>1$, then $H>1$;
- if $F_{31}=1$, then $H=G=1$.

2. Let $G<H$. Then the configuration III exists if $G<H<1$. It can be proved in the same way that this inequalities are incompatible.

So, only two double stream configurations, I and II, exist in the main sector.

3.3.10. STRUCTURAL CONDITIONS FOR VARIOUS DOUBLE CONFIGURATIONS

General relation (3.24) gets the following forms for two independent double stream configurations written for the functions $\varphi^{(i)}(\eta_1, \eta_2)$:

$$\text{Configuration} \quad I: \quad \begin{cases} \varphi^{(3)}(-l,-l)=\varphi^{(2)}(l,l) \\ \varphi^{(1)}(-l,l)=\varphi^{(4)} \end{cases} \tag{3.29a}$$

$$\text{Configuration } II-a: \begin{cases} \varphi^{(4)}(-l,-l)=\varphi^{(3)}(-l,-l) \\ \varphi^{(3)}(-l,-l)=\varphi^{(2)}(l,l) \end{cases} \tag{3.29b}$$

$$\text{Configuration } II-b: \begin{cases} \varphi^{(2)}(l,-l)+1/2=\varphi^{(1)}(-l,l)-1/2, \\ \varphi^{(1)}(-l,l)=\varphi^{(4)}(-l,l) \end{cases} \tag{3.29c}$$

In case I each independent node has two output streams, therefore each node gives one structural condition. In case II one node has only one output stream, that does not yield any condition, whereas another independent node has three output streams, that gives two structural conditions.

In all cases, we obtain two structural conditions necessary to close the problem (3.14), (3.21a).

3.4. SOLUTION OF THE CELL PROBLEM

3.4.1. CORRECT FORMULATION OF THE CELL PROBLEM

Finally, the closed problem on the cell gets the form:

$$\begin{cases} v^{(1)}\frac{d\varphi^{(1)}}{d\eta_1}=C_0-v^{(1)}\lambda_1, \quad \eta\in Y^1 \\ -v^{(2)}\frac{d\varphi^{(2)}}{d\eta_2}=C_0-v^{(2)}\lambda_2, \quad \eta\in Y^2 \\ v^{(3)}\frac{d\varphi^{(3)}}{d\eta_1}=C_0-v^{(3)}\lambda_1, \quad \eta\in Y^3 \\ -v^{(4)}\frac{d\varphi^{(4)}}{d\eta_2}=C_0-v^{(4)}\lambda_2, \quad \eta\in Y^4 \\ \varphi^{(1)}(l,l)q^{(1)}+\varphi^{(4)}(-l,-l)q^{(4)}=\varphi^{(3)}(-l,-l)q^{(3)}+\varphi^{(2)}(l,l)q_2 \\ \int\limits_{-l}^{l}(h^{(1)}\varphi^{(1)}+h^{(3)}\varphi^{(3)})d\eta_1+\int_{-l}^{l}(h^{(2)}\varphi^{(2)}+h^{(4)}\varphi^{(4)})d\eta_2=0 \\ \text{and one pair of conditions from (3.29)} \end{cases} \tag{3.30}$$

Herein relations (3.14) and (3.21a) have been used.

The problem has analytical solution in the form of (3.17)-(3.18):

$$\varphi^{(1)}(\eta_1)=\beta^{(1)}\eta_1+\alpha^{(1)}, \qquad \varphi^{(2)}(\eta_2)=\beta_2\eta_2+\alpha_2$$

$$\varphi^{(3)}(\eta_1)=\beta^{(3)}\eta_1+\alpha^{(3)}, \qquad \varphi^{(4)}(\eta_2)=\beta^{(4)}\eta_2+\alpha^{(4)}$$

where $-l<\eta_i<l$, $i=1,2$.

Coefficients $\beta^{(i)}$ are defined according to (3.18).

For $\alpha^{(i)}$ the following system results from (3.30):

$$\alpha^{(1)}q^{(1)}+\alpha^{(4)}q^{(4)}=\alpha^{(2)}q_2+\alpha^{(3)}q^{(3)}, \ \alpha^{(1)}h^{(1)}+\alpha^{(2)}h_2+\alpha^{(3)}h^{(3)}+\alpha^{(4)}h^{(4)}=0$$

with structural conditions (3.29):

– for the non-reflexive configuration I:

$$\alpha^{(3)}-\beta^{(3)}l=\alpha^{(2)}+\beta^{(2)}l, \quad \alpha^{(1)}-\beta^{(1)}l=\alpha^{(4)}+\beta^{(4)}l$$

– for the reflexive configuration II-a:

$$\alpha^{(4)}-\beta^{(4)}l=\alpha^{(3)}-\beta^{(3)}l, \quad \alpha^{(3)}-\beta^{(3)}l=\alpha^{(2)}+\beta^{(2)}l$$

– for the reflexive configuration II-b:

$$\alpha^{(2)}-\beta^{(2)}l+1=\alpha^{(1)}-\beta^{(1)}l+1, \quad \alpha^{(1)}-\beta^{(1)}l=\alpha^{(4)}+\beta^{(4)}l$$

Finally, it is easy to obtain for $\alpha^{(i)}$:

Configuration I:

$$\begin{cases} \alpha^{(1)}=\frac{2l^2}{B^2}\left[(h^{(4)}\beta_{14}-h^{(3)}\beta_{23})\cos\gamma_w+(h^{(2)}+h^{(3)})(\beta_{14}Z^{(4)}+\beta_{32}Z^{(3)})\right] \\ \alpha^{(2)}=\alpha^{(1)}-\frac{1}{B}(\beta_{14}Z^{(4)}+\beta_{32}Z^{(3)}), \\ \alpha^{(3)}=\alpha^{(2)}+\beta_{32}\beta_0, \\ \alpha^{(4)}=\alpha^{(1)}-\beta_{14}\beta_0 \end{cases} \tag{3.31a}$$

Configuration II-a:

$$\begin{cases} \alpha^{(2)}=-\frac{2l^2\cos\gamma_w}{B^2Z^{(1)}}\left[h^{(1)}(Z^{(3)}\beta_{32}-Z^{(4)}\beta_{24})+Z^{(1)}(h^{(3)}\beta_{32}+h^{(4)}\beta_{24})\right], \\ \alpha^{(1)}=\alpha^{(2)}+\frac{\beta_0}{Z^{(1)}}(Z^{(3)}\beta_{32}-Z^{(4)}\beta_{24}), \\ \alpha^{(3)}=\alpha^{(2)}+\beta_{32}\beta_0, \\ \alpha^{(4)}=\alpha^{(2)}+\beta_{24}\beta_0; \end{cases} \tag{3.31b}$$

Configuration II-b:

$$\begin{cases} \alpha^{(1)}=\frac{2l^2\cos\gamma_w}{B^2Z^{(3)}}\left[h^{(3)}(Z^{(2)}\beta_{21}+Z^{(4)}\beta_{14})+Z^{(3)}(h^{(4)}\beta_{14}-h^{(2)}\beta_{21})\right], \\ \alpha^{(3)}=\alpha^{(1)}-\frac{\beta_0}{Z^{(3)}}(Z^{(2)}\beta_{21}+Z^{(4)}\beta_{14})-\frac{Z^{(3)}}{Z^{(2)}}, \\ \alpha^{(2)}=\alpha^{(1)}+\beta_{21}\beta_0+1, \\ \alpha^{(4)}=\alpha^{(1)}-\beta_{14}\beta_0 \end{cases} \tag{3.31c}$$

Herein

$$\beta_{14}\equiv\frac{h^{(1)}}{Z^{(1)}}-\frac{h^{(4)}}{Z^{(4)}}; \quad \beta_{32}\equiv\frac{h^{(3)}}{Z^{(3)}}-\frac{h^{(2)}}{Z^{(2)}}; \quad \beta_{24}\equiv-\left(\frac{h^{(2)}}{Z^{(2)}}+\frac{h^{(4)}}{Z^{(4)}}\right)+\omega_0;$$

$$\beta_{21}\equiv-\left(\frac{h^{(2)}}{Z^{(2)}}+\frac{h^{(1)}}{Z^{(1)}}\right)+\omega_0;\quad \omega_0\equiv\frac{\sqrt{2}B}{\cos\gamma_w};\quad \beta_0\equiv\frac{l\cos\gamma_w}{B}.$$

The function $\varphi_I\equiv\varphi(\eta_1,\eta_2)$ is completely defined.

Solution (3.31) is defined in the main sector ($0\le\gamma<\pi/4$). Solution for other magnitude of γ_w results from (3.31) by means of a change of variables, (3.25).

3.4.2. SOLUTION OF THE SECOND CELL PROBLEM

The second cell function in (2.140), $\varphi_2(y)$, may be determined in analogous way.

Let us introduce the notation: $\phi\equiv\varphi_2$. Then it is the solution of the following problem, analogous to (3.14):

$$\begin{cases} u_1\frac{\partial\phi}{\partial\eta_1}+u_2\frac{\partial\phi}{\partial\eta_2}=C_0^*-\lambda_1u_1-\lambda_2u_2 \\ \langle\phi\rangle=0,\qquad C_0^*\equiv W_2/B=W\sin\gamma_w/B \end{cases}\tag{3.32}$$

As earlier, the function ϕ in the segment Y^i is designated as $\phi^{(i)}$.

Conservation mass condition in the nodes has the same form as (3.19) and (3.21):

$$q^{(1)}\phi^{(1)}(l,l)+q^{(4)}\phi^{(4)}(-l,-l)=q^{(3)}\phi^{(3)}(-l,-l)+q^{(2)}\phi^{(2)}(l,l)$$

Structural conditions are written in general form (3.24) as before, but the function $\Phi^{(i)}$ is now defined as $\Phi^{(i)}\equiv\phi^{(i)}+y_2$. For various double stream configurations we obtain:

$$\begin{cases} \text{Configuration I}: & \\ \qquad \phi^{(1)}(-l,l) = \phi^{(4)}(-l,l), & \\ \qquad \phi^{(3)}(-l,-l) = \phi^{(2)}(l,l)+1 & \\ \text{Configuration II-a}: & \\ \qquad \phi^{(4)}(-l,-l) = \phi^{(2)}(l,l)+1, & \\ \qquad \phi^{(3)}(-l,-l) = \phi^{(2)}(l,l)+1; & \\ \text{Configuration II-b}: & \\ \qquad \phi^{(4)}(-l,l) = \phi^{(1)}(-l,l), & \\ \qquad \phi^{(2)}(l,-l) = \phi^{(1)}(-l,l) & \end{cases}\tag{3.33}$$

As earlier, the solution has the piecewise linear form:

$$\phi^{(1)}=\mu^{(1)}\eta_1+\nu^{(1)};\ \phi^{(2)}=\mu^{(2)}\eta_2+\nu^{(2)};\ \phi^{(3)}=\mu^{(3)}\eta_1+\nu^{(3)};\ \phi^{(4)}=\mu^{(4)}\eta_2+v^{(4)}$$

where parameters $\mu^{(i)}$ are defined as:

$$\begin{aligned}
&\mu^{(1)}\equiv C_0^*/v^{(1)}-\lambda_1=\rho_0^* h^{(1)}/Z^{(1)}-\lambda_1,\\
&\mu^{(2)}\equiv C_0^*/v^{(2)}-\lambda_2=-\rho_0^* h^{(2)}/Z^{(2)}-\lambda_2,\\
&\mu^{(3)}\equiv C_0^*/v^{(3)}-\lambda_1=\rho_0^* h^{(3)}/Z^{(3)}-\lambda_1,\\
&\mu^{(4)}\equiv C_0^*/v^{(4)}-\lambda_2=\rho_0^* h^{(4)}/Z^{(4)}-\lambda_2; \quad \rho_0^*\equiv \sin\gamma_w/B
\end{aligned}$$

Coefficients $\nu^{(i)}$, $i=1-4$ are defined in various ways for various double configurations:

Configuration I:

$$\left\{\begin{aligned}
\nu^{(1)} &= \tfrac{2l^2}{B^2}\left[(h^{(4)}\mu_{14}-h^{(3)}\mu_{32})\cos\gamma_w+(h^{(2)}+h^{(3)})(\mu_{14}Z^{(4)}+\right.\\
&\left.\mu_{32}Z^{(3)})\right]\operatorname{tg}\gamma_w+\left[\tfrac{Z^{(3)}(h^{(2)}+h^{(3)})}{\cos\gamma_w}-h^{(3)}\right]\tfrac{2l^2}{B};\\
\nu^{(2)} &= \nu^{(1)}-\tfrac{\mu_0}{\cos\gamma_w}(\mu_{14}Z^{(4)}+\mu_{32}Z^{(3)})+\tfrac{Z^{(3)}}{\cos\gamma_w},\\
\nu^{(3)} &= \nu^{(2)}+\mu_{32}\mu_0+1,\\
\nu^{(4)} &= \nu^{(1)}-\mu_{14}\mu_0
\end{aligned}\right. \tag{3.34a}$$

Configuration II-a:

$$\left\{\begin{aligned}
\nu^{(2)} &= -\tfrac{2l^2\sin\gamma_w}{B^2Z^{(1)}}\left[h^{(1)}(Z^{(3)}\mu_{32}-Z^{(4)}\mu_{24})+Z^{(1)}(h^{(3)}\mu_{32}+h^{(4)}\mu_{24})\right]-\\
&\left[h^{(3)}+h^{(4)}+\tfrac{h^{(2)}\sin\gamma_w}{Z^{(1)}}\right]\tfrac{2l}{B};\\
\nu^{(1)} &= \nu^{(2)}+\tfrac{\mu_0}{Z^{(1)}}(\mu_{32}Z^{(3)}-\mu_{24}Z^{(4)})+\tfrac{\sin\gamma_w}{Z^{(1)}};\\
\nu^{(3)} &= \nu^{(2)}+\mu_{32}\mu_0+1;\\
\nu^{(4)} &= \nu^{(2)}+\mu_{24}\mu_0+1
\end{aligned}\right. \tag{3.34b}$$

Configuration II-b:

$$\begin{cases} \nu^{(1)} &= \frac{2l^2\cos\gamma_w}{B^2Z^{(3)}}\left[h^{(3)}(Z^{(2)}\mu_{21}+Z^{(4)}\mu_{14})+Z^{(3)}(\mu_{14}h^{(4)}-\mu_{21}h^{(2)})\right]; \\ \nu^{(3)} &= \nu^{(1)}-\frac{\mu_0}{Z^{(3)}}(\mu_{21}Z^{(2)}+\mu_{14}Z^{(4)}); \\ \nu^{(2)} &= \nu^{(1)}+\mu_{21}\mu_0; \\ \nu^{(4)} &= \nu^{(1)}-\mu_{14}\mu_0 \end{cases} \tag{3.34c}$$

Where

$$\mu_{14}\equiv\beta_{14}-\omega_0^*; \quad \mu_{32}\equiv\beta_{32}-\omega_0^*; \quad \mu_{24}\equiv\beta_{24}-\omega_0^*(1+\operatorname{tg}\gamma_w);$$

$$\mu_{21}\equiv\beta_{21}-\omega_0; \quad \omega_0^*\equiv\frac{\sqrt{2}B}{\sin\gamma_w}, \quad \omega_0\equiv\frac{\sqrt{2}B}{\cos\gamma_w}, \quad \mu_0\equiv\frac{l\sin\gamma_w}{B}$$

The second cell function $\varphi_2\equiv\phi(\eta_1,\eta_2)$ is completely defined.

3.5. EFFECTIVE DISPERSION TENSOR IN ANISOTROPIC LATTICES

3.5.1. CALCULATION OF THE DISPERSION TENSOR IN LATTICE MEDIUM

Let us use general correlation (2.141) defining the tensor $\mathbf{D}_{ij}$, and let us take into account Eq. (2.142):

$$D_{ij} \equiv \varepsilon \mathbf{D}_{ij}^{(0)}, \quad \text{where} \quad \mathbf{D}_{ij}^{(0)} = -\langle v_i \varphi_j^0 \rangle_1 \tag{3.35}$$

It is assumed that $\langle b\varphi^{(0)} \rangle = \langle \varphi^{(0)} \rangle = 0$, because $b \equiv 1$.

The expression for tensor components can be derived in the following way:

$$\mathbf{D}_{11}^{(0)} = -\langle v_1 \varphi_1^{(0)} \rangle_1 = -\sum_{i=1}^{4} \left(\int_{Y^i} v^{(i)} \cos \gamma_w \varphi^{(i)}(y) dy \right) =$$

$$= -\cos \gamma_w \sum_{i=1}^{4} h^{(i)} \frac{q^{(i)}}{h^{(i)}} \int_{-l}^{l} \varphi^{(i)}(\eta) d\eta = -2lW \cos \gamma_w \cos \gamma \sum_{i=1}^{4} Z^{(i)} \alpha^{(i)} =$$

$$= -\sqrt{2} W l \cos \gamma_w \sum_{i=1}^{4} Z^{(i)} \alpha^{(i)}; \qquad \text{Etc.}$$

As the result:

$$\begin{aligned}
\mathbf{D}_{11} &= -\sqrt{2}\varepsilon l W \cos \gamma_w \textstyle\sum_{i=1}^{4} Z^{(i)} \alpha^{(i)}, \\
\mathbf{D}_{12} &= -\sqrt{2}\varepsilon l W \cos \gamma_w \textstyle\sum_{i=2}^{4} Z^{(i)} \alpha^{(i)} (-1)^{i-1}, \\
\mathbf{D}_{21} &= -\sqrt{2}\varepsilon l W \cos \gamma_w \textstyle\sum_{i=1}^{4} Z^{(i)} v^{(i)}, \\
\mathbf{D}_{22} &= -\sqrt{2}\varepsilon l W \cos \gamma_w \textstyle\sum_{i=1}^{4} Z^{(i)} v^{(i)} (-1)^{i-1}
\end{aligned} \tag{3.36}$$

3.5.2. LONGITUDINAL AND TRANSVERSE DISPERSION

Usually the dispersion tensor is written in coordinate system rigidly connected with velocity vector W. Let (ξ_1, ξ_2) be the Cartesian system in a cell turned on the angle γ_w counter clockwise respectively to the coordinate system (y_1, y_2), so the axis ξ_1 coincides with the cell center (Fig. 3.13). The direct and inverse transformations of the basis become written in the form:

$$\xi_i = R_{ij} y_j, \quad y_k = R'_{km} \xi_m, \quad i,j,k,m=1,2 \tag{3.37}$$

where $R = \begin{pmatrix} \rho_1 & \rho_2 \\ -\rho_2 & \rho_1 \end{pmatrix}$, $R = \begin{pmatrix} \rho_1 & -\rho_2 \\ \rho_2 & \rho_1 \end{pmatrix}$, $\rho_1 = \cos\gamma_w$, $\rho_2 = \sin\gamma_w$.

Let D_{ij} be the coordinates of tensor (3.36) in a new system (ξ_1, ξ_2). It is easy to deduce from this, using the definition of a tensor:

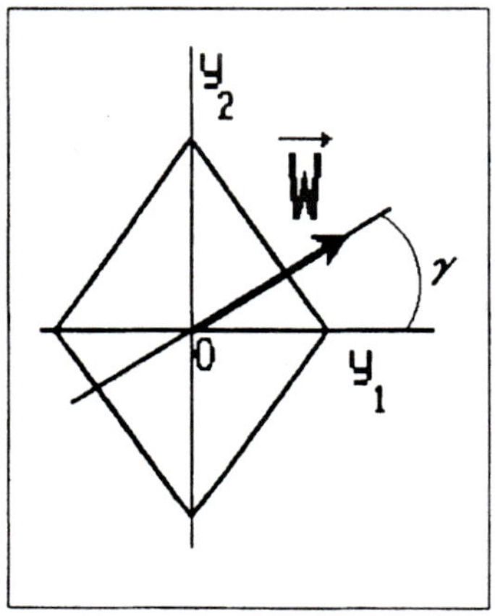

Figure 3.13. On transformation of dispersion tensor coordinates

$$\mathbf{D}'_{ij} = R_{ik} R'_{mj} \mathbf{D}_{km}, \qquad i,j,k,m=1,2 \tag{3.38}$$

$$\begin{cases} \mathbf{D}'_{11} = \rho_1 \rho^{(3)} (\mathbf{D}_{11} + \mathbf{D}_{23}) + \rho_1^2 \mathbf{D}_{12} + \rho_2^2 \mathbf{D}_{21} \\ \mathbf{D}'_{22} = -\rho_1 \rho_2 (\mathbf{D}_{11} + \mathbf{D}_{23}) + \rho_2^2 \mathbf{D}_{12} + \rho_1^2 \mathbf{D}_{21} \\ \mathbf{D}'_{12} = \rho_1^2 \mathbf{D}_{11} - \rho_3^2 \mathbf{D}_{22} - \rho_1 \rho_2 (\mathbf{D}_{12} - \mathbf{D}_{21}) \\ \mathbf{D}'_{21} = -\rho_2^2 \mathbf{D}_{11} + \rho_1^2 \mathbf{D}_{22} - \rho_1 \rho_1 (\mathbf{D}_{12} - \mathbf{D}_{21}) \end{cases}$$

Finally we have:

$$\mathbf{D}'_{11} = -\sqrt{2}\varepsilon l W \rho_0 \Big[Z^{(1)} \left(\alpha^{(1)} \chi_1 + v^{(1)} eta_1 \right) - Z^{(2)} \left(\alpha^{(2)} \chi_2 - v^{(2)} eta_2 \right) + Z^{(3)} \left(\alpha^{(3)} \chi_1 + v^{(3)} eta_1 \right) - Z^{(4)} \left(\alpha^{(4)} \chi_2 - v^{(4)} eta_2 \right) \Big],$$

$$\mathbf{D}'_{22} = -\sqrt{2}\varepsilon l W \rho_0 \Big[-Z^{(1)} \left(\alpha^{(1)} \eta_2 + v^{(1)} \chi_2 \right) + Z^{(2)} \left(\alpha^{(2)} \eta_1 - v^{(2)} \chi_1 \right) -$$

$$Z^{(3)}\left(\alpha^{(3)}\eta_2+v^{(3)}\chi_2\right)+Z^{(4)}\left(\alpha^{(4)}\eta_1-v^{(4)}\chi_1\right)\Big]$$

$$\mathbf{D}'_{12}=-\sqrt{2}\varepsilon lW\rho_0\Big[-Z^{(1)}\left(\alpha^{(1)}\chi_2-v^{(1)}\eta_2\right)-Z^{(2)}\left(\alpha^{(2)}\chi_1+v^{(2)}\eta_1\right)-$$
$$Z^{(3)}\left(\alpha^{(3)}\chi_2-v^{(3)}\eta_2\right)\Big],Z^{(4)}\left(\alpha^{(4)}\chi_1+v^{(4)}\eta_1\right)\Big],$$

$$\mathbf{D}'_{21}=-\sqrt{2}\varepsilon lW\rho_0\Big[-Z^{(1)}\left(\alpha^{(1)}\eta_1-v^{(1)}\chi_1\right)-Z^{(2)}\left(\alpha^{(2)}\eta_2+v^{(2)}\chi_2\right)-$$
$$Z^{(3)}\left(\alpha^{(3)}\eta_1-v^{(3)}\chi_1\right)-Z^{(4)}\left(\alpha^{(4)}\eta_2+v^{(4)}\chi_2\right)\Big]$$

where $\rho_0\equiv\cos\gamma_w\sin\gamma_w$; $\eta_1\equiv 1+\mathrm{tg}\,\gamma_w$; $\eta_2\equiv 1-\mathrm{tg}\,\gamma_w$; $\chi_1\equiv 1+\mathrm{ctg}\,\gamma_w$, $\chi_2\equiv 1-\mathrm{ctg}\,\gamma_w$.

Factors Z are determined according to (3.7), (3.9), factors $\alpha^{(i)}$ - according to (3.31), parameters $v^{(i)}$ - according to (3.34).

The component $\mathbf{D}'_{11}$ is the longitudinal dispersion, while $\mathbf{D}'_{22}$ is the transverse dispersion.

For examined lattices the non-diagonal elements $\mathbf{D}'_{12}$, $\mathbf{D}'_{21}$ are not equal to zero and $\mathbf{D}'_{12}\neq\mathbf{D}'_{21}$ due to lattice anisotropy and the lattice asymmetry with respect to any direction.

3.5.3. CRITICAL PHENOMENA IN CONVECTIVE DISPERSION PROCESSES

Analysis of tensor $\mathbf{D}$ variations at the monotonous change of averaged velocity vector orientation angle γ_w leads to the following results. As before, the main sector is considered.

Two following types of lattices exist when relationship $\mathbf{D}(\gamma_w)$ differs qualitatively.

a) Non-critical lattices, they are the lattices with $\pi_1>1$, where π_1 is defined according to (3.29). In these lattices the double stream configuration in nodes is not changed and corresponds to the type I at any magnitudes of γ_w in the main sector. The reverse flows in throats are always missing. Variation of the components of the tensor $\mathbf{D}$ depending on γ_w is shown in Fig. 3.14. This dependence is continuous and monotonous.

b) Critical lattices, where $0<\pi_1<1$. In these lattices, a critical direction γ_w^* exists, which corresponds to the following condition:

$$\mathrm{tg}\,\gamma_w^*=\pi_1 \tag{3.39}$$

If the respective orientation of the lattice and the flow direction is such that the angle γ is critical, then the tensor $\mathbf{D}$ increases infinitely, as seen from Fig. 3.15. In the critical point, the derivative $\partial D_{ij}/\partial\gamma_w$ has also a discontinuity and changes the sign.

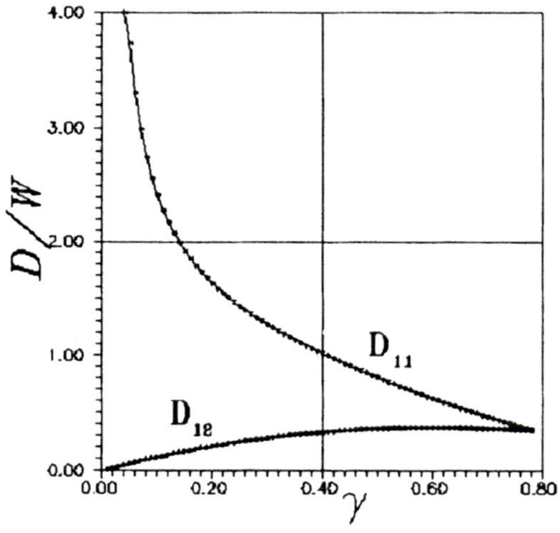

Figure 3.14. Dependence of effective dispersion on the angle of velocity vector orientation in noncritical lattices

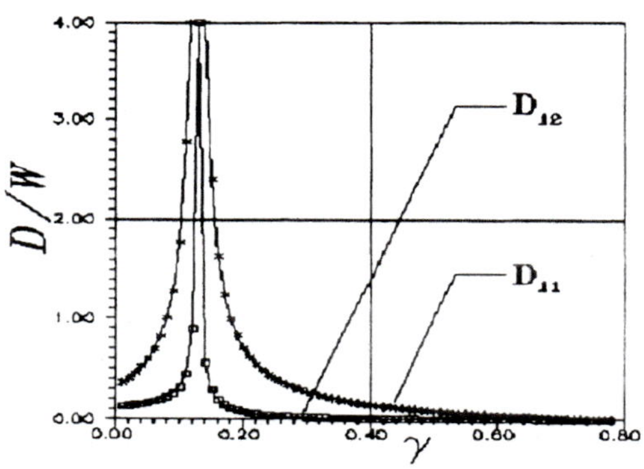

Figure 3.15. Dependence of effective dispersion on the angle γ between the lattice orientation and the flow direction, for a critical lattice

If the range of variation of γ_w is increased, several critical points can be detected.

It is obvious that the dispersion tensor is linear with respect to the averaged transport velocity W, that coincides with traditional representations about behavior of dispersion parameter without diffusion in the system.

3.5.4. ORIGIN OF CRITICAL BEHAVIOR. ARISING OF A NEW PHASE

Physical sense of arising the dispersion tensor jumps in critical points can be completely explained. Dissipation of substance in the lattice considerably depends on the direction of the microscale flow. A change of the configuration leads to formation of a stagnant zone in some throats of the cell.

Hence, the critical direction corresponds to the instance of double configuration replacement: configuration I is replaced by II or vice versa. In fact, during this transformation the flow direction is reversed in one throat of the cell. In the critical point any flow does not happen in this throat, and a stagnant zone is formed there. When the configuration I is replaced by II-a then the stagnant zone arises in the throat 4 (Fig. 3.10), when the configurations I is replaced by II-b the throat 2 becomes stagnant.

Due to reproduction of flow picture within each cell, appearance of traps leads to increasing two times the heterogeneity scale along the direction orthogonal to the trap axis, as shown in Fig. 3.16.

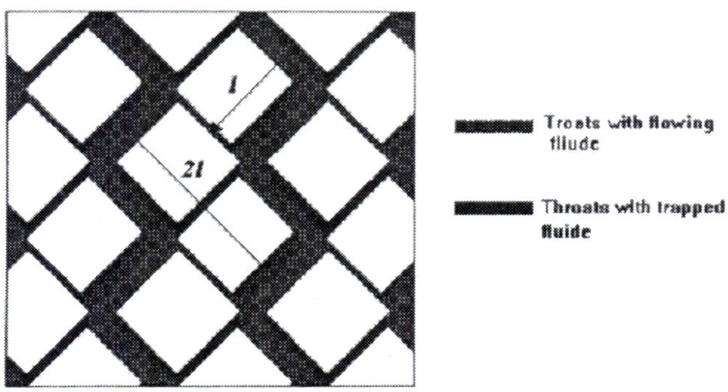

Figure 3.16. Increasing of the heterogeneity scale after trap formation

So far as the dispersion coefficient is proportional to the square of heterogeneity length, the corresponding component of dispersion tensor increases four times. Thus, the origin of dispersion jumps consists in formation of the second phase (the trapped phase).

While tending to the critical point the traps in the stagnant channels obtain new properties different to the properties of flowing phase. In particular, the continuity condition of concentration at the interface between the stagnant channels and the flowing channels is broken down.

It is known that near the phase transition point several phase properties undergo sudden changes. Vice versa, if some characteristic undergoes a sudden change in some point, this point corresponds to phase transition in some specified sense.

3.5.5. ORIGIN OF INFINITIES

The fact of infiniteness of the components of the tensor **D** has mathematical origin. It is caused by the fact that flow velocity beforehand the derivative of the cell function in (3.13) becomes equal to zero in stagnant throats. However the right-hand part is of order 1, then the cell function φ becomes unbounded. As far as the dispersion tensor is proportional to this function, it becomes unbounded too.

Presented solutions allows to identify critical points and to describe the tensor **D** behavior near them. However, they become incorrect strictly in these points. Really, the process in stagnant throat becomes non stationary and doesn't correspond to steady-state equation (3.13).

Near this state, one-phase description becomes invalid. The stagnant zones play the role of a second hydrodynamic phase.

3.5.6. MODIFICATION OF THE AVERAGING TECHNIQUE NEAR CRITICAL POINTS

A modification of averaging technique near critical points can be constructed on the idea of nonuniform averaging analogous to that proposed in the theory of highly heterogeneous media. The meaning of modifications is the following.

The modification of method should provide an automatic elimination of stagnant throats in critical point. This can be achieved by replacing the operation of averaging with nonuniform averaging:

$$\langle \chi_\omega \varphi \rangle \equiv \omega \langle \varphi \rangle_1 + \langle \varphi \rangle_2 \tag{3.40}$$

where φ is an arbitrary cell function; ω is the ratio of velocity module in the stagnant throat over the mean velocity in other throats; the symbols $\langle \cdot \rangle_1$, $\langle \cdot \rangle_2$ denote integration over subdomains Y_1 and Y_2 on the cell, where Y_1 is occupied by stagnant throat.

One can see that $\omega=0$ in critical point and the operation (3.40) entirely eliminates the stagnant throat. Near the critical point, influence of stagnant throat is manifested in high order approximations. Far from the critical point $\omega \sim 1$ and operation (3.40) coincides with classical averaging.

TWO-PHASE FLOW IN DOUBLE POROSITY MEDIA

4.1. INTRODUCTORY NOTES

Problem of upscaling two-phase flow through highly heterogeneous media deals with a more complex case of transport equations, characterized by nonlinearity and degeneration of relative permeabilities, as well as by specific capillary phenomena which determine discontinuous behavior of the saturation, that has not anything similar in linear case.

As usual in all this book, the porous medium consisting of two types of rocks with contrast permeabilities will be considered. In the basic part of this chapter, the highly permeable sub-domain will be assumed to be connected. The "inverse" case, when the low permeable medium forms a connected domain, whereas the highly permeable blocks are disconnected one from other, will be examined in the last chapter of this book.

The classical form of two-phase equations will be considered, which corresponds to non compressible flow through non deformable porous medium, with Darcy's law governing the motion of both the fluids:

$$\begin{aligned} m\left(\frac{x}{\varepsilon}\right)\frac{\partial s}{\partial t} &= \operatorname{div}\left(\frac{K\left(\frac{x}{\varepsilon}\right)k_\alpha\left(s,\frac{x}{\varepsilon}\right)}{\mu_\alpha}\operatorname{grad}P_\alpha\right) \\ -m\left(\frac{x}{\varepsilon}\right)\frac{\partial s}{\partial t} &= \operatorname{div}\left(\frac{K\left(\frac{x}{\varepsilon}\right)k_\beta\left(s,\frac{x}{\varepsilon}\right)}{\mu_\beta}\operatorname{grad}P_\beta\right) \qquad (4.1) \\ P_\beta - P_\alpha &= P_c\left(s,\frac{x}{\varepsilon}\right) \end{aligned}$$

where α and β are the indexes of fluid phases; s is the saturation of the α phase; m and K are the porosity and the permeability of the medium; k is the relative phase permeability; P_c is the capillary pressure; μ is the fluid viscosity.

This system of equations is widely used in petroleum science to describe the flow of water (phase α) and oil (phase β), or in hydrogeology to describe the flow of groundwater (phase α) and air (phase β) with supplemented simplification that P_β=*const* in last case.

In general case both the relative permeabilities and the capillary pressure are distinct in various components of the porous medium. So, we deal with five types of oscillating functions:

- porosity $m\left(\frac{x}{\varepsilon}\right)$;

- permeability $K\left(\frac{x}{\varepsilon}\right)$, which is a tensor of the second rank (9 functions in 3D);

- function of capillary pressure versus saturation $P_c\left(s, \frac{x}{\varepsilon}\right)$;

- tensor function of relative permeability for the α-phase versus saturation, $k_\alpha\left(\frac{x}{\varepsilon}\right)$ (9 functions in 3D);

- tensor function of relative permeability for the β-phase versus saturation, $k_\beta\left(\frac{x}{\varepsilon}\right)$ (9 functions in 3D).

The problem is to deduce the equations with non-oscillating functions, which would describe the flow in some homogenized medium, remaining equivalent to initial equations (4.1) in some sense.

Equation (4.1) will be named as microscale, while the averaged equations corresponds to the macroscale.

4.1.1. FEATURES OF THE PROBLEM

A priori analysis allows to detect the following features of the problem and the following expected results.

Incompressibility of Fluids

Fluid compressibility being not taken into account, the system behavior in the time is expected to be considerably different from that examined in chapter 2. In particular, summing two first equations (4.1) we get the same system in an "elliptic-parabolic" form:

$$-m\left(\frac{x}{\varepsilon}\right)\frac{\partial s}{\partial t} - \mathrm{div}\left(\frac{Kk_\beta}{\mu_\beta}\mathrm{grad}P_\alpha\right) - \mathrm{div}\left(\frac{Kk_\beta}{\mu_\beta}\mathrm{grad}P_c\Big|_{s_{fixed}}\right) = \mathrm{div}\left(\frac{Kk_\beta}{\mu_\beta}\frac{\partial P_c}{\partial s}\mathrm{grad}s\right), \tag{4.2a}$$

$$\mathrm{div}\left[\left(\frac{Kk_\alpha}{\mu_\alpha}+\frac{Kk_\beta}{\mu_\beta}\right)\mathrm{grad}P_\alpha\right] = -\mathrm{div}\left[\frac{Kk_\beta}{\mu_\beta}\mathrm{grad}P_c\right] \tag{4.2b}$$

The first equation is parabolic respectively to saturation s, while the second is elliptic with respect to pressure P_α.

Thus, the pressure field being quasi steady-state, any flow structures are excluded, if they are associated to the fluid compressibility, as for instance the C-S type of flow described in chapter 2.

Nonstationarity and Fluid Redistribution in Space

Nonstationarity of (4.1) is caused by the saturation variation in the time. In fact, the saturation variation can be reached simultaneously with fluid redistribution in space. Therefore our porous medium, made of two rock types, constitutes a simplest model, which allows to study the nonstationary phenomena of fluid redistribution.

On the other hand, it is known from laboratory experiments that the characteristic time of stabilization for two-phase flow is large even in homogeneous medium. So, this time is expected to be of high order in media with high degree of heterogeneity.

Hence, we can expect that such a medium model will lead to definition of the characteristic eigen times of fluid redistribution in the porous space.

From this point of view, the double porosity model may be examined as a tool to solve one of the most important problems of the theory of two-phase flow. It consists in improvement of classical equations of two-phase flow in porous media. One of the recognized defects of the classical equations of two-phase flow through porous media is the hypothesis of local capillary equilibrium, this theory is based on. As the result, the attempt to introduce the time dependent functions as relative permeabilities constitutes a basic way how to modify the conventional two-phase equations. It was applied in [39], [35], [36], [37], [38].

About the Role of Boundary Conditions

The equation for P_α being elliptic, it is necessary to be careful with choosing the domain of flow, Ω. In particular, the domain should be preferably bounded, because the fundamental solution of an elliptic equation (in linear case) tends to infinity when $x\to\infty$, in 1D and 2D cases. But working in a bounded domain we have to examine some boundary conditions. Thus, it is impossible to deduce the homogenized equations for (4.1) independently of the boundary conditions for pressure. We have always to deduce the homogenized boundary-value problem for (4.1).

For the sake of determinacy, we will examine only the Dirichlet problem for the pressure P_α, or for equation (4.2b), assuming that the pressure $P_\alpha^*(x)$ at the boundary is known. The principle of maximum for the elliptic equations determines that both the maximum and the minimum values of the pressure P_α are reached on the domain boundary. Therefore the function $P_\alpha^*(x)$ must give us two characteristic values of the pressure field: the minimal and the maximal pressures, P_A and P_B. Equation (4.2b) being invariant respectively to the pressure shift, only the difference $\Delta P = P_A - P_B$ is of interest. This value gives a scale of variation for fluid pressure.

From the point of view of the physics, only the pressure scale ΔP is important for the system behavior, but not the complete boundary function $P^*_\alpha(x)$. Therefore we can expect that homogenization of the mentioned boundary-value problem will lead to rather a universal macroscale model, whose qualitative structure does not depend on the entire function $P^*_\alpha(x)$, but probably depends on the pressure scale ΔP.

4.1.2. OVERVIEW OF THE EXISTING RESULTS

To describe the heterogeneity degree a parameter ω_K is introduced as earlier, which is equal to the ratio of the low and the high permeabilitiy. A medium is moderately heterogeneous, if $\omega_K \sim 1$. When $\omega_K \to 0$, the medium is highly heterogeneous.

Phenomenological Approach. Exchange Models

The most known attempts to construct averaged equations for two-phase flow correspond to papers [17], where the ideas of Barenblatt-Zheltov model were used. The macroscopic system has been represented in form of two continuous media with an exchange term between them.

Within the framework of this approach the basic problems consist in assuming some form of the exchange term with respect to saturation, and in postulating some form of the effective phase permeabilities. Certainly, such "postulation" is usually based on some physical ideas rich in content.

In [17], the volume of matrix ("fracture") was assumed small, while the blocks were assumed to be operating only as the sources. The flow inside the blocks was governed by capillary forces (counter-current imbibition). The macroscale model is:

$$\begin{cases} \operatorname{div} \vec{V}_\alpha^{II} = -\tilde{q}; \\ \\ m \dfrac{\partial S^I}{\partial t} = \tilde{q} \end{cases} \tag{4.3}$$

where the upper indexes I and II correspond to the tight blocks and the highly permeable matrix; V_α is the flow velocity of the α-phase (water); S^I is the averaged block saturation; $\tilde{q}$ is the rate of exchange by water volume between the blocks and the matrix.

Skvortsov [17] suggested the following approximation for the exchange rate q:

$$\tilde{q}(x,t) \simeq \begin{cases} 0, & t<t_0(x) \\ N_0 \dfrac{exp\left(-\lambda_0[t-t_0(x)]\right)}{\sqrt{t-t_0(x)}}, & t>t_0(x) \end{cases}$$

with empirical constant parameters N_0, λ_0. These formulae have been deduced using the approximate analytical solution of a 1D problem of capillary imbibition into a porous medium.

The function $t_0(x)$ is the time moment, when the displacement front reaches the point x. This function may be obtained as the solution of a Bacley-Leverett problem in a homogeneous medium without any exchange processes. It is clear, that this model does not take into account influence of the exchange processes on the velocity of the front motion.

Phenomenological Approach. Exchange+Dispersivity Models

The similar model has been developed in a series of papers [27], [28], but an important supplement has been introduced there. Flow of the α-phase through the highly permeable matrix II is described as:

$$\frac{\partial}{\partial t}\left(m^{II}S^{II}\rho_\alpha\right)+\text{div}\left(\rho_\alpha\vec{V}_\alpha\right) = -q+\text{div}\left(m^{II}S^{II}\widehat{D}\text{grad}\rho_\alpha\right) \tag{4.4}$$

where m is the porosity; q is the rate of exchange by water mass between the blocks and the matrix, which plays the same role as the term $\tilde{q}$ in Eq. (4.3) ; $\widehat{D}$ being the macroscale coefficient of hydraulic dispersion.

The following correlations have been proposed for the exchange term:

- correlation of [10]:

$$q = R_0\lambda_1 e^{-\lambda_1 t}, \quad R_0 \equiv m^I(s^* - s_*); \tag{*}$$

where s_* and $(1-s^*)$ are residual water and oil saturations respectively;

- correlation of [40] and [27]:

$$q = R_0\frac{\lambda_1\lambda_2}{\lambda_2-\lambda_1}\left(e^{-\lambda_1 t}-e^{-\lambda_2 t}\right) \tag{**}$$

- correlation of [28]:

$$q = R_0\frac{\lambda_1\lambda_2}{\lambda_2-\lambda_1}\int\limits_0^t \left[e^{-\lambda_1(t-\tau)}-e^{-\lambda_2(t-\tau)}\right]\frac{\partial S^{II}}{\partial\tau}d\tau \tag{***}$$

where the parameters λ_1^{-1} is the characteristic time to transfer the cumulative oil mass from the block towards the interface, while λ_2^{-1} is the characteristic time to transfer the same oil mass from the interface into the matrix.

Relation (*) results from (**), when $\lambda_2 \gg \lambda_1$. Relation (***) may be obtained from (**) as a convolution with the saturation derivative, according to general relation proposed in [30].

Exponential terms in all these relations follow from postulation of a linear dependence between the rate om mass exchange q and the difference of masses in blocks and in matrix. Such a linear dependence is not proved.

As to the convolution (***), it was proposed in [30] by the analogy with linear case of one-phase flow, where arising of similar convolution with the pressure derivative is an exact result (see chapter 1 of this book). In two-phase case, which is described by nonlinear equations respectively to saturation, arising of the convolution operator with a kernel dependent only on the difference $(t-\tau)$ seems to be a very coarse approximation.

Note that a new important term is introduced in (4.4) , which describes the macroscale dispersivity of the saturation field caused by spatial heterogeneity. Regretfully the proposed description of two-phase dispersivity can not be used on practice, because the parameter $\widehat{D}$ is undefined. Moreover, the dispersivity in (4.4) is caused by gradients of fluid density and becomes zero when the fluid is incompressible. Really, the two-phase dispersivity should be caused by gradients of saturation.

Up-scaling Relative Permeability

In all the models mentioned above the macroscale flow velocity is defined using the macroscale analogy to Darcy's law:

$$\vec{V}_\alpha = -\frac{\widehat{K}\widehat{k}_\alpha(S)}{\mu_\alpha}\mathrm{grad}P_\alpha$$

where $\widehat{K}$ is the macroscale absolute permeability; $\widehat{k}_\alpha(S)$ is the effective relative permeability; μ_α is the viscosity.

The problem is how to recalculate the macroscale permeabilities $\widehat{K}$ and $\widehat{k}_\alpha(S)$, knowing the microscale absolute permeability $K(x)$ and the microscale relative permeability $k_\alpha(s)$.

Known procedures of scale-up are based on the exact analytical formula of harmonic averaging for the effective permeability of 1D medium:

$$\widehat{K} = \frac{1}{\left\langle \frac{1}{K(x)} \right\rangle}$$

where the symbol $< >$ denotes spatial averaging.

Then, decomposing the real multi-dimensional flow pattern onto a system of 1D streams, we can use this formula along each one-dimensional stream. The art consists in that how to perform such a decomposition. All classical scale-up methods, as [45], [60], [34], [41], [43], [46], [18], [57] and [29], were based on this simple idea. Two-phase flow is treated in analogous way, but the saturation variation along each streams should be taken into account.

For example, in the method [45] and [60], the medium is assumed to be layered, and the flow is assumed to be directed along the medium layers. Then the averaged permeability may be calculated as a harmonic mean along each layer, on the first step of averaging, and consequently as an arithmetic mean along the direction perpendicular to layering.

Other methods developed later, as [41], are more general, and do not require some special structure of the medium to be applied. Selection of 1D streams is performed in such a way, that all the streams form a regular periodic lattice, which replaces the original continuous flow field. Moreover, the pressure distribution in such a lattice is considerably simplified, assuming that the lattice constitutes a system of hydrodynamically independent "blocks". Each block consists in 2^n adjacent smallest elements of the lattice, where n is the space dimension. Pressure field inside each block is assumed to be independent of other blocks. Therefore averaging can be made within each block independently of other blocks. Such a transition from the smallest scale l of the lattice up to the scale of the block $2l$ is an elementary scale-up. Performing the next elementary scale-up to the scale $4l$ we may improve the results obtained. The method was proved by multiple numerical experiments.

It is necessary to note, that such an algorithm is implicitly based on a hypothesis of self-similarity of flow on each spatial scale. In fact, the system of streams is selected in a similar way after each elementary scale-up. Moreover all other simplified hypothesis concerning the pressure field remain similar too.

The described algorithms show that the problem of upscaling phase permeabilities and deducing the form of exchange terms are independent one of other. This circumstance seems to be a mistake rather than an advantage. Results obtained by homogenization have shown the effective relative permeabilities dependent on the parameters of the exchange process.

Concluding Ideas of Phenomenological Approach

Completing this section, it is possible to conclude, that four basic features of the desired two-phase macroscale model have been emphasized within the framework of phenomenological approach:

1. The macroscale model of two-phase flow in highly heterogeneous medium should be presented in a "splitted form", i.e., in the form of equations

each of them corresponds to the matrix or to the blocks.

Further, it can be shown that this hypothesis is usually satisfied, but is not necessary. Moreover, in case of small capillary forces, the hypothesis of splitting becomes a basic obstacle to deduce the effective macroscale model.

2. Equations for blocks and matrix must be related by an exchange term, whose form remains unknown. Approximations (*) - (***) shown above will not be confirmed by the homogenization theory.

3. Exchange process can cause some memory effects, associated to the saturation redistribution within the heterogeneous medium. Description of such phenomena is undefined in nonlinear case.

4. Heterogeneity changes the structure of absolute permeability and relative permeability. Recalculation of relative permeabilities based on renormalization technique is well developed in case of moderately heterogeneous media. But validity of such procedures for highly heterogeneous media is not proved.

5. Heterogeneity should stipulate the dispersion phenomenon. However this effect is not satisfactory described.

Homogenized Models for Moderately Heterogeneous Media

The first results obtained by various averaging procedures, have shown the problem to be more complicated, than the form of the exchange term and the independent scale-up of relative permeabilities. We will examine these results in this section.

The simplest situation corresponds to the case of a moderately heterogeneous medium, with phase permeability and capillary pressure being identical in all parts of the medium:

$$\omega \sim 1 : \qquad k_i(s), P_c(s) \ \text{ do not depend on } \ x$$

The solution for the problem of averaging two-phase flow in the medium of such a type has been investigated by Amaziane and Bourgeat [3]. It was then shown, that the homogenized model has the same type as the initial one.

The case where the functions are distinct at various points of the medium was investigated by the method of volume averaging [56], [14], [1], by asymptotic expansion [58], and by two-scale convergence [4], [12], [21].

The classical result of [58] and [56] consists in the following macroscopic model respectively to microscale equations (4.1):

$$\langle m \rangle \frac{\partial S}{\partial t} - \frac{\partial}{\partial x_i} \left(\frac{\widehat{K_\alpha^{ij}}(S)}{\mu_\alpha} \frac{\partial P_\alpha}{\partial x_j} \right) = 0 \tag{4.5a}$$

$$-\langle m\rangle\frac{\partial S}{\partial t}-\frac{\partial}{\partial x_i}\left(\frac{\widehat{K}^{ij}_{\beta}(S)}{\mu_\beta}\frac{\partial\left[P_\alpha+\widehat{P}_c(S)\right]}{\partial x_j}\right)=0 \tag{4.5b}$$

$$\widehat{P}_c(S)=\widehat{P}^{I}_c(S^I)=\widehat{P}^{II}_c(S^{II}) \tag{4.5c}$$

$$\widehat{P}^{I}_c(S^I)\equiv P^{I}_c(S^I),\quad \widehat{P}^{II}_c(S^{II})\equiv P^{II}_c(S^{II}) \tag{4.5d}$$

$$S\equiv\frac{\langle m\rangle_1 S^I+\langle m\rangle_2 S^{II}}{\langle m\rangle} \tag{4.5e}$$

Where S^I, S^{II} and S are the averaged saturation for the blocks, the matrix and the whole medium; P_α is the macroscale pressure in the α-phase; $\widehat{P}^{II}_c(S^I)$, $\widehat{P}^{II}_c(S^{II})$ and $\widehat{P}_c(S)$ are the macroscale capillary pressure for the blocks, matrix and the whole cell. The symbols $\langle\cdot\rangle$, $\langle\cdot\rangle_1$ and $\langle\cdot\rangle_2$ mean the integration over the whole cell Y, the sub-domain Y^I and the sub-domain Y^{II} respectively.

Parameters $\widehat{K}^{ij}_\alpha(S)$, $\widehat{K}^{ij}_\beta(S)$ and $\widehat{P}_c(S)$ are determined via the cell functions $\psi_{\gamma,i}$ for each fluid γ, $\gamma=\alpha,\beta$, $i=1,2,3$:

$$\widehat{K}^{ij}_\gamma(S)=\left\langle K_{ij}(y)k^I_\gamma(S^I)\left(\frac{\psi_{\gamma,i}}{y_j}+\delta_{ij}\right)\right\rangle_1+$$
$$\left\langle K_{ij}(y)k^{II}_\gamma(S^{II})\left(\frac{\psi_{\gamma,i}}{y_j}+\delta_{ij}\right)\right\rangle_2 \tag{4.6}$$

where $\psi_{\gamma,k}$ is defined as solution of the cell problem:

$$\begin{cases}\dfrac{\partial}{\partial y_i}\left(K_{ij}(y)k_\gamma(y,S^I,S^{II})\left(\dfrac{\partial\psi_{\gamma,k}}{\partial y_j}+\delta_{kj}\right)\right)=0, & y\in Y\\[2ex] \langle\psi_{\gamma,k}\rangle=0, \qquad \psi_{\gamma,k}\text{ is } y\text{-periodic}\end{cases} \tag{4.7}$$

where

$$k_\gamma(y,S^I,S^{II})\equiv\begin{cases}k_\gamma(S^I), & y\in Y^I\\ k_\gamma(S^{II}), & y\in Y^{II}\end{cases}$$

Physical sense of the model (4.5) is the following. Eqs. (4.5a) and (4.5a) describe mass balance for each fluid phase, α and β respectively, in the averaged medium. Eqs. (4.5c) are the macroscale capillary equilibrium equations. Two relations (4.5d) comprise three saturations, therefore only one of them is independent. Hence, two equations (4.5d) define two functions:

$$S^I=S^I(S),\qquad S^{II}=S^{II}(S) \tag{4.8}$$

where Eqs. (4.5d) and (4.5e) are taken into account.

Due to this fact, the effective permeabilities (4.6) depend on one saturation S, although two saturations S^I and S^{II} enter in the local problem (4.7).

The plot 4.1 shows a graphical solution of the sub-system (4.5c).

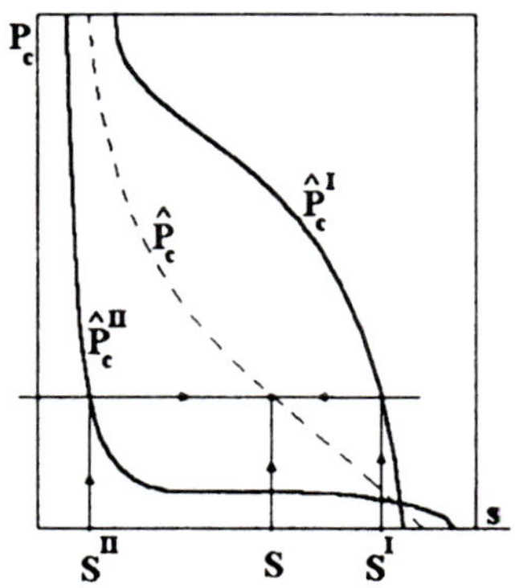

Figure 4.1. Graphical solution of the system (4.5c)

Knowing the value S we can calculate the functions (4.8), i.e. $S^I(S)$ and $S^{II}(S)$, according to (4.5c), and the capillary pressures $\hat{P}_c^I(S^I)$ and $\hat{P}_c^{II}(S^{II})$ which should be equivalent to the effective capillary pressure in the point S. Using this algorithm we can construct all the function $\hat{P}_c(S)$ (dashed curve in Fig. 4.1).

We see that the global block pressure and the global high conductivity pressure are equivalent.

Models of Flow in Highly Heterogeneous Media

Homogenization of two-phase flow in highly heterogeneous porous media, i.e. media whose component parts show a great difference for permeabilities, is practically still an open problem.

The linear case with respect to the saturation represents a simple convection-diffusion transport equation with highly oscillating convective flow velocity, as well as dispersion parameter. This case was studied below, in chapter 2. For some classes of probable ratios between determining parameters it was studied in [44], [50] and [55].

Nonlinear equations of two phase flow were examined in [31], [22], [24], [51]. In [22] the two-scale limit of the homogenization was obtained for a ratio between permeabilities of blocks and fractures equal to ε^2. Two-scale convergence is strictly proved, but the macroscopic model is not less complex, than the micro-scale equations, because micro- and macro-scale variables are not splitted. The similar problem is appropriated to the paper [31].

In paper [51] the totality of systems was shown to be nonuniform. Four classes of two-phase systems in highly heterogeneous media leading to various upscaling limits were constructed. A macroscopic model showing capillary non-equilibrium effects was proposed for one class of systems. More general case of correspondence between parameters was presented in [24].

In this chapter, the model for various ratios between parameters will be deduced.

4.2. BASIC PROBLEMS

Basing on the preliminary analysis of existing results stated above, the most important theoretical problems of averaging two-phase flow can be formulated.

4.2.1. CHANGE OF THE EQUATION TYPE AFTER AVERAGING

As we mentioned earlier, the problem of up-scaling has been formulated according to the needs of petroleum engineers to obtain a tools allowing to use the coarse grids for numerical modeling of the flow in underground reservoirs. All the classical method of scale-up are based on the hypothesis of *invariance of the flow equations with respect to the procedure of up-scaling.* This hypothesis says that mathematical type of macroscale equations is the same as for microscale model. This hypothesis allows to reduce the problem of up-scaling to a simple renormalization of the transport parameters.

Highly heterogeneous media constitute a basic case, where the hypothesis of invariance respectively to up-scaling becomes not valid. Even in the linear case of one-phase flow through highly heterogeneous medium, arising of an integral-differential operator has been detected in the macroscopic model, after up-scaling the diffusion equation [11], [9], [49]. Because the type of the macroscale and the microscale equations is distinct for highly heterogeneous media, up-scaling can not be reduced to the simple parameter renormalization. The first step of up-scaling consists in deducing the macroscale model of flow, which is not known a priori.

In case of two-phase flow in highly heterogeneous medium, the type of macroscale equations is not clear. Operator of convolution arisen in linear case can be formally transmitted to nonlinear case. On the other hand, averaging of convection-diffusion linear equations in chapter 2 has shown arising of another objects like equations with delaying argument for some classes of flow. It is natural to wait appearance of some combined operators in non-linear case of two-phase flow, however these operators are expected to have a very complicate, not explicit structure.

4.2.2. NONLOCALITY AND NONLINEARITY

Long-Memory and Exchange Phenomena

It is expected that equations for blocks and matrix should be related by an exchange term, whose form is unknown, but some previous results mentioned in chapter 1, or approximate equation (***) which is a closure relation for phenomenological equation (4.4) , say that the exchange term can introduce a long-term memory in the system.

In fact, the basic reason causing the type change of the motion equations in highly heterogeneous media are long-term correlations in time originating between events happening on the blocks and on the matrix. On the one hand, the flow structure inside the matrix and blocks is sufficiently distinct. On the other hand, these flows are coupled at the block/matrix interface. That coupling is non-local in time, because the block flow and the matrix flow are evolving in various time scales.

In case of two-phase flow, non-locality is completed by non-linearity, that leads to such macroscale models which are not more simple than the microscale equations [21], [22]. However, a replacement of the nonlinear nonlocal complicated procedures by some coarse approximations like (***) in (4.4) is impossible too.

In this chapter other way is developed which allows to obtain useful results It constitutes in studying the "weak non-locality".

Weak Nonlocality

It is known, that the convolution operator $\int_0^t \frac{\partial u}{\partial \tau} R(t-\tau)d\tau$, which is nonlocal in time, may be transformed into a local operator, when the integral kernel can be simplified. Let the kernel $R(t)$ has a form similar to the exponential function with a length of exponential decay equal to t^*. When this length t^* is small, the kernel can be converted into the delta-function, such that

$$R(t) \quad \rightarrow \quad t^* \delta(t)$$

Then the integral operator may be converted into the simple operator of differentiation:

$$\int_0^t \frac{\partial u}{\partial \tau} R(t-\tau)d\tau \;\simeq\; t^* \frac{\partial u}{\partial t} \tag{4.9}$$

The idea is to use the similar formula in case, when the time t^* is not small, as an approximate relation.

Recognizing that the exact description of nonlocal phenomena in nonlinear case is a very complicated problem, we can try to replace "strong" nonlocality by a *"weak nonlocality"* basing on the equation like (4.9) . Equation (4.9) is a transition from a system with long-term memory to another system with a short memory. This short memory effects are described by local operators, but they keep some information about nonlocal behavior. Due to that property, short memory terms may be called as "weakly nonlocal".

Studying the weak nonlocality is a fruitful way to examine two-phase flow. In this case we can wait to obtain closed macroscale equations in the explicit and very effective form.

To ensure weakly nonlocal macroscale behavior, we can select the parameters of the process in such a way, that the expected relaxation times would have a rather small magnitude, in order to the transition (4.9) could be possible. For example, basing on the model for Translation-Source type of one flow explained in chapter 1, where the relaxation effects are important, we can wait that the relaxation time is of order $\varepsilon_p \sim \varepsilon^2/\omega_K$. Then, in order to the time t^* be small, the heterogeneity degree must be no very large, i.e., ω_K must be larger than ε^2.

Basing of these ideas, we will examine two-phase flow through such media, where

$$\omega_K > \varepsilon^2 \tag{4.10}$$

Importance of First Terms in Asymptotic Expansion

It follows from the previous text that to keep weak nonlocality in desired macroscale equations, we need to keep the effects of order of the relaxation time $t^* \sim \varepsilon_p$. Thus, deduction of the zero term of asymptotic expansion is not a sufficient goal of this book. Important new effects can be obtained, if the first terms with respect to ε_p will be kept.

Thus, by an a priori choose of parameters according to (4.10) we simplify the system analysis, but to keep basic features of system behavior we need to complicate the analysis by keeping first-order terms in asymptotic expansions.

4.2.3. MACROSCALE RELATIVE PERMEABILITY: DEPENDENCE ON THE EXCHANGE PROCESS

Dependence of the effective relative permeability on two saturations

Results of previous chapters show that up-scaling of flow through highly heterogeneous media leads to splitting of the pressure (or concentration) field onto two macroscale fields appropriated to the blocks and the matrix respectively. In is natural to wait that two macroscale saturation fields will arise after up-scaling the two-phase flow. Thus, macroscale relative permeabilities should depend on both the saturations. Such a dependence on two arguments is not traditional, but may be analyzed before performing any up-scaling.

When these saturations are related by some equation, which allows to exclude one of them, only one saturation is independent, and the relative permeabilities depend on one variable only, as in classical case. The typical case of such a functional relation is demonstrated in section 4.1.2, Eq. (4.5c) , which is devoted to moderately heterogeneous media.

However, even in this simplest case, the effective relative permeability "feels" the equation establishing connection between blocks and matrix therefore the scale-up procedure should be dependent on the exchange process.

Dynamic Effective Relative Permeability
Such a simple functional relation between two saturations seems to be something nonrealistic, that can be reached only in case of complete capillary equilibrium in the whole medium. But the equilibrium in a highly heterogeneous medium is not very typical state. Thus, we can wait that two saturations are independent, therefore the effective relative permeabilities depend on two variables. Moreover, we can wait that effective relative permeabilities display a dynamic behavior.

4.2.4. "EXTRACTION" OF THE SATURATION FROM CELL PROBLEM. SPLITTING THE SATURATION AND NETEROGENEITY

One of the basic problems of up-scaling two-phase flow is associated to development of powerful numerical algorithms to compute effective relative permeabilities. More strictly speaking, the problem is as follows.

In classical cases, the effective relative permeability is determined via cell functions, which are solutions of the boundary-value problems in a heterogeneity cell. These functions parametrically depend on saturation, as seen in (4.6) and (4.7) . One solution of the cell problem yields only one point of the function effective relative permeability versus saturation. The problem is how to eliminate saturation from this algorithm, and how to obtain whole the function of effective relative permeability basing on a single solution of the cell problem.

This problem will be solved in section 4.8.3.

4.2.5. AVERAGING OF CONVECTION DOMINATED FLOW. TWO-PHASE DISPERSION

Capillary forces being weak, the saturation transport equation becomes hyperbolic, and up-scaling of such an object is more complicated problem.

Averaging the Hyperbolic Equations
The first obstacle to obtain constructive results is that the solution of the hyperbolic equation can be discontinuous. Averaged behavior near the jumps is expected to be very different from that happening far from them. So the averaged model qualitatively depend on the amount of saturation jumps, i.e., on the initial and boundary conditions. This problem is common in the theory of averaging hyperbolic equations.

Basic mathematical results in averaging hyperbolic equation of first order are published in [5], [6] and [7] for 1D case. Nonlinear 1D equation of two-phase flow without capillary forces was averaged in [23], [20] and [29].

Dependence of the Cell Problem on Macroscale Variable

The second problem of averaging the hyperbolic equation is that the similarity of solution in various cells is disturbed, when the transport velocity is non-periodic, or non-ergodic in stochastic case. It means that the cell problem depends on macroscale variable, which can not be eliminated. Thus the cell problem must be solved in each macroscale point. Such an algorithm is useless. In two-phase case, the transport velocity is a nonlinear function of saturation, thus it may be periodic only in casual way. The problem is how to eliminate the macroscale variable from the cell problem.

Macroscale Two-Phase Dispersion

The third problem concerns the following. As shown in chapter 2, splitting onto two macroscale saturations becomes non-fruitful when diffusion effects are weak on the microscale. In this case we proposed to describe the system using one concentration, averaged over the whole cell. This approach leads to arising a phenomenon of the macroscale dispersion.

In two-phase, nonlinear, case, the similar approach should be applied. An example of such an attempt to introduce formally dispersivity into the macroscale models of two-phase flow was shown in section 4.1.2, Eq. (4.4) , where some critical remarks were said.

Another approach to deduce thoroughly two-phase equations containing the two-phase dispersion may be proposed. Only one macroscale saturation should be introduced, that will give a nonlinear macroscale saturation transport equation comprising the effective nonlinear capillary dispersion, i.e. a diffusion term with nonlinear diffusion parameter dependent on the convective transport velocity.

Because in two-phase flow models, diffusion and dispersion are associated to capillary forces, arising of such a capillary dispersion will give description of the dynamic macroscale capillary pressure dependent on the transport velocity.

Actually this approach is in development.

Some close ideas are developing in papers [33] and [47].

4.3. DEFINITION OF THE PROBLEM

4.3.1. PHYSICAL SETTING

Structure of the Medium

Consider the flow of two incompressible phases (the α−phase and the β−phase) in a medium $\Omega \subset I\!R^3$ made of a highly permeable connected "matrix", Ω^{II}, and of a system of periodically located lowly permeable "inclusions" or "blocks", Ω^I, such that $\Omega = \Omega^I \cup \Omega^{II}$, as shown on Fig.4.2. Let Γ_x be the interface separating two media, Ω^I and Ω^{II}.

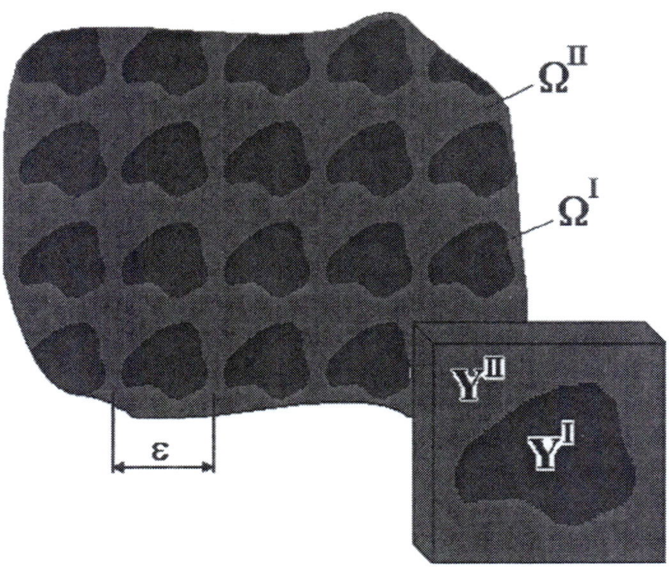

Figure 4.2. Structure of the medium with highly permeable connected matrix II and lowly permeable inclusions I

Each of such sub-domains Ω^i, $i = I, II$, is characterized by a rock permeability tensor K^i and a porosity m^i, by the relative phase permeability functions $k^i_\alpha(s)$, $k^i_\beta(s)$, and the capillary pressure function $P^i_c(s)$. The capillary pressure is the difference between the pressures in two fluids : $P_c = P_\beta - P_\alpha$. The viscosity μ of each fluid phase is supposed to be constant. The volume fractions of both media are of the same order.

We assume first the heterogeneity is fast oscillating, such that the heterogeneity scale l is more small than the macroscopic scale L:

$$\varepsilon \equiv l/L << 1$$

Secondly, we assume the medium is highly heterogeneous with respect to the permeability and to the capillary pressure, such that:

1) the permeability of the inclusions I is must lower than the permeability of the matrix II;

2) the order of capillary forces in the inclusions is must higher than in the matrix.

In third, we assume the medium is periodic, such that the heterogeneity scale ε is equal to the heterogeneity period.

Assume the rock permeabilities K^I and K^{II}, as well as the porosities m^I and m^{II} to be constant.

Fluid-Medium Properties

Let P be the pressure and s be the saturation of the α–phase.

Let s_*, s^* ($s_* < s^*$) denote the residual saturations correspondent to the percolation threshold for each phase.

Assume both the relative permeabilities $k_i(s)$ and the capillary pressure $P_c(s)$ are distinct in the media I and II, as shown on Fig.4.3.

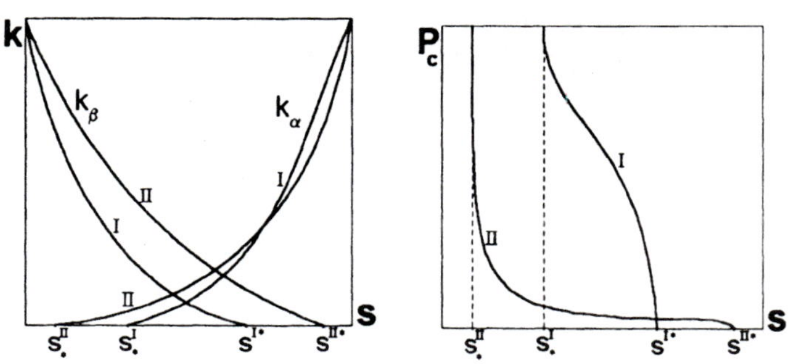

Figure 4.3. Micro-scale relative permeabilities (a) and capillary pressures (b) for blocks I and matrix II

The relative ratio between s_*^I and s_*^{II}, as well as between s^{I*} and s^{II*} is any.

4.3.2. MATHEMATICAL SETTING

Properties of Nonlinear Functions
Assume the following properties for the functions, where $i{=}I, II$:

$$P_c^i(s^i)\geq 0,\ \frac{dP_c^i}{ds^i}<0,\ \text{when } s_*^i\leq s^i\leq s^{i*};\ \left|\int_{s_*^i}^{s^{i*}} P_c^i(s)ds\right|<\infty,\ P_c^i(s^{i*})=0 \quad (4.11)$$

$$k_\alpha^i(s^i)\in[0,1],\ 0<\frac{dk_\alpha^i}{ds^i}<\infty,\ \text{when } s_*^i\leq s^i\leq s^{i*};\ \ k_\alpha^i(s_*^i)=0 \quad (4.12)$$

$$k_\beta^i(s^i)\in[0,1],\ -\infty<\frac{dk_\beta^i}{ds^i}<0,\ \text{when } s_*^i\leq s^i\leq s^{i*};\ \ k_\beta^i(s^{i*})=0 \quad (4.13)$$

where s_*^i, s^{i*} are the residual saturations corresponding to the percolation threshold for each phase in the medium i ($i{=}I, II$);

In inequalities (4.12) and (4.13), the principal fact consists in that the relative permeability has a non-zero derivative at the percolation threshold saturation, that follows from basic concepts of percolation theory. The properties (4.11) define singular behavior of the capillary pressure curves, but in such a manner that these functions remain integrable.

Characteristic Magnitudes of the Process
Let us introduce the following characteristic magnitudes:

- The minimum and the maximum pressures P_A and P_B. As far as the pressure field is described by elliptic equation (4.2b), these two characteristic pressures must be equivalent to a minimum and a maximum boundary-values, according to the principle of maximum for the elliptic systems.

- Averaged capillary pressures:

$$P_c^{i,0}\equiv\frac{1}{s^{i*}-s_*^i}\int_{s_*^i}^{s^{i*}} P_c^i(s)ds, \quad i{=}I, II \quad (4.14)$$

These values exist due to property (4.11).

- Time of complete filling the matrix by one of the fluids (β-fluid, for instance), due to the pressure drop $\Delta P{=}P_A{-}P_B$:

$$t_*\equiv\frac{m^{II}\mu_\beta L^2}{K^{II}\Delta P} \quad (4.15)$$

Flow Equations

Equations describing two-phase flow in Ω with respect to the α-phase saturation s and to the α-phase dimensionless pressure p may be written in the following form, for any $\tau>0$:

In Ω^I:

$$\operatorname{div}_x\Big(\omega_K\lambda^I_\alpha(s)\operatorname{grad}_x p\Big)=\omega_m\frac{\partial s}{\partial\tau}, \tag{4.16a}$$

$$\operatorname{div}_x\Big(\omega_K\lambda^I_\beta(s)\operatorname{grad}_x(p+p_c(s)/Pe^I)\Big)=-\omega_m\frac{\partial s}{\partial\tau} \tag{4.16b}$$

In Ω^{II}:

$$\operatorname{div}_x\Big(\lambda^{II}_\alpha(s)\operatorname{grad}_x p\Big)=\frac{\partial s}{\partial\tau}, \tag{4.17a}$$

$$\operatorname{div}_x\Big(\lambda^{II}(s)\operatorname{grad}_x(p+p_c(s)/Pe^{II})\Big)=-\frac{\partial s}{\partial\tau} \tag{4.17b}$$

On Γ_x:

$$[p]_{\Gamma_x}=0\ , \tag{4.18a}$$

$$p^I_c|_{\Gamma_x}=\omega_c p^{II}_c|_{\Gamma_x}\ , \tag{4.18b}$$

$$\omega_K\lambda^I_\alpha\frac{\partial p^I}{\partial x_i}n_i|_{\Gamma_x}=\lambda^{II}_\alpha\frac{\partial p^{II}}{\partial x_i}n_i|_{\Gamma_x}\ , \tag{4.18c}$$

$$\omega_K\left(\lambda^I\frac{\partial p^I}{\partial x_i}+\frac{\lambda^I_\beta}{Pe^I}\frac{\partial p^I_c}{\partial x_i}\right)n_i|_{\Gamma_x}=\left(\lambda^{II}\frac{\partial p^{II}}{\partial x_i}+\frac{\lambda^{II}_\beta}{Pe^{II}}\frac{\partial p^{II}_c}{\partial x_i}\right)n_i|_{\Gamma_x} \tag{4.18d}$$

The initial condition :

$$s(x,y,0)=s^0(x). \tag{4.19}$$

The boundary conditions require only that

$$p|_{\partial\Omega}=O(1) \tag{4.20a}$$

and

$$s|_{\partial\Omega}=s_{\partial\Omega}(x,t) \tag{4.20b}$$

where $s_{\partial\Omega}(x,t)$ is a given function.

According to remarks in section 4.1.1, the macroscale model is expected to be independent on the form of boundary condition (4.20b) .

The following notations have been introduced for the dimensionless variables with $i=I,II$:

$$p^i\equiv\frac{P^i-P_B}{P_A-P_B},\ p^i_c(s)\equiv\frac{P^i_c(s)}{P^{i,0}_c},\ \tau\equiv\frac{t}{t_*},\ x\equiv\frac{X}{L},$$

$$\lambda_j^i(s) \equiv \frac{k_j^i(s)\mu_\beta}{\mu_j}, \quad j=\alpha,\beta; \quad \lambda^i(s) \equiv \lambda_\alpha^i(s) + \lambda_\beta^i(s)$$

$$\omega_K \equiv \frac{K^I}{K^{II}}, \quad \omega_m \equiv \frac{m^I}{m^{II}}, \quad \omega_c \equiv \frac{P_c^{II,0}}{P_c^{I,0}}, \quad Pe^i \equiv \frac{P_A - P_B}{P_c^{i,0}}$$

Discontinuity of the dimensionless capillary pressure p_c in (4.18) is due to the rescaling procedure and corresponds to a continuity condition for capillary pressure with dimension:

$$[P_c]_{\Gamma_x} = 0 \tag{4.21}$$

Conditions on Γ_x mean the continuity for the flow rate and the pressure in each phase.

Relation (4.20) means that the dimension pressure is between the minimum and the maximum boundary-values P_A and P_B. More detailed definition of the boundary-value condition is not necessary, because its detailed structure does not influence the homogenized behavior.

The initial function $s^0(x)$ and its influence on the macroscale behavior will be discussed furthermore, in section 4.5.3.

The parameters $\lambda(s; x/\varepsilon)$, $\lambda_\alpha(s; x/\varepsilon)$, and $\lambda_\beta(s; x/\varepsilon)$ are ε-periodic with respect to their second argument.

As to the capillary pressure, along with the functions $p_c^I(x,\tau)$ and $p_c^{II}(x,\tau)$ we use another notations for two capillary pressures as functions of saturation: $\tilde{p}_c^I(s^I)$ and $\tilde{p}_c^{II}(s^{II})$. It is naturally, that

$$p_c^i(x,\tau) = \tilde{p}_c^i(s(x,\tau)), \qquad i = I, II \tag{4.22}$$

Two functions $\tilde{p}_c^I(s)$ and $\tilde{p}_c^{II}(s)$ are known a priori.

Parameters of Heterogeneity

The parameters ω_K, ω_c and ω_m measure the degree of heterogeneity of the porous medium and constitute the ratio between the parameters shown on Fig. 4.4.

The capillary pressure function $P_c(s)$ has the form

$$P_c(s) = \frac{\sigma \cos\Theta}{\sqrt{K/m}} J(s) \tag{4.23}$$

with σ the surface tension, Θ the wettability angle and $J(s)$ the dimensionless Leverett-function. Then, in the simplest case, when the wettability characteristics are uniform in the medium, we would obtain the relationship

$$\omega_c \sim \sqrt{\omega_K / \omega_m} \tag{4.24}$$

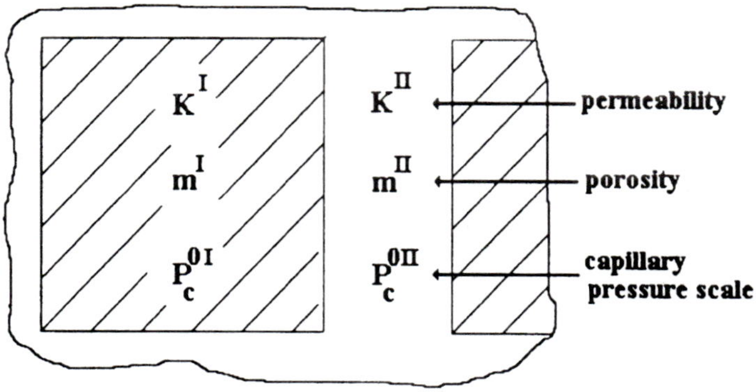

Figure 4.4. Structure of the medium and its parameters

However this relation between ω_c and ω_K is not true in more general case when the medium is heterogeneous with respect to the wettability characteristics, i.e. the function $J(s)$ is different in any component of the medium, as well as the angle Θ.

In order to take into account the more general case, we have to assume that the parameter ω_c is rather independent on ω_K or ω_m, but relation (4.24) gives some limitations for the ratio between ω_c and ω_K.

According to definition of a dual porosity medium we impose the following assumptions on the range of parameters :

$$0<\omega_K\ll 1, \qquad \omega_c\ll 1 \tag{4.25}$$

We exclude a case when $\omega_c\gg 1$ corresponding also to the dual-porosity medium. The such a case stands for the media with inverse wettability of medium I and medium II and will be considered in future.

Contrarily to last relations we assume for ω_m the following:

$$\omega_m\sim 1 \tag{4.26}$$

relation (4.26) corresponds to most situations occurring in oil and gas reservoirs.

According to these assumptions we introduce a new system of independent parameters, more convenient for further analysis :

$$\varepsilon; \qquad \varepsilon_p\equiv\frac{\varepsilon^2\omega_m}{\omega_K}; \qquad Pe^I; \qquad Pe^{II}$$

The parameter ε_p characterizes the ratio between t^* and t_*, time of the saturation convective wave propagation through a dense block and through the entire highly permeable system correspondingly:

$$\varepsilon_p = \frac{t^*}{t_*} = \frac{\varepsilon L}{\left(\frac{K^I}{m^I \mu}\right)\frac{\Delta p}{\varepsilon L}} \Bigg/ \frac{L}{\left(\frac{K^{II}}{m^{II}\mu}\right)\frac{\Delta p}{L}}$$

The Peclet numbers Pe^i (or "capillary numbers") measure ratio between the macroscopic hydrodynamic pressure drop and capillary pressure in the tight blocks ($i{=}I$) or in highly permeable matrix ($i{=}II$). Therefore Pe^i are the macroscale Peclet numbers. The microscale Peclet numbers should be introduced as

$$Pe^i_{mic} \equiv \frac{(P_A - P_B)\, l}{P_c^{i,0} L} \tag{4.27}$$

where l is the size of a cell.

Then

$$Pe^i_{mic} = \varepsilon Pe^i, \qquad i{=}I, II \tag{4.28}$$

4.3.3. TWO-SCALE FORMULATION

Extension of functions and operators

Before the asymptotic expansions be begun, it is necessary to give the two-scale reformulation for our problem.

Let any function $u(x)$ be the solution of the differential equation

$$\frac{du}{dx} = f(u, x) \tag{4.29}$$

with some given function $f(u, x)$.

Let an extended function $\tilde{u}(x, y)$ be introduced in such a way that

$$\tilde{u}(x, y)\Big|_{y=x/\varepsilon} = u(x) \tag{4.30}$$

Then we get the basic identity for differentiation operator:

$$\frac{du(x)}{dx} \equiv \frac{1}{\varepsilon}\frac{\partial \tilde{u}(x, y)}{\partial y}\Bigg|_{y=x/\varepsilon} + \frac{\partial \tilde{u}(x, y)}{\partial x}\Bigg|_{y=x/\varepsilon} \tag{4.31}$$

which allows to examine the two-scale differential equation

$$\frac{1}{\varepsilon}\frac{\partial \tilde{u}(x, y)}{\partial y} + \frac{\partial \tilde{u}(x, y)}{\partial x} = \tilde{f}(\tilde{u}, x, y) \tag{4.32}$$

instead of (4.29). As both equations (4.29) and (4.32) are equivalent (when $y{=}x/\varepsilon$), then their solution must be equivalent too when $y{=}x/\varepsilon$.

Equation (4.32) does not contain fast oscillating functions, that is its advantage.

Fast Variables and Unit Cell

Effecting a change of variable $x \to (x, y{=}x/\varepsilon)$, we are now considering the pressure and the saturation depending on both the large and the small scale x and y : $p(x,t) \to \tilde{p}(x,y,t)$, $s(x,t) \to \tilde{s}(x,y,t)$, assuming identity (4.30). The symbol of tilde will be ignored anywhere.

Taking into account relation (4.31), we have new operators :

$$\partial_k^\varepsilon \equiv \partial_k + \varepsilon \partial_{\mathrm{x}k} \qquad \partial_k \equiv \frac{\partial}{\partial y_k}, \quad \partial_{\mathrm{x}k} \equiv \frac{\partial}{\partial x_k} \ ; \quad \partial_\tau \equiv \frac{\partial}{\partial \tau}$$

To any point x in the domain Ω we associate the image $y{=}x/\varepsilon$ in the cell $Y{=}\{-1/2{<}y_i{<}1/2, \ i{=}1,2,3\}$, which is emphasized on Fig.4.2.

The image of the surface Γ_x, the surface $\Gamma{=}\partial Y^I$ divides the cell into two open sets Y^I (the dense block) and Y^{II} (the highly permeable matrix), and Γ is assumed to be not intersecting the external boundary of Y : $\Gamma \cap \partial Y{=}\emptyset$.

Two-scale Formulation of the Problem

Then the initial problem (4.16a) — (4.20) becomes :
$x{\in}\Omega, \quad y{\in}Y^I$:

$$\varepsilon_p \partial_\tau s = \partial_k^\varepsilon (\lambda_\alpha^I \partial_k^\varepsilon p) \tag{4.33a}$$

$$\mathrm{Pe} \partial_k^\varepsilon (\lambda^I \partial_k^\varepsilon p) = \partial_k^\varepsilon (\lambda_\beta^I \partial_k^\varepsilon p_c) \tag{4.33b}$$

$x{\in}\Omega, \quad y{\in}Y^{II}$:

$$\varepsilon^2 \partial_\tau s = \partial_k^\varepsilon (\lambda_\alpha^{II} \partial_k^\varepsilon p) \tag{4.33c}$$

$$\partial_k^\varepsilon (\lambda^{II} \partial_k^\varepsilon p) = \partial_k^\varepsilon (\lambda_\beta^{II} \partial_k^\varepsilon p_c) \tag{4.33d}$$

The conditions on the interface Γ keep the same form as (4.18), with $\partial/\partial x_k$ replaced by ∂_k^ε:

$$[p]_\Gamma = 0 \ , \tag{4.34a}$$

$$p_c^I\big|_\Gamma = \mathrm{Pe}{\cdot}p_c^{II}\big|_\Gamma \ , \tag{4.34b}$$

$$\varepsilon^2 \omega_m \lambda_\alpha^I (\partial_i^\varepsilon p^I) n_i\big|_\Gamma = \varepsilon_p \lambda_\alpha^{II} (\partial_i^\varepsilon p^{II}) n_i\big|_\Gamma, \tag{4.34c}$$

$$\varepsilon^2 \omega_m \left(\lambda^I \partial_i^\varepsilon p^I + \frac{1}{\mathrm{Pe}} \lambda_\beta^I \partial_i^\varepsilon p_c^I \right) n_i\big|_\Gamma = \varepsilon_p \left(\lambda^{II} \partial_i^\varepsilon p^{II} + \lambda_\beta^{II} \partial_i^\varepsilon p_c^{II} \right) n_i\big|_\Gamma \tag{4.34d}$$

The initial and boundary conditions are kept in the same form as (4.19) and (4.20):

$$s(x,y,0){=}s^0(x,y)$$

$$p\big|_{\partial\Omega} = O(1) \tag{4.35}$$

$$s\big|_{\partial\Omega} = s_{\partial\Omega}(x,t)$$

All the functions dependent on y are $y-$periodic.

The necessary conditions of existence for a y-periodic solution of (4.33a)-(4.33d) are :

$$\varepsilon\varepsilon_p[\omega_m\langle\partial_\tau s\rangle_1+\langle\partial_\tau s\rangle_2]=\varepsilon^2\omega_m\langle\partial_{xi}(\lambda_\alpha^I\partial_i^\varepsilon p)\rangle_1+\varepsilon_p\langle\partial_{xi}(\lambda_\alpha^{II}\partial_i^\varepsilon p)\rangle_2, \tag{4.36a}$$

$$\varepsilon^2\omega_m\langle\partial_{xi}(\lambda^I\partial_i^\varepsilon p)\rangle_1+\varepsilon_p\langle\partial_{xi}(\lambda^{II}\partial_i^\varepsilon p)\rangle_2= \frac{\varepsilon^2\omega_m}{\text{Pe}}\langle\partial_{xi}(\lambda_\beta^I\partial_i^\varepsilon p_c)\rangle_1+\epsilon_p\langle\partial_{xi}(\lambda_\beta^{II}\partial_i^\varepsilon p_c)\rangle_2 \tag{4.36b}$$

Where we used the following notations

$$\langle\cdot\rangle\equiv\int_Y(\cdot)dy,\quad \langle\cdot\rangle_1\equiv\int_{Y^I}(\cdot)dy,\quad \langle\cdot\rangle_2\equiv\int_{Y^{II}}(\cdot)dy \tag{4.37}$$

Conditions (4.36a), (4.36b) were obtained by integrating equations (4.33a) — (4.33d) over the entire cell Y and by using the Green formula, with the conditions at the interface (4.34c) - (4.34d) and with:

$$\int_Y(\partial f/\partial y_i)dy=0,\quad \forall i=1,2,3$$

for any y-periodic function $f(y)$.

Condition of Nonuniform Averaging

As described earlier in chapter 1, the nonuniform homogenization will be applied, which requires some restrictions for the structure of asymptotic expansions. Any function $f(x,y,t)$ may be presented as a sum of two components, one of them depends on the fast variable y, while another component is not dependent on y:

$$f=u(x,y,t)+w(x,t) \tag{4.38a}$$

Then the non-oscillating component should be equivalent to the non-uniformly averaged value, i.e.:

$$\omega_K\langle f\rangle_1+\langle f\rangle_2=w$$

or:

$$\omega_K\langle u\rangle_1+\langle u\rangle_2=0 \tag{4.38b}$$

4.4. CLASSIFICATION OF TWO-PHASE FLOWS

4.4.1. ANALYSIS OF THE MICROSCALE EQUATIONS

Some A Priori Properties of the Microscale Equations

Equations (4.33c) and (4.33d) in the matrix contain only one parameter ε, which is located beforehand the time derivative, that defines existence of a time boundary layer of a scale ε^2. Within this time scale the perturbation is propagating along the matrix. Such a boundary layer is typical for homogenization and is usually ignored being very short in time.

As to equations on the block (4.33a) and (4.33b), they include two other parameters ε_p and Pe, which may be of various orders, as smaller as greater than 1. Variations of ε_p determine delaying of the saturation transport through a block respectively to the matrix. Such a delaying leads to various geometry of streamlines inside a block. The second parameter, Pe, defines predomination of convective or diffusive type of transport.

Reduced Form of the Saturation Transport Equation

We are looking now at what happens for the flow in the blocks (medium I). We start by physical considerations deduced from the following transport equation for saturation, derived from equation (4.33a) — (4.33d).

Replacing λ_α in (4.33a) by $\lambda_\alpha = \lambda F$, multiplying relation (4.33a) by Pe, multiplying (4.33b) by $(-F)$, and adding to (4.33a), we finally obtain the associated transport equation :

$$\begin{aligned}\mathrm{Pe}\cdot\varepsilon_p \partial_\tau s = {} & \mathrm{div}_y\left(\mathbf{V}_y^D\right) + \mathbf{V}_y^C\cdot \mathrm{grad}_y F(s) + \\ & \varepsilon\left[\mathbf{V}_y^C\cdot \mathrm{grad}_x F(s) + \mathbf{V}_x^C\cdot \mathrm{grad}_y F(s)\right] + \\ & \varepsilon\left[\nabla_y\cdot \mathbf{V}_x^D + \nabla_x\cdot \mathbf{V}_y^D\right] + \\ & \varepsilon^2\left[\mathbf{V}_x^C\cdot \mathrm{grad}_x F(s) + \nabla_x\cdot \mathbf{V}_x^D\right], \qquad y\in Y^I,\ x\in\Omega \end{aligned} \quad (4.39)$$

where $\mathbf{V}^D$ is the vector velocity of the diffusion flux, caused by capillary forces ; $\mathbf{V}^C$ is a velocity of the total convective transport;

$$\mathbf{V}_y^D = \lambda_\beta F \mathrm{grad}_y p_c \quad , \quad \mathbf{V}_x^D = \lambda_\beta F \mathrm{grad}_x p_c,$$

$$\mathbf{V}_y^C = \mathrm{Pe}\lambda \mathrm{grad}_y p - \lambda_\beta \mathrm{grad}_y p_c \quad , \quad \mathbf{V}_x^C = \mathrm{Pe}\lambda \mathrm{grad}_x p - \lambda_\beta \mathrm{grad}_x p_c$$

In the last relations using

$$\lambda \mathrm{grad} p - \frac{1}{\mathrm{Pe}}\lambda_\beta \mathrm{grad} p_c = \lambda_\alpha \mathrm{grad} p_\alpha + \lambda_\beta \mathrm{grad} p_\beta,$$

we see that $\mathbf{V}^C$ is the total flow rate.

4.4.2. ELEMENTARY FLOWS IN A CELL

The classes of two-phase flow are the same as in case of the miscible mixture transport studied in chapter 2. However, some specific properties arise, in particular, the diagram of flow becomes another.

In (4.39), the variations in saturation values are caused by five groups of terms in the right-hand side. The last group of terms corresponds to a purely macroscopic flow, and the first four groups of terms describe four classes of elementary flows on the micro-scale. These four types of elementary flows could be classified according to their physical meaning.

First, two distinct geometrical types of flow introduced in chapter 1 as a source and a translation flow, are kept for a two-phase system too. They are shown in Fig. 1.5, section 1.2.2. The source flow is formed by a radial system of streamlines in the block, while the translation flow is formed by a system of streamlines on a block, which are oriented in the same direction as in the matrix.

Another two types of elementary flow corresponds to two physical types of saturation transport, such as convection or diffusion. Note, that the diffusion term in transport equation (4.39) is related with capillary phenomena ("capillary diffusion").

Any of two geometrical types of flows can be produced either by diffusion, or by convection forces, leading either to a Diffusion-Source and a Convection-Source or to a Diffusion-Translation and a Convection-Translation.

Diffusion-Source Flow (D-S)

D-S flow is produced by the first term in (4.39). This first term, only expressed by micro-scale operators, is describing the capillary diffusion into a block. Because this process does not depend on macroscopic saturation gradients, it can be only caused by local difference between saturation in the block and in the highly permeable medium. On the global model this term will lead to a source-type of flow.

Fig.4.5 illustrates the flow structure within D-S mode. The dark fluid begins to invade the porous medium saturated by a white fluid. The displacement is directed to the right from the left. In the connected matrix the displacement takes a form of a dissipated front. Each of blocks is being filled with invading dark fluid from the block boundary into the block center. These interior block waves are propagating with a dissipated front too, that is caused by capillary diffusion. Flow pattern is not identical on all the cells, but is similar.

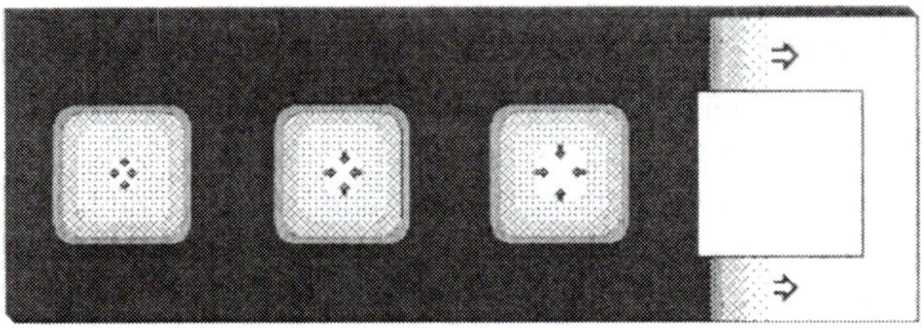

Figure 4.5. Illustration for the Capillary Diffusion-Source Flow

Dispersion-Translation Flow (D-T)
D-T flow is related to the fourth group of terms in Eq. (4.39). It describes the translation of substance between neighboring cells by means of capillary diffusion flux. More precisely, there is a capillary diffusion in a cell $\nabla_y \cdot \mathbf{V}_x^D$ coming from the microscopic oscillations of the macroscopic flux $\mathbf{V}_x^D$, and a diffusion spreading the flux between neighboring cells, $\nabla_x \cdot \mathbf{V}_y^D$.

This type of flow is illustrated by Fig.4.6. The displacement is performed in the form of two waves with dissipated fronts, one of them is propagating through the matrix, while another one comes more slowly in the same direction along the series of blocks.

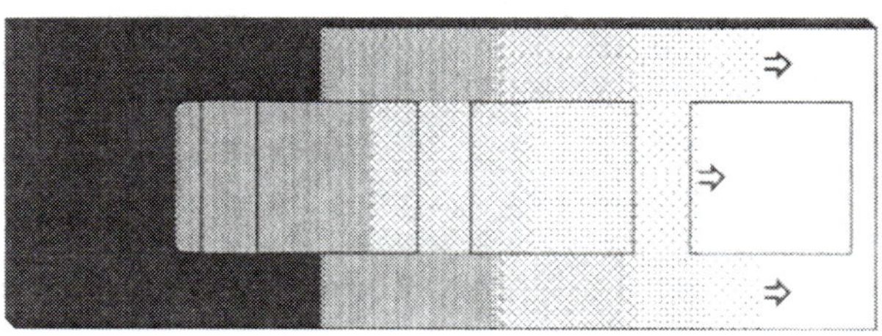

Figure 4.6. Flow pattern for Diffusion-Translation Mode

It was shown in [50] that this last type of flow does not give any new term, but leads only to some additional transport coefficients in the averaged equations.

Convection-Translation Flow (C-T)
C-T flow is related to the third group of terms in (4.39). This group describes the saturation convective transport, on the micro-level, in directions

influenced by macroscopic flow.

The flow pattern shown on Fig. 4.7 is similar to that appropriated to the D-T mode, but both the displacement fronts are sharp, without any dissipation.

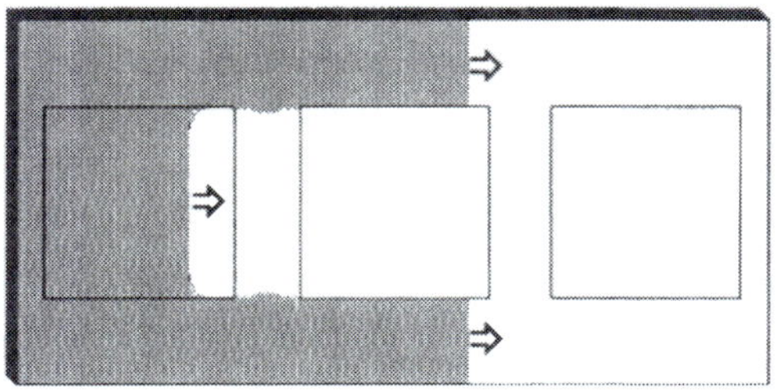

Figure 4.7. Flow pattern for Convection-Translation Mode

Convection-Source Flow (C-S)

C-S flow is related to the second term in (4.39). This second term is also expressed only by micro-scale operators, but they are describing a pure convective saturation transport into a block. Because it does not depend on the directions of macroscopic flows, it can only be caused by the local difference between the fluid pressures in the block and in the highly permeable medium. On the global level this term will correspond to a source-type of flow too.

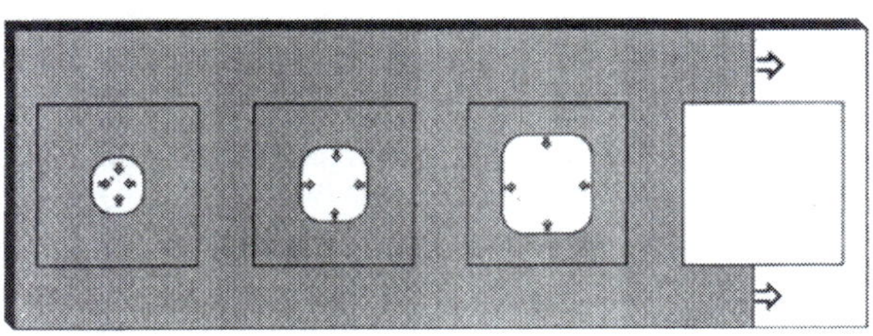

Figure 4.8. Flow pattern for Convection-Source Mode

The flow pattern is shown in Fig. 4.8. It is similar to that appropriated to the D-S mode, but all the displacement fronts are sharp without any dissipation, as in the matrix, as in the blocks.

This class of flow can exist only, when the fluids are compressible. It appears to be impossible for incompressible fluids, because there is no difference between the pressures in medium I and II.

4.4.3. DIAGRAM OF ELEMENTARY FLOWS FOR MEDIA WITH HETEROGENEOUS WETTABILITY

Let us examine a rather general case, when the capillary pressure is not correlated to permeability, so the media I and II differ not only in permeability, but in wettability too.

In order to reduce the total spectrum of probable systems, we assume, within the framework of this section only, the equilibrium between hydrodynamics and capillary forces in the highly permeable matrix Ω^{II} on macroscale:

$$Pe_c^{II,0} \sim P_A - P_B; \quad \text{then } Pe^I \sim \frac{P_A - P_B}{P_c^{I,0}} \sim \frac{P_c^{II,0}}{P_c^{I,0}} = \omega_c$$

such that all the terms in Eqs. (4.16a), (4.16b) for the highly permeable medium are of same order 1.

Various ratios between parameters ε, ε_p, Pe^I will determine the predominance of a flow type on the other ones. Asymptotic analysis of (4.39) be done in the same way as in chapter 2 and in [50], leads to a full classification of all possible systems, which is presented on Fig. 4.9.

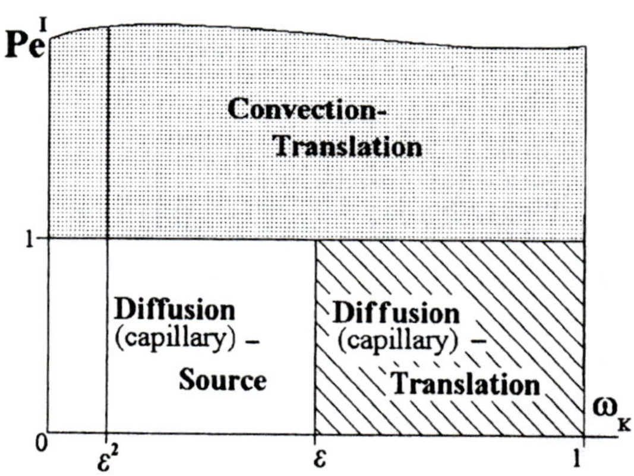

Figure 4.9. Diagram Pe^I vs. ω_K of elementary flows for two-phase fluid in media with heterogeneous wettability, when $Pe^{II} \sim 1$

The picture in Fig. 4.9 differs from the analogous diagram obtained for compressible fluids ; herein the fluids are incompressible, and only one kind of convection flow can occur.

However one more important difference can be found. In the upper area, where convection-translation type of flow is dominating, the following is true:

$$Pe^I > 1 \qquad \Longrightarrow \qquad P_c^I < P_c^{II}$$

but

$$\omega_K < 1 \qquad \Longrightarrow \qquad K^I < K^{II}$$

Such a situation can occur only when the wettability angle Θ in the medium I is considerably larger than in the medium II, as shown in Fig. 4.10-b. Therefore the medium should be heterogeneous with respect to wettability. When the wettability angle is uniform in the whole medium, the capillary pressure is greater in low permeable subdomain I, as shown in Fig. 4.10-a.

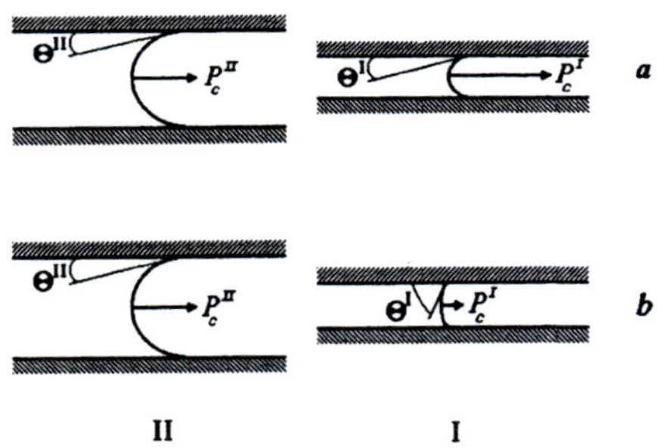

Figure 4.10. Relative ratio between capillary forces P_c^I and P_c^{II} in a heterogeneous medium when:

a) $\Theta^I = \Theta^{II}$ (homogeneous wettability);
b) $\Theta^I \gg \Theta^{II}$ (heterogeneous wettability)

4.4.4. DIAGRAM OF ELEMENTARY FLOWS FOR MEDIA WITH HOMOGENEOUS WETTABILITY

When the media I and II are equivalent with respect to wettability, then capillary pressure is well correlated with permeability, such that:

$$P_c \sim \sqrt{\frac{m}{K}}$$

Then two capillary numbers, Pe^I and Pe^{II}, are not independent, because:

$$Pe^{II} = Pe^I \frac{P_c^{I,0}}{P_c^{II,0}} = Pe^I \sqrt{\frac{m^I K^{II}}{m^{II} K^I}} = Pe^I \sqrt{\frac{\omega_m}{\omega_K}} \sim \frac{Pe^I}{\sqrt{\omega_K}} \tag{4.40}$$

and

$$\omega_c = \sqrt{\frac{\omega_K}{\omega_m}} \sim \sqrt{\omega_K} \tag{4.41}$$

Among four parameters, Pe^I, Pe^{II}, ω_c and ω_K, only two are independent; for example, ω_K and Pe^{II}. In this case we do not need to fix $Pe^{II} \sim 1$ as in the precedent section.

The diagram of flow is shown in Fig. 4.11.

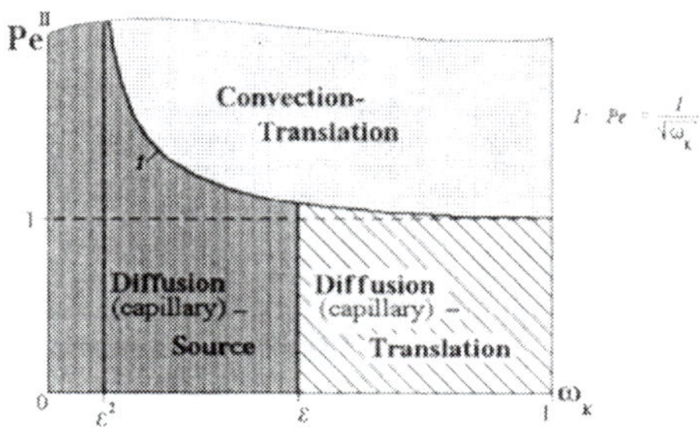

Figure 4.11. Diagram Pe^{II} vs. ω_K of elementary flows for two-phase fluid in media with homogeneous wettability

The same diagram may be presented via coordinates Pe^I vs. ω_K, then we will obtain the picture shown in Fig. 4.9.

4.5. EQUILIBRIUM MODES OF CAPILLARY DOMINATED FLOW

First of all the capillary dominated flow will be examined.

The classification given above emphasizes two modes of such a flow, namely D-S and D-T modes, which differ in geometry of streamline. However the streamline geometry is not a unique criterion defining the system class.

Another classification may be suggested basing on some nonlinear properties of the capillary pressure curves. The basis to emphasize the classes is the degree of their difference from the case of a linear capillary pressure function.

On the other hand, capillary forces, playing a stabilizing role, determine some equilibrium behavior. Any two-phase system is tending to this capillary equilibrium state at large times, according to the properties of parabolic equations (respectively to the saturation). In case of a linear capillary pressure versus saturation curve, only one capillary equilibrium mode exists. When this function is nonlinear, four various equilibrium modes may be detected from the analysis of the equilibrium condition.

4.5.1. FOUR EQUILIBRIUM MODES OF TWO-PHASE BEHAVIOR WITH DOMINATING CAPILLARY FORCES

Being governed by capillary forces, any solution of Eqs.(4.16a) — (4.18d) tends to some capillary equilibrium limit at large times, which is determined by condition (4.18b). This condition requires equality of the block and the matrix capillary pressures at the interface. The last is contrasted to the condition of high heterogeneity of the medium respectively to the capillary forces. This contradiction becomes more considerable when the parameter ω_c becomes more small. Hence, the capillary equilibrium behavior of the system is not clear, when the medium is highly heterogeneous ($\omega_c \to 0$).

To understand it, an analysis of limit behavior of condition (4.18b) when $\omega_c \to 0$ is needed.

If the functions $\tilde{p}_c^i(s^i)$ are analytical and monotonously decreasing, according to the properties (4.11), then the limit of Eq. (4.18b) is unique: $s^I \to s^{I*}$. In fact, using the regular asymptotic expansions for relation (4.18b), we get:

$$\tilde{p}_c^I(s_0^I + \omega_c s_1^I + ...) \sim \omega_c \tilde{p}_c^{II}(s_0^{II} + \omega_c s_1^{II} + ...) \qquad (*)$$

The following relation results from this equation: $\tilde{p}_c^I(s_0^I)=0$, or $s_0^I = s^{I*}$.

Moreover, the parameter ω_c enters into equation (4.18b) in regular way.

However, the realistic functions $\tilde{p}_c^i(s^i)$ are singular, that leads to arising of a singular dependence on the parameter ω_c too. Therefore other situations, singular with respect to the parameter ω_c, appear to be possible, when equilibrium condition (4.18b) is satisfied.

The analysis yields four various equilibrium modes.

Quasilinear Mode (Regular)
It corresponds to the case:

$$\begin{cases} \tilde{p}_c^I(s^I) \sim \omega_c \to 0, \\ \tilde{p}_c^{II}(s^I) \sim 1 \end{cases} \tag{4.42a}$$

or for the saturations:

$$\begin{cases} s^I \to s^{I*} \\ s^{II} = O(1) \end{cases} \tag{4.42b}$$

This mode is called "quasi-linear", because the system behaves like in the case, when the capillary pressure versus saturation function would be linear.

It is very important that the dependence of capillary pressures and saturations on ω_c is regular, as it follows from (*).

Nonlinear Mode
It corresponds to the case:

$$\begin{cases} \tilde{p}_c^I(s^I) \sim \omega_c \Phi(\omega_c) \to 0, \\ \tilde{p}_c^{II}(s^{II}) \sim \Phi(\omega_c) \to \infty \end{cases} \tag{4.43a}$$

where $\Phi(\omega_c)$ is some nonlinear function of ω_c.

The last relationships (4.43a) gives for saturations:

$$\begin{cases} s^I \to s^{I*}, \\ s^{II} \to s_*^{II} \end{cases} \tag{4.43b}$$

Arising of such a mode is due to nonlinear and singular properties of the capillary pressure function.

As shown, the dependence of capillary pressures and saturations on ω_c is singular with logarithmic character. Respectively to two other modes, which will be explained below, this mode may be characterized as "weakly" singular.

Partially Frozen Mode
It corresponds to the case:

$$\begin{cases} \tilde{p}_c^I(s^I) \sim 1 \\ \tilde{p}_c^{II}(s^{II}) \sim \dfrac{1}{\omega_c} \to \infty \end{cases} \tag{4.44a}$$

or, for saturations:

$$\begin{cases} s^I = O(1) \\ s^{II} = s_*^{II} + O(e^{-1/\omega_c}) \to s_*^{II} \end{cases} \tag{4.44b}$$

The saturation in highly permeable domain is equal to the lower percolation threshold, which is the higher saturation when the α-phase remains discontinuous. Therefore, only the β-phase flows through the medium II. The medium II may be said as frozen respectively to the α-phase.

However, the flow in the medium I is two-phase.

The dependence of matrix capillary pressure on ω_c is singular.

Entirely Frozen Mode
It corresponds to the case:

$$\begin{cases} \tilde{p}_c^I(s^I) \sim \dfrac{1}{\omega_c} \to \infty \\ \tilde{p}_c^{II}(s^{II}) \sim \dfrac{1}{\omega_c^2} \to \infty \end{cases} \tag{4.45a}$$

or, for saturations:

$$\begin{cases} s^I = s_*^I + O(e^{-1/\omega_c}) \to s_*^I \\ s^{II} = s_*^{II} + O(e^{-1/\omega_c^2}) \to s_*^{II} \end{cases} \tag{4.45b}$$

The α-phase is immobile in both the media. The flow is entirely single-phase.

Both the matrix and the block capillary pressure singularly depend on ω_c.

All the limit modes are shown on Fig.4.12 on the diagram "Capillary pressure vs. saturation".

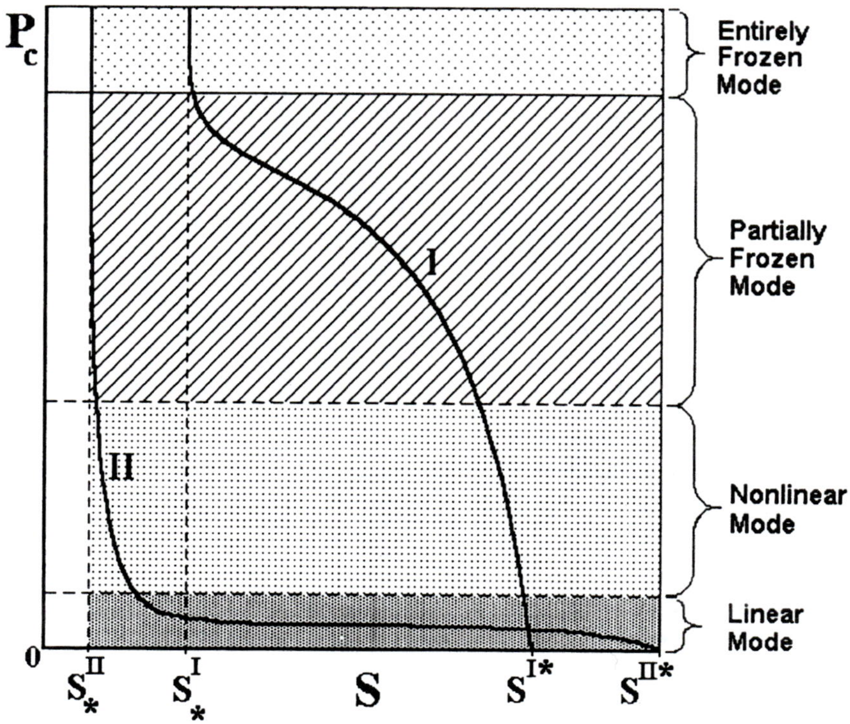

Figure 4.12. Capillary pressure vs. saturation. Four modes of capillary equilibrium for a two-phase mixture in double porosity medium

4.5.2. EXAMPLES

Let the capillary pressure function can be described as

$$P_c(s) = -P_c^0 \ln \sigma(s), \quad 0<\sigma<1 \tag{4.46}$$

where

$$\sigma(s) = \frac{s - s_*}{s^* - s_*}, \quad s_* < s < s^* \tag{4.47}$$

The property (4.11) is satisfied.

It is easy to show that P_c^0 is equal to the integral:

$$P_c^0 = \int_{s_*}^{s^*} P_c(s) ds = \int_0^1 P_c(\sigma) d\sigma$$

In the dimensionless form we get:

$$p_c^i(\sigma^i) = -\ln \sigma^i, \quad 0<\sigma^i<1, \quad i=I, II \tag{4.48}$$

The equilibrium condition when $\omega_c \to 0$ takes the form:

$$\ln \sigma^I \sim \omega_c \ln \sigma^{II}, \quad \text{or} \quad \sigma^I \sim \left(\sigma^{II}\right)^{\omega_c}$$

Then the classification of flow modes mentioned above becomes as follows.

1. Quasi-linear Mode:

$$\begin{cases} p_c^I \sim \omega_c, \\ p_c^{II} \sim 1 \end{cases} \Longrightarrow \begin{cases} \sigma^I \sim 1-\omega_c \sigma_1^I, \quad \sigma_1^I = O(1) \\ \sigma^{II} = O(1) \end{cases} \tag{4.49a}$$

2. Nonlinear Mode:

$$\begin{cases} p_c^I \sim \omega_c \ln \omega_c \to 0, \\ p_c^{II} \sim \ln \omega_c \to \infty \end{cases} \Longrightarrow \begin{cases} \sigma^I \sim \omega_c^{\omega_c} \to 1 \\ \sigma^{II} \sim \omega_c \end{cases} \tag{4.49b}$$

3. Partially Frozen Mode:

$$\begin{cases} p_c^I \sim 1 \\ p_c^{II} \sim \dfrac{1}{\omega_c} \to \infty \end{cases} \Longrightarrow \begin{cases} \sigma^I = O(1) \\ \sigma^{II} \sim e^{-1/\omega_c} \to 0 \end{cases} \tag{4.49c}$$

4. Entirely Frozen Mode:

$$\begin{cases} p_c^I \sim \dfrac{1}{\omega_c} \to \infty \\ p_c^I \sim \dfrac{1}{\omega_c^2} \to \infty \end{cases} \Longrightarrow \begin{cases} \sigma^I \sim e^{-1/\omega_c} \to 0, \\ \sigma^{II} \sim e^{-1/\omega_c^2} \to 0 \end{cases} \tag{4.49d}$$

4.5.3. ON EXISTENCE OF ALL THE EQUILIBRIUM MODES

The entirely frozen mode (4.49d) is rather trivial, being correspondent to a simple single-phase flow in the whole medium. Therefore this mode will be ignored in this book.

The sense of all the remaining modes, except the entirely frozen, consists in that the α-phase tends to filling the low permeable blocks, independently of the filling degree of the matrix. However such an "independence" seems questionable in practice, because both the block and the matrix saturations are more strongly linked in reality. Hence, the problem is, whether all these modes may be reached actually.

In fact, when we consider the medium with disconnected low permeable blocks and connected matrix, it is necessary to take into account that the matrix is the basic transport channel. Any perturbation can reach the blocks only through the matrix. Thus, the block saturation considerably depends on the matrix saturation.

Let the process be considered when the saturation is monotonously increasing in the matrix, which corresponds to the invasion of the α-phase. Because invasion happens through the medium II, saturation S^{II} is higher than the threshold S_*^{II}, therefore only nonlinear and linear modes can be reached. Saturation growth is accompanied with transition top-down on Fig.4.12, and with passing from the nonlinear to the quasi-linear mode. Thus, existence of both these modes may be considered as a temporary event only.

Entirely frozen and partially frozen mode can not exist within such a process.

All these laws refer only to a process of monotonous variation of the saturation in time. Other processes may lead to another conclusions, when anyone from these modes may exist as a time limit.

Summing up all previous reasonings, we have to conclude that for rather generalized form of the initial and the boundary-value conditions, the solution of two-phase problem (4.16a) — (4.20) passes through all the equilibrium modes detected below.

From this point of view, the generalized behavior of the capillary pressure and the saturation, when $\omega_c \to 0$ can be described in the following way:

$$\begin{aligned} &\tilde{p}_c^i(s) \to \tilde{p}_{c0}^i(s;\omega_c), \\ &s^i \to s_0^i(\omega_c), \quad i{=}I,II \end{aligned} \tag{4.50}$$

when the functions $\tilde{p}_{c0}^i(s;\omega_c)$ are singularly dependent on ω_c.

That conclusion will be used to develop asymptotic expansions for the examined problem of two-phase flow.

It is necessary to note that the system evolution may not correspond to capillary equilibrium conditions. When the variation of the matrix saturation in time happens more rapidly than the equilibrium saturation in the blocks will be established, the system behavior is non-equilibrium. It is clear, that the relaxation time is not a unique constant value, but should depend on the corresponding limiting equilibrium mode. In other words, the relaxation time is expected to be dependent on the matrix or block saturation value.

4.6. DEDUCTION OF THE MACROSCALE MODEL FOR D-S FLOW

In this section we examine the case of ratio between parameters corresponding to the Diffusion-Source type, according to a classification given in [50], [51], and in section 4.4.

The case of independent parameters ω_c and ω_K will be assumed, such that the medium can be heterogeneous with respect to wettability. The diagram 4.10 is then true.

Let the parameters are as follows:

$$Pe^{II}{\sim}1, \qquad Pe^{I}{\sim}\sqrt{\varepsilon}, \qquad \varepsilon_p{\sim}\sqrt{\varepsilon}$$

4.6.1. FORM OF ASYMPTOTIC EXPANSIONS

We will seek the asymptotic expansion for the problem (4.33a)–(4.36b) according to a nonuniform two-scale expansion method explained above. The zero term of expansions is a "nonuniform homogenized value":

$$p=\begin{cases} p_0(x,\tau)+\sqrt{\varepsilon}p^I_{1/2}(x,y,\tau)+\varepsilon p^I_1(x,y,\tau)+\varepsilon\sqrt{\varepsilon}..., & y\in Y^I \\ p_0(x,\tau)+\sqrt{\varepsilon}p^{II}_{1/2}(x,y,\tau)+\varepsilon p^{II}_1(x,y,\tau)+\varepsilon\sqrt{\varepsilon}p^{II}_{3/2}(x,y,\tau)+ \\ \varepsilon^2 p^{II}_2(x,y,\tau)+..., \quad y\in Y^{II} \end{cases} \tag{4.51a}$$

Due to the definition of the capillary pressure and from condition (4.34b) we have the expansion

$$p_c=\begin{cases} \sqrt{\varepsilon}\Big(p_{c0}(x,\tau)+\sqrt{\varepsilon}p^I_{c,1/2}(x,y,\tau)+\varepsilon p^I_{c1}(x,y,\tau)+\varepsilon\sqrt{\varepsilon}...\Big), & y\in Y^I \\ p_{c0}(x,\tau)+\sqrt{\varepsilon}p^{II}_{c,1/2}(x,y,\tau)+\varepsilon p^{II}_{c1}(x,y,\tau)+\varepsilon\sqrt{\varepsilon}p^{II}_{c,3/2}(x,y,\tau)+ \\ \varepsilon^2..., \quad y\in Y^{II} \end{cases} \tag{4.51b}$$

The independence of the first terms p_0, p_{c0} on y is due to the contact conditions at the interface (4.34a), (4.34b).

The relationship between the capillary pressure p_c and the saturation being non linear, we use a Taylor expansion of the saturation in (4.51b) and we get the following expansions for the saturation

$$s=\begin{cases} s^I_0(x,\tau)+\sqrt{\varepsilon}s^I_{1/2}(x,y,\tau)+\varepsilon s^I_1(x,y,\tau)+\varepsilon\sqrt{\varepsilon}\cdots, & y\in Y^I \\ s^{II}_0(x,\tau)+\sqrt{\varepsilon}s^{II}_{1/2}(x,y,\tau)+\varepsilon^{II}_1(x,y,\tau)+\varepsilon\sqrt{\varepsilon}\cdots, & y\in Y^{II} \end{cases} \tag{4.51c}$$

Then we obtain the Taylor's expansion for both functions p_c^I and p_c^{II}, according to (4.51b) :

$$\tilde{p}_c^I(s^I) = \tilde{p}_c^I(s_0^I)+\sqrt{\varepsilon}\tilde{p}_c^{I\prime}(s_0^I)s_{1/2}^I+\varepsilon\dots \tag{4.52a}$$

$$\tilde{p}_c^{II}(s^{II}) = \tilde{p}_c^{II}(s_0^{II})+\sqrt{\varepsilon}\tilde{p}_c^{II\prime}(s_0^{II})s_{1/2}^{II}+\varepsilon\dots \tag{4.52b}$$

According to nonuniform homogenization we seek the expansion in such a way, that the non-oscillating component of the solution, which does not depend on the fast variable y, is equal to the nonuniformly averaged value (4.38). Therefore we obtain the following restrictions for the structure of expansions :

$$\varepsilon\sqrt{\varepsilon}\omega_m\Big(\langle p_0\rangle_1+\sqrt{\varepsilon}\langle p_{1/2}^I\rangle_1+\varepsilon\cdots\Big)+\sqrt{\varepsilon}\langle p_{1/2}^{II}\rangle_2+\varepsilon\langle p_1^{II}\rangle_2+$$
$$+\,\varepsilon\sqrt{\varepsilon}\langle p_{3/2}^{II}\rangle_2+\varepsilon^2\cdots = 0, \tag{4.53a}$$

$$\varepsilon\sqrt{\varepsilon}\omega_m\Big(\langle p_{c0}\rangle_1+\sqrt{\varepsilon}\langle p_{1/2}^I\rangle_1+\varepsilon\dots\Big)+\sqrt{\varepsilon}\langle p_{c,1/2}^{II}\rangle_2+\varepsilon\langle p_{c1}^{II}\rangle_2+$$
$$+\,\varepsilon\sqrt{\varepsilon}\langle p_{c,3/2}^{II}\rangle_2+\varepsilon^2\cdots = 0 \tag{4.53b}$$

The macroscopic model will be then deduced from averaged equations (4.36a), (4.36b) in which the terms associated to the sub-domain Y^I have a more higher order. Therefore, for obtaining the macroscopic model it is necessary to define less terms in Y^I, than in Y^{II}. For instance, one can see that in order to obtain the equations of order $O(\sqrt{\varepsilon}+\varepsilon+)$ it is sufficient to determine p_0, and the following terms : $p_{1/2}^I$; s_0^I, $s_{1/2}^I$ in Y^I and $p_{1/2}^{II}$, p_1^{II}, $p_{3/2}^{II}$; s_0^{II}, $s_{1/2}^{II}$, s_1^{II}, $s_{3/2}^{II}$ in Y^{II}.

Because identity (4.42) relates the functions $p_c^i(x,t)$ and $\tilde{p}_c^i(s)$, then equating the same power-value terms in asymptotic expansions (4.51b) and in Taylor's series (4.52a), (4.52b) we obtain finely :

$$\begin{aligned}&\tilde{p}_c^I(s_0^I)=\sqrt{\varepsilon}p_{c0}(x,\tau), \quad \tilde{p}_c^{II}(s_0^{II})=p_{c0}(x,\tau),\\ &\tilde{p}_c^{I\prime}(s_0^I)s_{1/2}^I=\sqrt{\varepsilon}p_{c,1/2}^I(x,y,\tau), \quad \tilde{p}_c^{II\prime}(s_0^{II})s_{1/2}^{II}=p_{c,1/2}^{II}(x,y,\tau)\end{aligned} \tag{4.54}$$

4.6.2. SEQUENCE OF LOCAL PROBLEMS IN Y^I

Substitution of expansions (4.51a) - (4.51c) in (4.33a) - (4.33d), (4.34a) - (4.34d) and equating same powers of $\sqrt{\varepsilon}$, ε and $\varepsilon\sqrt{\varepsilon}$ leads to a sequence of problems.

From (4.33a), (4.34a), (4.33b), (4.34b) we derive the boundary-value problems in Y_1 for $p_{1/2}^I$ and $p_{c,1/2}^I$:

$$\partial_\tau s_0^I=\partial_i(\lambda_{\alpha 0}^I\partial_i p_{1/2}^I), \qquad p_{1/2}^I|_\Gamma=p_{1/2}^{II}|_\Gamma \tag{4.55a}$$

$$\partial_i(\lambda_0^I\partial_i p_{1/2}^I)=\partial_i(\lambda_{\beta 0}^I\partial_i p_{c,1/2}^I), \qquad p_{c,1/2}^I|_\Gamma=p_{c,1/2}^{II}|_\Gamma \tag{4.55b}$$

4.6.3. SEQUENCE OF LOCAL PERIODIC PROBLEMS IN Y^{II}

First we can prove the independence of $p^{II}_{1/2}$ and $p^{II}_{c,1/2}$ on y:

$$p^{II}_{1/2}=p^{II}_{1/2}(x,\tau), \qquad p^{II}_{c,1/2}=p^{II}_{1/2}(x,\tau) \tag{4.56}$$

because we get in Y^{II} for terms of the order $\sim\sqrt{\varepsilon}$:

$$\partial_i(\lambda^{II}_{\alpha 0}\partial_i p^{II}_{1/2})=0,$$
$$\partial_i(\lambda^{II}_0\partial_i p^{II}_{1/2})-\lambda^{II}_{\beta 0}\partial_i p^{II}_{c,1/2})=0,$$

with homogeneous boundary-value conditions, e.t.c.

Then in Y^{II}, we obtain for the y–periodic functions, p^{II}_1, $p^{II}_{3/2}$ from (4.33c), (4.34c), (4.53a), and for p^{II}_{c1}, $p^{II}_{c,3/2}$ from (4.33d), (4.34d), (4.53b) :

$$\sim\varepsilon\ :\ \partial_i\Big(\lambda^{II}_{\alpha 0}(\partial_i p^{II}_1+\partial_{xi}p_0)\Big)=0,\quad \lambda^{II}_{\alpha 0}(\partial_i p^{II}_1+\partial_{xi}p_0)n_i|_\Gamma=0, \tag{4.57a}$$

$$\partial_i\Big(\lambda^{II}_0(\partial_i p^{II}_1+\partial_{xi}p_0)-\lambda^{II}_{\beta 0}(\partial_i p^{II}_{c1}+\partial_{xi}p_{c0})\Big),$$
$$\lambda^{II}_0(\partial_i p^{II}_1+\partial_{xi}p_0)-\lambda^{II}_{\beta 0}(\partial_i p^{II}_{c1}+\partial_{xi}p_{c0})n_i|_\Gamma=0, \tag{4.57b}$$

$$\sim\varepsilon\sqrt{\varepsilon}\ :\ \partial_i\Big(\lambda^{II}_{\alpha 0}(\partial_i p^{II}_{3/2}+\partial_{xi}p^{II}_{1/2})+\lambda^{II}_{\alpha,1/2}(\partial_i p^{II}_1+\partial_{xi}p_0)\Big)=0,$$
$$\lambda^{II}_{\alpha 0}(\partial_i p^{II}_{3/2}+\partial_{xi}p^{II}_{1/2})+\lambda^{II}_{\alpha,1/2}(\partial_i p^{II}_1+\partial_{xi}p_0)n_i|_\Gamma=0,, \tag{4.57c}$$

$$\partial_i\Big(\lambda^{II}_0(\partial_i p^{II}_{3/2}+\partial_{xi}p^{II}_{1/2})-\lambda^{II}_{\beta 0}(\partial_i p^{II}_{c,3/2}+\partial_{xi}p^{II}_{1/2})\Big)+$$
$$+\partial_i\Big(\lambda^{II}_{1/2}(\partial_i p^{II}_1+\partial_{xi}p_0)-\lambda^{II}_{\beta,1/2}(\partial_i p^{II}_{c1}+\partial_{xi}p_{c0})\Big)=0,$$
$$\Big(\lambda^{II}_0(\partial_i p^{II}_{3/2}+\partial_{xi}p^{II}_{1/2})-\lambda^{II}_{\beta 0}(\partial_i p^{II}_{c,3/2}+\partial_{xi}p^{II}_{1/2})+$$
$$+\lambda^{II}_{1/2}(\partial_i p^{II}_1+\partial_{xi}p_0)-\lambda^{II}_{\beta,1/2}(\partial_i p^{II}_{c1}+\partial_{xi}p_{c0})\Big)n_i|_\Gamma=0, \tag{4.57d}$$

with a condition that any function p^{II}_k is y–periodic.

From (4.53a) we have:

$$p^{II}_k=p^{II}_{k1}(x,y,\tau)+p^{II}_{k0}(x,\tau),\qquad \langle p^{II}_{k1}\rangle_2=0,\quad k=1;3/2 \tag{4.58a}$$
$$p^{II}_{c,k}=p^{II}_{c,k1}(x,y,\tau)+p^{II}_{c,k0}(x,\tau),\qquad \langle p^{II}_{c,k1}\rangle_2=0,\quad k=1;3/2 \tag{4.58b}$$

It is clear that the functions p^I_k may be examined as extensions of functions p^{II}_k to all Y. Hence, for these extended functions we return to the y–periodicity in the whole domain Y.

4.6.4. SEQUENCE OF PROBLEMS FOR NON-OSCILLATING COMPONENTS

Basing on (4.51a)-(4.51c) and (4.52a), (4.52b), averaged equations (4.36a) give the following problems :

$$\sim\varepsilon \ : \ \langle\partial_{xi}(\lambda^{II}_{\alpha 0}(\partial_i p^{II}_1+\partial_{xi}p^{II}_0))\rangle_2= \quad (4.59a)$$

$$=\langle\omega_m\partial_\tau s^I_0\rangle_1+\langle\partial_\tau s^{II}_0\rangle_2 \ , \quad (4.59b)$$

$$\sim\varepsilon\sqrt{\varepsilon} \ : \ \langle\partial_{xi}(\lambda^{II}_{\alpha 0}(\partial_i p^{II}_{3/2}+\partial_{xi}p^{II}_{1/2})+\lambda^{II}_{\alpha,1/2}(\partial_i p^{II}_1+\partial_{xi}p_0))\rangle_2=$$

$$=\langle\omega_m\partial_\tau s^I_{1/2}\rangle_1+\langle\partial_\tau s^{II}_{1/2}\rangle_2 \ , \quad (4.59c)$$

and (4.36b) gives :

$$\sim\varepsilon \ : \ \langle\partial_{xi}(\lambda^{II}_0(\partial_i p^{II}_1+\partial_{xi}p_0)-\lambda^{II}_{\beta 0}(\partial_i p^{II}_{c1}+\partial_{xi}p_{c0}))\rangle_2=0, \quad (4.60a)$$

$$\sim\varepsilon\sqrt{\varepsilon} \ : \ \langle\partial_{xi}\Big(\lambda^{II}_0(\partial_i p^{II}_{3/2}+\partial_{xi}p^{II}_{1/2})+\lambda^{II}_{1/2}(\partial_i p^{II}_1+\partial_{xi}p_0)-$$

$$-\lambda^{II}_{\beta 0}(\partial_i p^{II}_{c,3/2}+\partial_{xi}p^{II}_{c,1/2})-\lambda^{II}_{\beta,1/2}(\partial_i p^{II}_{c1}+\partial_{xi}p_{c0})\Big)\rangle_2=0 \quad (4.60b)$$

Where we have used the notations for the terms in Taylor series :

$$\lambda_0\equiv\lambda(s_0), \quad \lambda'_{s0}\equiv(d\lambda/ds)_{s=s_0}, \quad \lambda_{1/2}=\lambda'_{s0}s_{1/2} \quad (4.61)$$

4.6.5. CORRECTNESS OF THE PROBLEMS OF VARIOUS ORDERS

From (4.54) and (4.61) we conclude that the following terms do not depend on the variable y :

$$s^{II}_{1/2}=s^{II}_{1/2}(x,\tau), \ \lambda^{II}_{1/2}=\lambda^{II}_{1/2}(x,\tau), \ \lambda^{II}_{\alpha,1/2}=\lambda^{II}_{\alpha,1/2}(x,\tau), \ \lambda^{II}_{\beta,1/2}=\lambda^{II}_{\beta,1/2}(x,\tau) \quad (4.62)$$

From (4.55a), (4.55b) :

$$p^I_{1/2}=\frac{1}{\lambda^I_{\alpha 0}}\frac{\partial s^I_0}{\partial\tau}\varphi(y)+p^{II}_{1/2}(x,\tau) \ ; \quad (4.63a)$$

$$p^I_{c,1/2}=\frac{\lambda^I_0}{\lambda^I_{\alpha 0}\lambda^I_{\beta 0}}\frac{\partial s^I_0}{\partial\tau}\varphi(y)+p^{II}_{c,1/2}(x,\tau) \quad (4.63b)$$

where the function $\varphi(y)$ is defined by the cell problem :

$$\partial_i(\partial_i\varphi)=1, \qquad y\in Y^I \ ; \qquad \varphi|_\Gamma=0 \quad (4.64)$$

From (4.57a), (4.57b) we get :

$$p^{II}_1=\psi^{II}_k(y)\partial_{xk}p_0+p^{II}_{10}(x,\tau) \ ; \quad (4.65a)$$

$$p^{II}_{c1}=\psi^{II}_k(y)\partial_{xk}p_{c0}+p^{II}_{c1}(x,\tau) \ ; \quad (4.65b)$$

from (4.57c), (4.57d) :

$$p^{II}_{3/2} = \psi^{II}_k(y)\partial_{xk}p^{II}_{1/2} + p^{II}_{3/2,0}(x,\tau) \; ; \tag{4.66a}$$

$$p^{II}_{c,3/2} = \psi^{II}_k(y)\partial_{xk}p^{II}_{c,1/2} + p^{II}_{c,3/2,0}(x,\tau) \tag{4.66b}$$

where the functions ψ^i are periodic solutions of the auxiliary problems :

$$\begin{cases} \partial_i(\partial_i\psi^{II}_k + \delta_{ki}) = 0, & y \in Y^{II} \; ; \quad \psi^{II}_k \text{ is } y\text{-periodic} \\ (\partial_i\psi^{II}_k + \delta_{ki})n_i\big|_\Gamma = 0 \; ; & \langle\psi^{II}_k\rangle_2 = 0 \end{cases} \tag{4.67}$$

We have finally for the saturation, from (4.62) and (4.54) :

$$p^{I\prime}_c(s^I_0)s^I_{1/2}(x,\tau) = \sqrt{\varepsilon}\Big\{\frac{\lambda^I_0}{\lambda^I_{\alpha0}\lambda^I_{\beta0}}\frac{\partial s^I_0}{\partial\tau}\varphi(y) + p^{II}_{c,1/2}(x,\tau)\Big\} \tag{4.68}$$

and from (4.51b), (4.52a), (4.52b), (4.34b) :

$$p^I_c(s^I_0) = \sqrt{\varepsilon}p^{II}_c(s^{II}_0) \tag{4.69}$$

Each of systems (4.55a), (4.55b) is a linear Dirichlet problem for the Poisson's equation in a bounded domain Y^I, and under sufficient regularity assumptions has a unique solution in the Sobolev space $H^1(Y^I)$. Each system (4.67) is a Poisson's equation in a torus Y^{II} with Neumann conditions on the boundary Γ and periodicity conditions on the boundary ∂Y. These problems have unique solutions in the Sobolev space $H^1_{\text{per}}(Y^{II})$.

4.6.6. RESULTING EQUATIONS FOR NON-OSCILLATING COMPONENT

Relations (4.59b), (4.59c), (4.60a), (4.60b) give the following expressions for the first order terms :

$$\partial_{xi}(\lambda^{II}_{\alpha0}\langle\partial_i\psi^{II}_k + \delta_{ik}\rangle_2\partial_{xk}p_0) = \langle\omega_m\partial_\tau s^I_0\rangle_1 + \langle\partial_\tau s^{II}_0\rangle_2, \tag{4.70a}$$

$$\partial_{xi}(\langle\partial_i\psi^{II}_k + \delta_{ik}\rangle_2(\lambda^{II}_{\alpha0}\partial_{xk}p^{II}_{1/2} + \lambda^{II}_{\alpha,1/2}\partial_{xk}p_0)) =$$
$$= \langle\omega_m\partial_\tau s^I_{1/2}\rangle_1 + \langle\partial_\tau s^{II}_{1/2}\rangle_2 \tag{4.70b}$$

$$\partial_{xi}(\lambda^{II}_0\langle\partial_i\psi^{II}_k + \delta_{ik}\rangle_2\partial_{xk}p_0) = \partial_{xi}(\lambda^{II}_{\beta0}\langle\partial_i\psi^{II}_k + \delta_{ik}\rangle_2\partial_{xk}p_{c0}) \tag{4.71a}$$

$$\partial_{xi}(\langle\partial_i\psi^{II}_k + \delta_{ik}\rangle_2(\lambda^{II}_0\partial_{xk}p^{II}_{1/2} + \lambda^{II}_{1/2}\partial_{xk}p_0)) =$$
$$= \partial_{xi}(\langle\partial_i\psi^{II}_k + \delta_{ik}\rangle_2(\lambda^{II}_{\beta0}\partial_{xk}p^{II}_{c,1/2} + \lambda^{II}_{\beta,1/2}\partial_{xk}p_{c0})) \tag{4.71b}$$

4.6.7. MACROSCOPIC MODEL OF TWO-PHASE FLOW

We first present our final result giving a global homogenized problem. Let macroscopic quantities, as the global block saturation S^I, the global saturation of high conductivity medium S^{II}, the global block α−phase pressure $\widehat{p}^I$, the global α−phase pressure of high conductivity medium $\widehat{p}^{II}$, the global block capillary pressure $\widehat{p}^I_c$, the global capillary pressure of high conductivity medium $\widehat{p}^{II}_c$ be defined as:

$$S^I \equiv s_0^I + \sqrt{\varepsilon}\frac{1}{\theta}\langle s^I_{1/2}\rangle_1, \qquad S^{II} \equiv s_0^{II} + \sqrt{\varepsilon} s^{II}_{1/2}, \tag{4.72a}$$

$$\widehat{p}^I \equiv p_0 + \sqrt{\varepsilon}\frac{1}{\theta}\langle p^I_{1/2}\rangle_1, \qquad \widehat{p}^{II} \equiv p_0 + \sqrt{\varepsilon} p^{II}_{1/2}, \tag{4.72b}$$

$$\widehat{p}^I_c \equiv \sqrt{\varepsilon}\left(p_{c0} + \sqrt{\varepsilon}\frac{1}{\theta}\langle p^I_{c,1/2}\rangle_1\right), \quad \widehat{p}^{II}_c \equiv p_{c0} + \sqrt{\varepsilon} p^{II}_{c,1/2}, \tag{4.72c}$$

where θ is a volume fraction of blocs.

The global homogenized model has a following form, with the order of residual terms $o(\sqrt{\varepsilon})$, in all Ω for $\tau>0$:

$$(1-\theta)\frac{\partial S^{II}}{\partial \tau} - \frac{\partial}{\partial x_i}\left(\zeta_\alpha \Lambda^{ki}_\alpha(S^{II})\frac{\partial \widehat{p}^{II}}{\partial x_k}\right) = -\omega_m \theta \frac{\partial S^I}{\partial \tau} \tag{4.73a}$$

$$\frac{\partial}{\partial x_i}\left(\left(\zeta_\alpha \Lambda^{ki}_\alpha(S^{II}) + \zeta_\beta \Lambda^{ki}_\beta(S^{II})\right)\frac{\partial \widehat{p}^{II}}{\partial x_k}\right) =$$

$$= \frac{\partial}{\partial x_i}\left(\zeta_\beta \Lambda^{ki}_\beta(S^{II})\frac{\partial \widehat{p}^{II}_c}{\partial x_k}\right) \tag{4.73b}$$

$$\frac{1}{\sqrt{\varepsilon}} p^I_c(S^I) - p^{II}_c(S^{II}) = -\tau_*(S^I)\frac{\partial S^I}{\partial \tau} \tag{4.73c}$$

with the block pressure defined as:

$$\widehat{p}^I = \widehat{p}^{II} + \tau_*(S^I)(1 - F_\alpha(S^I))\frac{\partial S^I}{\partial \tau}, \quad F_\alpha(S^I) \equiv \frac{\lambda^I_\alpha(S^I)}{\lambda^I(S^I)} \tag{4.73d}$$

$$\widehat{p}^{II}_c = p^{II}_c(S^{II}), \quad \widehat{p}^I_c = p^I_c(S^I) \tag{4.73e}$$

with the initial conditions

$$S^I(x,0) = s^*(x), \qquad S^{II}(x,0) = s^*(x) \tag{4.73f}$$

The effective functions $\Lambda^{ki}_i(S)$ are determined by :

$$\Lambda^{ki}_\alpha(S) = \widehat{K}_{ki} k^{II}_\alpha(S) \quad , \quad \Lambda^{ki}_\beta(S) = \widehat{K}_{ki} k^{II}_\beta(S), \tag{4.74a}$$

$$\widehat{K}_{ki} = \langle \partial_i \psi^{II}_k + \delta_{ki}\rangle_2 \quad , \quad i,k = 1,2,3 \tag{4.74b}$$

and the capillary relaxation time in (4.73c) is defined as :

$$\tau_*(S^I) \equiv \sqrt{\varepsilon}\langle\varphi\rangle_1 \frac{\lambda^I(S^I)}{\lambda^I_\alpha(S^I)\lambda^I_\beta(S^I)} \tag{4.74c}$$

We see that the global block pressure and the global high conductivity pressure differ by the small terms.

Below, we show how this global homogenized model has been derived.

4.6.8. DEDUCTION OF MACROSCOPIC EQUATIONS FROM THE EQUATIONS FOR NON-OSCILLATING COMPONENTS OF FIRST ORDERS

Homogenized model (4.73a)-(4.73e) was deduced in the following way.
1. Multiplying equations (4.70b), (4.71b) by $\sqrt{\varepsilon}$, and adding them to (4.70a), (4.71a) correspondingly, we obtain two averaged equations :

$$\partial_{\mathrm{x}i}\Big(\langle\partial_i\psi^{II}_k+\delta_{ik}\rangle_2\Big[\lambda^{II}_{\alpha0}\partial_{\mathrm{x}k}p_0+\sqrt{\varepsilon}(\lambda^{II}_{\alpha0}\partial_{\mathrm{x}k}p^{II}_{1/2}+\lambda^{II}_{\alpha1/2}\partial_{\mathrm{x}k}p_0)\Big]\Big)=$$
$$= \langle\omega_m\partial_\tau(s^I_0+\sqrt{\varepsilon}s^I_{1/2})\rangle_1+\langle\partial_\tau(s^{II}_0+\sqrt{\varepsilon}s^{II}_{1/2})\rangle_2, \tag{4.75a}$$

$$\partial_{\mathrm{x}i}\Big(\langle\partial_i\psi^{II}_k+\delta_{ik}\rangle_2\Big[\lambda^{II}_0\partial_{\mathrm{x}k}p_0+\sqrt{\varepsilon}(\lambda^{II}_0\partial_{\mathrm{x}k}p^{II}_{1/2}+\lambda^{II}_{1/2}\partial_{\mathrm{x}k}p_0)\Big]\Big)=$$
$$= \partial_{\mathrm{x}i}\Big(\langle\partial_i\psi^{II}_k+\delta_{ik}\rangle_2\Big[\lambda^{II}_{\beta0}\partial_{\mathrm{x}k}p_{c0}+\sqrt{\varepsilon}(\lambda^{II}_{\beta0}\partial_{\mathrm{x}k}p^{II}_{c,1/2}+$$
$$\lambda^{II}_{c,1/2}\partial_{\mathrm{x}k}p_{c0})\Big]\Big), \tag{4.75b}$$

By regrouping the terms, we have the equivalence of order of accuracy for

$$\lambda^{II}_{\alpha0}\partial_{\mathrm{x}k}p_0+\sqrt{\varepsilon}(\lambda^{II}_{\alpha0}\partial_{\mathrm{x}k}p^{II}_{1/2}+\lambda^{II}_{\alpha,1/2}\partial_{\mathrm{x}k}p_0) \sim (\lambda^{II}_{\alpha0}+\sqrt{\varepsilon}\lambda^{II}_{\alpha,1/2})\partial_{\mathrm{x}k}(p_0+$$
$$\sqrt{\varepsilon}p^{II}_{1/2})=(\lambda^{II}_{\alpha0}+\sqrt{\varepsilon}\lambda^{II'}_{\alpha0}s^{II}_{1/2})\partial_{\mathrm{x}k}P^{II}=\lambda^{II}_\alpha(s^{II}_0+\sqrt{\varepsilon}s^{II}_{1/2})\partial_{\mathrm{x}k}P^{II}=$$
$$=\lambda^{II}_\alpha(S^{II})\partial_{\mathrm{x}k}P^{II} \tag{4.76}$$

Then we deduce equations (4.73a, 4.73b) from (4.75a, 4.75b) and (4.76).
2. Equation (4.73c) may be deduced by using relations (4.69) and (4.72a) :

$$0 = p^I_c\Big(S^I-\sqrt{\varepsilon}\frac{1}{\theta}\langle s^I_{1/2}\rangle_1\Big)-\sqrt{\varepsilon}p^{II}_c\Big(S^{II}-\sqrt{\varepsilon}s^{II}_{1/2}\Big)=p^I_c(S^I)-$$
$$-\sqrt{\varepsilon}\frac{1}{\theta}p^{I\prime}_c(S^I)\langle s^I_{1/2}\rangle_1-$$
$$-\sqrt{\varepsilon}\Big(p^{II}_c(S^{II})+\sqrt{\varepsilon}p^{II\prime}_c(S^{II})s^{II}_{1/2}\Big)+o(\varepsilon)\sim$$
$$\sim p^I_c(S^I)-\sqrt{\varepsilon}\frac{1}{\theta}p^{I\prime}_c(s^I_0)\langle s^I_{1/2}\rangle_1-\sqrt{\varepsilon}\Big(p^{II}_c(S^{II})+\sqrt{\varepsilon}p^{II\prime}_c(s^{II}_0)s^{II}_{1/2}+\Big)+\cdots$$

and by substituting in them relations (4.68) and (4.54) :

$$
\begin{aligned}
0 &= p_c^I(S^I) - \varepsilon\left[\frac{1}{\theta}\frac{\lambda_0^I}{\lambda_{\alpha 0}^I\lambda_{\beta 0}^I}\frac{\partial s_0^I}{\partial \tau}\langle\varphi(y)\rangle_1 + p_{c,1/2}^{II}(x,\tau)\right] \\
&- \sqrt{\varepsilon}p_c^{II}(S^{II}) + \varepsilon p_{c,1/2}^{II}(x,\tau) \\
&= p_c^I(S^I) - \sqrt{\varepsilon}p_c^{II}(S^{II}) - \varepsilon\frac{1}{\theta}\frac{\lambda_0^I}{\lambda_{\alpha 0}^I\lambda_{\beta 0}^I}\frac{\partial s_0^I}{\partial \tau}\langle\varphi(y)\rangle_1
\end{aligned}
$$

which coincides with (4.73c).
3. Relation (4.73e) results from (4.54) :

$$
\begin{aligned}
P_c^{II} &= p_{c0} + \sqrt{\varepsilon}p_{c,1/2}^{II} = p_c^{II}(s_0^{II}) + \sqrt{\varepsilon}p_c^{II\prime}(s_0^{II})s_{1/2}^{II} \\
&\sim p_c^{II}(s_0^{II} + \sqrt{\varepsilon}s_{1/2}^{II}) = p_c^{II}(S^{II})
\end{aligned}
$$

Other equations in (4.73a)-(4.73e) may be easily obtained.

4.7. MACROSCALE MODELS OF TWO-PHASE FLOW WITH CAPILLARY FORCES. CAPILLARY RELAXATION

4.7.1. GENERAL MACROSCALE MODEL

The D-T flow can be examined in the same way as in section 4.6. Combining both the results we can deduce a generalized model for flow with considerable capillary forces.

Using these results and returning back to the variable with dimensions, we can present them more convenient to applications:

$$m^{II}(1-\theta)\frac{\partial S^{II}}{\partial t} - \frac{\partial}{\partial X_i}\left(\frac{\widehat{K}^{ik}_\alpha(S^{II},t)}{\mu_\alpha}\frac{\partial \widehat{P}^{II}_\alpha}{\partial X_k}\right) = -m^I\theta\frac{\partial S^I}{\partial t} \tag{4.77a}$$

$$-m^{II}(1-\theta)\frac{\partial S^{II}}{\partial t} - \frac{\partial}{\partial X_i}\left(\frac{\widehat{K}^{ik}_\beta(S^{II},t)}{\mu_\beta}\frac{\partial\left(\widehat{P}^{II}_\alpha+\widehat{P}^{II}_c(S^{II})\right)}{\partial X_k}\right) = $$

$$=m^I\theta\frac{\partial S^I}{\partial t} \tag{4.77b}$$

$$\widehat{P}^I_c(S^I,t) = \widehat{P}^{II}_c(S^{II}) \tag{4.77c}$$

where

$$\widehat{P}^I_c(S^I,t)\equiv P^I_c(S^I) - t_*(S^I)\frac{\partial S^I}{\partial t}P^{I,0}_c, \tag{4.78a}$$

$$\widehat{P}^{II}_c(S^{II})\equiv P^{II}_c(S^{II}) \tag{4.78b}$$

The pressure in blocks is determined as:

$$\widehat{P}^I_\alpha=\widehat{P}^{II}_\alpha+t^*(S^I)\left(1-F_\alpha(S^I)\right)\frac{\partial S^I}{\partial t},\quad F_\alpha(S^I)\equiv\frac{k^I_\alpha(S^I)}{k^I(S^I)+k^I_\beta(S^I)\frac{\mu_\alpha}{\mu_\beta}} \tag{4.79}$$

where θ is the volume fraction of the blocks; P^i_c is the microscale capillary pressure (with dimension) in the medium i, $(i{=}I,II)$; $\widehat{P}^i_c$ is the macroscale (effective) capillary pressure; $\widehat{K}^{ij}_\alpha$ and $\widehat{K}^{ij}_\beta$ are the effective phase permeabilities; $\widehat{P}_\alpha$ is the macroscale pressure in the fluid α; S^I and S^{II} are the macroscale block and matrix saturations.

Effective phase permeabilities $\widehat{K}^{ij}_\alpha$ and $\widehat{K}^{ij}_\beta$, and capillary relaxation time t^* are defined in the next sections.

Equation (4.77c) defines a function:

$$S^I = S^I(S^{II},t) \tag{4.80}$$

which is the solution of the following ordinary differential equation:

$$t_*(S^I)\frac{\partial S^I}{\partial t}P_c^{I,0} + \P_c^I(S^I) = P_c^{II}(S^{II}) \tag{4.81}$$

4.7.2. CAPILLARY RELAXATION PHENOMENON

Capillary relaxation time

In (4.78), the capillary relaxation time t^* determines the characteristic time, after which the saturation on blocks reaches the equilibrium state. It is defined explicitly as

$$t^*(S^I) = \frac{l^2 m^I \mu_\beta}{P_c^{I,0} K^I}\langle\varphi\rangle_1 \frac{\left[k_\alpha^I(S^I)+k_\beta^I(S^I)\frac{\mu_\alpha}{\mu_\beta}\right]}{k_\alpha^I(S^I)k_\beta^I(S^I)} \tag{4.82}$$

where φ is defined as the solution of the Dirichlet cell problem:

$$\begin{cases} \dfrac{\partial}{\partial y_i}\left(K_{ik}^{II}\dfrac{\partial\varphi}{\partial y_k}\right) = -1; \quad y\in Y^I \\ \varphi|_\Gamma = 0 \end{cases} \tag{4.83}$$

l is the dimensional period of heterogeneity, $k_\alpha(s)$ and $k_\beta(s)$ are the micro-scale relative permeabilities.

Solution of Cell Problem

If the block is a sphere of radius ξ and K^{II} is a scalar, the problem (4.83) has an analytical solution: $\varphi(r)=(r^2-\xi^2)/6$, and we obtain then an explicit expression for the value $\langle\varphi\rangle_1$ in (4.82):

$$\langle\varphi\rangle_1 = \frac{2\xi^2}{15}$$

Then:

$$t^*(S^I) = \frac{2\xi^2 l^2 m^I \mu_\beta}{15 P_c^{I,0} K^I}\frac{\left[k_\alpha^I(S^I)+k_\beta^I(S^I)\frac{\mu_\alpha}{\mu_\beta}\right]}{k_\alpha^I(S^I)k_\beta^I(S^I)}, \quad \xi=(3\theta/4\pi)^{1/3} \tag{4.84}$$

In Fig.4.13 the dependence $\tau_*(S)\equiv t^*(S)/t_*$ is drown, with t_* determined according to (4.15), for relative permeabilities $k_\alpha(S)=S^2$, $k_\beta(S)=(1-S)^3$ and $\mu_\alpha/\mu_\beta=1$.

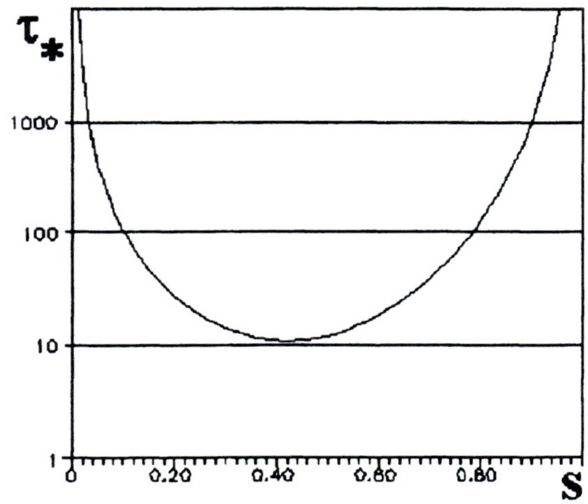

Figure 4.13. Capillary relaxation time vs. block saturation

As seen, the relaxation rate is very small, when the system becomes single phase ($s \to 0$, or $s \to 1$). In these cases any relaxation is stopped (frozen). Nonlinear behavior of the relaxation time shows, that although the formal order of τ_* is $\sqrt{\varepsilon}$, its real order can by sufficiently more large. Hence, the phenomena of phase redistribution in space can be very slow, that corresponds to experimental observation.

Dynamic Capillary Pressure on Blocks

The saturation field in the matrix II is in equilibrium state according to (4.78b), while the block saturation S^I defined by Eqs. (4.77c) and (4.78a) varies in time and follows the relaxation law.

This relaxation is equivalent to a global process of saturation distribution in space caused by capillary forces. According to experimental data, this relaxation is very slow and can proceed some days in a porous sample of a small volume less than 100 cm^3.

Due to this non instantaneous relaxation the macroscale block capillary pressure, $\widehat{P}_c^I$, becomes dependent not only on the block saturation, but on the time also, as seen from (4.78a).

So, the effective capillary pressure curve in dense blocks displays dynamic behavior and is varying in time in such a way that it tends to the microscale curve, $P_c^I(s)$, at $t \to \infty$, as shown in Fig. 4.14.

Note that the capillary pressure in high conductive matrix is in equilibrium state, and is equal to the microscale function $P_c^{II}(s)$, according to (4.78b).

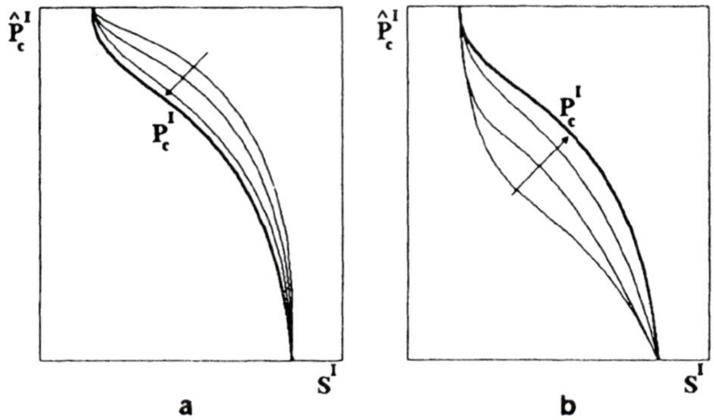

Figure 4.14. Dynamic capillary pressure in blocks when the block saturation grows (a), or decreases (b) in time

Contrarily to moderately heterogeneous media (4.18), the concept of a mean effective capillary pressure for the whole medium becomes useless when the medium is highly heterogeneous. Any unique capillary pressure curve is not able to describe behavior of this system.

Difference Between Matrix and Block Pressures

Relation (4.79) shows that a difference between the matrix and block pressures arises due to capillary phenomena.

It is necessary to note, that in a moderately heterogeneous medium, no difference exists in the matrix and the block pressures for each fluid phase.

The difference $\widehat{P}_\alpha^I - \widehat{P}_\alpha^{II}$ is governed by a kinetic law (4.79) with the same order of relaxation time as in (4.77c).

However, it is necessary to note, that the difference in the phase pressures is small enough, whereas two saturations S^I and S^{II} are considerably distinct.

4.7.3. EXPLANATION OF NON-EQUILIBRIUM BEHAVIOR

To explain the results obtained concerning to capillary relaxation, let us primary remind, the saturation field in the blocks may be considered as quasi steady-state, while the blocks saturation is in non-equilibrium state.

Dependence of the Relaxation Time on the Saturation

A relation of the same type as (4.77c) was first proposed by G.Barenblatt [15], [17], [16] as a general approach to describe non-equilibrium capillary processes, with a constant relaxation time. In our model, the relaxation time is not constant, it depends on the saturation in the blocks as described in

(4.82). To explain the physical meaning of this formula, we, first, note, that the relaxation process happens in the form of an invasion of the α-phase into the blocks. During this invasion, the β-phase has to leave the blocks, hence a counter-flow is forming. The relaxation time is equal to the characteristic length of a block divided by the velocity of the invading phase flow during such a two-phase counter-motion. If the relative permeability for one of the fluids is very small, invasion and flowing out are complicated, therefore filling of the blocks is very slow, and the relaxation time tends to infinity, as shown in Fig. 4.13.

Sense of Capillary Non-Equilibrium Equation. Limit cases

To understand the sense of equation (4.77c), or its dimensionless analog (4.73c), we may examine limiting cases, taking into account the following limitation for the capillary number in blocks Pe^I and for the heterogeneity degree ω_K which are true within the framework of the model (4.77a) - (4.78):

$$0 \leq Pe^I \leq 1, \qquad \varepsilon^2 \leq \omega_K \leq 1$$

The first condition says that capillary forces are important, the second one determines that the medium is highly heterogeneous but not too highly.

Examine formal limiting cases: $Pe^I \to 0$, $Pe^I \to 1$ and $\varepsilon^2/\omega_K \to 0$ assuming the flow structure remains invariant.

1. When $Pe^I=0$, i.e. the capillary imbibition in the blocks, the formal limiting equation resulting from (4.77c) is then:

$$\widehat{P}^I_c(s^I)=0 \tag{4.85}$$

That is an effective condition of capillary equilibrium in the medium when the capillary forces in the blocks are infinitely greater than in the matrix. On practice, highly contrasted capillary properties are always associated to a highly heterogeneous field of permeability.

It means that the block saturation remains close to one at any time, according to the property of the capillary pressure curve. The blocks are always totally filled by the α-phase. If the porous medium was saturated by the β-phase at the initial state, the blocks must be instantly filled by another fluid. Such a system is oversimplified, thus, the problem is to take in account the transient process. It is clear, to keep transient process we should assume the value of Pe^I is small, although it differs from zero. The corresponding result is the equation (4.77c) or its dimensionless analog (4.73c).

2. When we formally assume $Pe=1$, equation (4.73c) will be not changed, with keeping relaxation term, without any changes in the definition of the relaxation time $\tau_* \sim \varepsilon_p$.

3. When only we assume the medium is moderately or weakly heterogeneous, i.e. the second parameter $\varepsilon_p \rightarrow 0$, or $\varepsilon^2/\omega_K \rightarrow 0$, a new capillary equilibrium equation may be obtained from (4.73c):

$$\frac{1}{Pe} p_c^I(S^I) - p_c^{II}(S^{II}) = 0$$

or, in dimensional form:

$$P_c^I(S^I) = P_c^{II}(S^{II}) \tag{4.86}$$

that means a standard situation examined by various authors, for instance [58], [3], [21], when the blocks and the matrix permeabilities are not contrasted.

Thus, the deduced macroscale equations describe transient behavior of a system which is tending to one of two probable equilibrium states defined by limiting equations (4.85), or (4.86), which are appropriated to a permeability field of very high heterogeneity, or of moderate heterogeneity, depending on the order of the parameters Pe and ε_p.

It is necessary to note, that the value of the parameter ε_p is not very important for the goals of our research. In mathematical literature the case $\varepsilon_p \sim 1$ has been selected in [9] as a special case of the double porosity media, leading to a macroscale model corresponding to the zero approximation with respect to ε. However, this result concerns only the compressible non-stationary flow (parabolic equations with respect to the fluid pressure), when the macro-scale behavior appears to be unlocal in time. When the fluid is non-compressible (elliptic equation respectively to pressure), the pressure field is quasi steady-state, then the case $\varepsilon_p \sim 1$ does not give any special. Herein we were studying non-compressible flow and the exact value of the parameter ε_p is not really important as soon as it has been chosen in the range satisfying the general condition of double porosity media.

4.7.4. NONLINEAR RELAXATION

Due to the chose of parameters the formal order of the capillary relaxation time is rather small. In a linear system this would mean the weak non-equilibrium, or the equilibrium state is quickly reached. In non-linear case the situation is more complicated.

Let some process of water penetration into the blocks be examined, such that the water saturation S^I monotonously increases in time. If the process is governed by law (4.77c) – (4.78a) , then variation of S^I happens in such a way as shown in Fig. 4.15.

The "quasi-linear stage" corresponds to the central part of the function $t^*(S^I)$ imaged in Fig. 4.16, when the relaxation time is most small. However

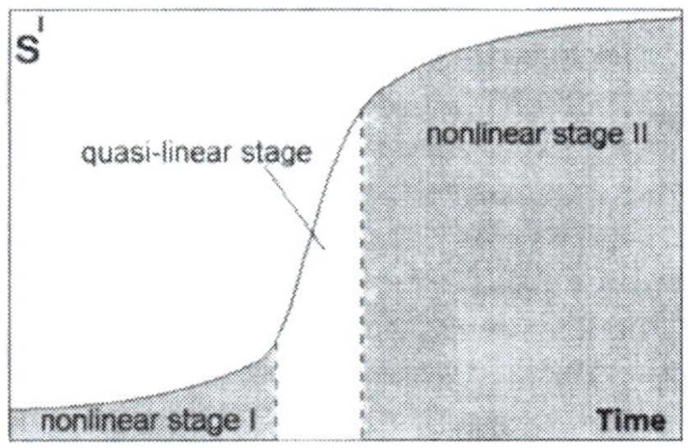

Figure 4.15. Nonlinear relaxation of the block saturation

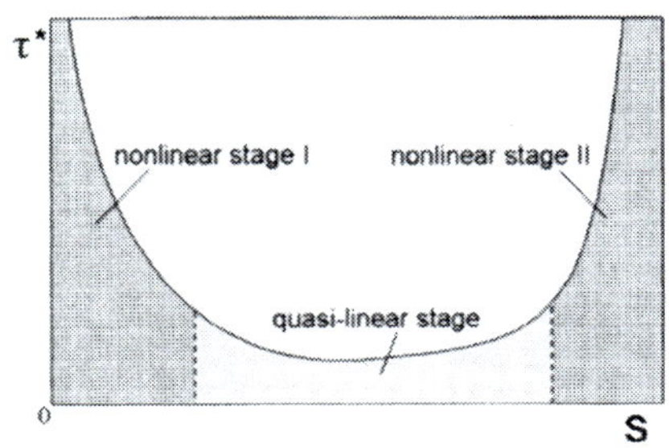

Figure 4.16. Quasi-linear and nonlinear parts of capillary relaxation time

the relaxation of the sunction S^I includes also two nonlinear stages which correspond to those parts of the function $t^*(S^I)$ where the relaxation time is fast and considerably varying. Due to these nonlinear stages the total relaxation process appears very long.

4.8. EFFECTIVE PHASE AND RELATIVE PERMEABILITIES

4.8.1. EFFECTIVE PHASE PERMEABILITIES

General Definition

For each phase γ, with $\gamma=\alpha,\beta$, the effective phase permeability tensor $\widehat{K}^{ij}_{\gamma}$ is determined in the following way:

$$\widehat{K}^{ij}_{\gamma}\left(S^{II},S^{I}\right)=\widehat{K}^{II}_{ij}k^{II}_{\gamma}\left(S^{II}\right)+\widehat{K}^{I}_{ij}k^{I}_{\gamma}\left(S^{I}\right)+\widehat{K}^{II,I}_{ij}k^{II}_{\gamma}\left(S^{II}\right) \quad (4.87a)$$

where three constant tensors are:

$$\widehat{K}^{II}_{ij}=\left\langle K^{II}_{ik}\left(\frac{\partial\psi^{II}_{j}}{\partial y_k}+\delta_{kj}\right)\right\rangle_2 \quad (4.87b)$$

$$\widehat{K}^{I}_{ij}=\left\langle K^{I}_{ik}\left(\frac{\partial\psi^{I}_{j}}{\partial y_k}+\delta_{kj}\right)\right\rangle_1 \quad (4.87c)$$

$$\widehat{K}^{II,I}_{ij}=\left\langle K^{II}_{ik}\left(\frac{\partial\psi^{II,I}_{j}}{\partial y_k}\right)\right\rangle_2 \quad (4.87d)$$

and three types of cell functions $\psi^{II}_k(y)$, $\psi^{I}_k(y)$ and $\psi^{II,I}_k(y)$ are solutions of the following cell problems:

$$\begin{cases} \dfrac{\partial}{\partial y_i}\left(K^{II}_{ij}\left(\dfrac{\partial\psi^{II}_k}{\partial y_j}+\delta_{kj}\right)\right)=0, \quad y\in Y^{II} \\[2ex] K^{II}_{ij}\left(\dfrac{\partial\psi^{II}_k}{\partial y_j}+\delta_{kj}\right)n_i\Big|_{\Gamma}=0 \\[2ex] \langle\psi^{II}_k\rangle_2=0, \qquad \psi^{II}_k \text{ is } y\text{-periodic} \end{cases} \quad (4.88)$$

$$\begin{cases} \dfrac{\partial}{\partial y_i}\left(K^{I}_{ij}\left(\dfrac{\partial\psi^{I}_k}{\partial y_j}+\delta_{kj}\right)\right)=0, \quad y\in Y^{I} \\[2ex] \psi^{I}_k\big|_{\Gamma}=\psi^{II}_k\big|_{\Gamma} \end{cases} \quad (4.89)$$

$$\begin{cases} \dfrac{\partial}{\partial y_i}\left(K_{ij}^{II}\dfrac{\partial \psi_k^{II,I}}{\partial y_j}\right) = 0, \quad y \in Y^{II} \\ K_{ij}^{II}\dfrac{\partial \psi_k^{II,I}}{\partial y_j} n_i|_\Gamma = K_{ij}^{I}\left(\dfrac{\partial \psi_k^{I}}{\partial y_j} + \delta_{kj}\right) n_i|_\Gamma \\ \langle \psi_k^{II,I} \rangle_2 = 0, \qquad \psi_k^{II,I} \text{ is } y\text{-periodic} \end{cases} \tag{4.90}$$

Note, when both the saturations are equal to 0 or 1 we obtain an analogous relation for the effective global permeability from (4.87a) :

$$\widehat{K}_{ij} = \widehat{K}_{ij}^{II} + \widehat{K}_{ij}^{I} + \widehat{K}_{ij}^{II,I} \tag{4.91}$$

Physical Sense of Eq. (4.87)

Eq. (4.87) presents the effective phase permeability in a splitted form, where some constant coefficients are multiplied by some functions of saturation. The physical sense of this factors and functions is easy to be explained.

The constant tensor $\widehat{K}_{ij}^{II}$ is the effective global permeability of the highly permeable matrix II only, i.e., it corresponds to the case when the blocks are assumed to be impenetrable.

The constant tensor $\widehat{K}_{ij}^{I}$ is an additive contribution to the effective global permeability of the medium caused by translation type of flow through the blocks. Note that $\widehat{K}_{ij}^{I}$ is not equal to the effective permeability of blocks, as the averaged permeability can not be obtained by summation of the permeabilities of all parts.

The constant tensor $\widehat{K}_{ij}^{II,I}$ is an additional contribution to the effective global permeability caused by the secondary influence of the translation flow through the blocks on the flow through the matrix II.

These values are completely equivalent to A_{ij}^{II}, A_{ij}^{I} and $A_{ij}^{II,I}$ in the definition of the effective permeability introduced in chapter 1.

Functions k_γ^I and k_γ^{II} are the microscale relative permeabilities for the fluid γ.

Tensor properties of effective phase permeabilities

The effective phase permeabilities $\widehat{K_\alpha^{ij}}$ and $\widehat{K_\beta^{ij}}$ keep the same tensor properties as the effective global permeability $\widehat{K}_{ij}$ (4.91) . In particular, the following is easy to be proved.

Diagonal case 1

Let three conditions be satisfied:

- the block shape is uniformly symmetric respectively to all the axis;

- the microscale permeability tensor is diagonal, i.e. $K^{II}_{ij}(y)\equiv 0$, $K^{I}_{ij}(y)\equiv 0$, $\forall i\neq j$;

- the microscale permeabilities $K_{ii}(y)$ are even with respect to each axis.

Then the effective phase permeabilities are diagonal:

$$\widehat{K^{ij}_{\alpha}}\equiv 0, \qquad \widehat{K^{ij}_{\beta}}\equiv 0; \qquad \forall i\neq j \tag{4.92}$$

Certainly, the effective global permeability $\widehat{K_{ij}}$ is also diagonal.

Diagonal case 2

Let three conditions be satisfied:

- the block shape is symmetric respectively to each axis;

- the microscale permeability is scalar, i.e. $K^{II}_{ij}(y)=K^{II}(y)\delta_{ij}$, $K^{I}_{ij}(y)=K^{I}(y)\delta_{ij}$, $\forall i,j$;

- the microscale permeabilities $K^{II}(y)$ and $K^{I}(y)$ are even with respect to each axis.

Then the effective phase permeabilities are diagonal, i.e., (4.92) is true too.

Scalar case

Let in the last case the block shape be uniformly symmetric respectively to all the axis.

Then the effective phase permeabilities are scalar:

$$\widehat{K^{ij}_{\alpha}}(S^{II},t)=\widehat{K_{\alpha}}(S^{II},t)\delta_{ij}, \qquad \widehat{K^{ij}_{\beta}}(S^{II},t)=\widehat{K_{\beta}}(S^{II},t)\delta_{ij}; \quad \forall i,j \tag{4.93}$$

The "Diagonal case 1" shows that the anisotropy of the microscale permeability field and the asymmetric form of the blocks are not sufficient to arising of non-diagonal terms in the macroscale permeability tensor.

The "Diagonal case 2" shows that the anisotropy of the macroscale permeability field may be produced only by an asymmetric shape of the blocks, although the microscale permeability is isotropic.

The "Scalar case", i.e. the scalar macroscale field may be obtained only when the microscale permeability is isotrope, and the block shape is uniformly symmetric.

4.8.2. EFFECTIVE RELATIVE PERMEABILITIES

On practice of reservoir engineering, the relative permeabilities are more widely used than the phase permeabilities $\widehat{K^{ij}_{\alpha}}$ and $\widehat{K^{ij}_{\beta}}$ introduced in (4.87) . In scalar case, the relative permeability for the phase α (or β) is determined as $k_{\alpha}(s)=K_{\alpha}(s)/K$, where K is the global permeability. When the effective phase permeability $\widehat{K^{ij}_{\alpha}}$ and the effective global permeability $\widehat{K_{ij}}$ are tensors, definition of relative permeability becomes not obvious.

Definition of the Tensor Effective Relative Permeability
In general tensor case, the effective relative permeability $\widehat{k}_\alpha$ or $\widehat{k}_\beta$ may be introduced only via a convolution of two tensors, i.e.

$$\widehat{K}_\gamma^{ij}(S^{II},t)=\widehat{K}_{ik}\widehat{k}_\gamma^{kj}(S^{II},t), \quad \gamma=\alpha,\beta; \quad i,j=1,2,3 \tag{4.94}$$

In R^n, for a fixed fluid phase γ, Eqs. (4.94) constitute a system of n^2 equations for n^2 components $\left\{\widehat{k}_\gamma^{kj}\right\}_{k,j=1}^{n}$ of a tensor function of second rank. This system is closed and the relative permeabilities are completely defined. For instance, in R^2 this system takes the form:

$$\begin{aligned} \widehat{K}_{11}\widehat{k}_\gamma^{11} + \widehat{K}_{12}\widehat{k}_\gamma^{21} &= \widehat{K}_\gamma^{11} \\ \widehat{K}_{11}\widehat{k}_\gamma^{12} + \widehat{K}_{12}\widehat{k}_\gamma^{22} &= \widehat{K}_\gamma^{12} \\ \widehat{K}_{21}\widehat{k}_\gamma^{11} + \widehat{K}_{22}\widehat{k}_\gamma^{21} &= \widehat{K}_\gamma^{21} \\ \widehat{K}_{21}\widehat{k}_\gamma^{12} + \widehat{K}_{22}\widehat{k}_\gamma^{22} &= \widehat{K}_\gamma^{22} \end{aligned} \tag{4.95}$$

When the medium is highly heterogeneous, the system (4.94) has an analytical solution. It may be deduced using decomposition (4.87) , (4.91) for the effective phase and global permeabilities, and taking into account that all the values $\widehat{K}_{ij}^{I}$ and $\widehat{K}_{ij}^{II,I}$ are of order ω_K. Then at $\omega_K<<1$ we can obtain from (4.94) and (4.87) for any phase $\gamma=\alpha,\beta$:

$$\widehat{k}_\gamma^{ij}(S^{II},S^{I}) \simeq k_\gamma^{II}(S^{II}) + \xi_\gamma^{ij}(S^{II},S^{I}), \quad i,j=1,2,3 \tag{4.96}$$

where the components of the tensor ξ_γ^{ij} are solutions of a linear algebraic system:

$$\widehat{K}_{ik}^{II}\xi_\gamma^{kj}(S^{II},S^{I}) = \widehat{K}_{ij}^{I}\left[k_\gamma^{I}(S^{I}) - k_\gamma^{II}(S^{II})\right] \tag{4.97}$$

It follows that the macroscale relative permeabilities do not depend on the tensor $\widehat{K}_{ij}^{II,I}$, although the phase permeabilities $\widehat{K}_\gamma^{ij}$ depend on it.

Diagonal Effective Relative Permeabilities
As it follows from the "diagonal case 1" and the "diagonal case 2", if the effective phase permeabilities are diagonal, then the microscale global permeability is diagonal too. In this case it follows from (4.94) that the effective relative permeabilities are also diagonal:

$$\widehat{K}_\alpha^{ii} = \widehat{K}_{ii}\widehat{k}_\alpha^{ii}, \qquad \widehat{K}_\beta^{ii} = \widehat{K}_{ii}\widehat{k}_\beta^{ii}; \quad i=1,2,3 \tag{4.98}$$

In these formulas no of summation is made over i.

Using (4.87) we can represent the relative permeability in the explicit form for each fluid γ:

$$\widehat{k}_\gamma^{ii}(S^{II},t) = \frac{\widehat{K}_{ii}^{II} k_\gamma^{II}(S^{II}) + \widehat{K}_{ii}^{I} k_\gamma^{I}(S^{I}) + \widehat{K}_{ii}^{II,I} k_\gamma^{II}(S^{II})}{\widehat{K}_{ii}^{II} + \widehat{K}_{ii}^{I} + \widehat{K}_{ii}^{II,I}}, \quad i=1,2,3 \quad (4.99)$$

(no of summation is made over i).

This relation may be simplified, if we take into account that $\widehat{K}_{ii}^{II}$ and $\widehat{K}_{ii}^{I}$ are of order ω_K. Using asymptotic expansion of (4.99) at $\omega_K \rightarrow 0$, we obtain for each fluid γ:

$$\widehat{k}_\gamma^{ii}(S^{II},t) \simeq k_\gamma^{II}(S^{II}) + \frac{\widehat{K}_{ii}^{I}}{\widehat{K}_{ii}^{II}} \left[k_\gamma^{I}(S^{I}) - k_\gamma^{II}(S^{II}) \right], \quad i=1,2,3 \quad (4.100)$$

(no of summation over i).

This relation can be deduced directly from Eqs. (4.96) and (4.97) .

Scalar Effective Relative Permeability

Scalar effective relative permeability can be obtained in two cases.

The first one is rather trivial. When the effective global permeability field is scalar, then the effective relative permeability is scalar too. In other words, if

$$\widehat{K}_{ij}^{II} = \widehat{K}^{II} \delta_{ij}, \quad \widehat{K}_{ij}^{I} = \widehat{K}^{I} \delta_{ij}$$

then

$$\widehat{k}_\gamma^{ij}(S^{II}, S^{I}) = \widehat{k}_\gamma(S^{II}, S^{I}) \delta_{ij}$$

where

$$\widehat{k}_\gamma(S^{II}, S^{I}) \simeq k_\gamma^{II}(S^{II}) + \frac{\widehat{K}^{I}}{\widehat{K}^{II}} \left[k_\gamma^{I}(S^{I}) - k_\gamma^{II}(S^{II}) \right] \quad (4.101)$$

In the second case, the effective global permeability may be a tensor, but all the ratios $\widehat{K}_{ij}^{I}/\widehat{K}_{ij}^{II}$ for all $i,j=1,2,3$ must be equivalent. Thus, if

$$\frac{\widehat{K}_{ij}^{I}}{\widehat{K}_{ij}^{II}} = \nu = const, \quad \forall i,j \quad (4.102)$$

(no of summation is made over repeating indexes), then:

$$\widehat{k}_\gamma(S^{II}, S^{I}) \simeq k_\gamma^{II}(S^{II}) + \nu \left[k_\gamma^{I}(S^{I}) - k_\gamma^{II}(S^{II}) \right] \quad (4.103)$$

Property (4.102) will be called further as the *"uniform anisotropy"*.

4.8.3. SPLITTING THE SATURATION AND HETEROGENEITY

One of the most important problems of up-scaling two-phase flow is that the standard algorithms performing calculation of phase permeability is very time expensive. More strictly speaking, in the classical models [58], [56] the cell problem defines some intermediary functions like $\psi_k^\alpha(y;S)$ and $\psi_k^\beta(y;S)$ in (4.6) , (4.7) , which depend on the saturation S as the parameter. It means that the cell problem has to be computed for each value of saturation S. The cell problem being solved once gives only one point of the effective relative permeability curve. The problem is how to extract the saturation from the algorithm of phase permeability calculation.

In other words, this fundamental problem of scale-up theory is formulated as a problem of "splitting saturation and heterogeneity geometry", or excluding the saturation from the cell problem.

Eqs. (4.87) , (4.96) , and (4.103) show that the problem is automatically solved in case of highly heterogeneous medium. All cell problems (4.88) - (4.90) do not include saturation. One solution of all the cell problems gives the whole effective phase permeability vs. saturation curve.

This is reached due to splitting the functions dependent on saturation, as microscale relative permeabilities, $k_\gamma^I(s)$ and $k_\gamma^{II}(s)$, $(\gamma{=}\alpha,\beta)$, and the constant factors dependent on heterogeneity geometry, as $\widehat{K}_{ij}^{II}$ and $\widehat{K}_{ij}^{I}$.

Splitting of phase permeability (4.87) remains true only for highly heterogeneous media, i.e. when $\omega_K{<<}1$, because it can be considered as an asymptotic expansion of the problem (4.19) - (4.20) over ω_K when $\omega_K{\to}0$.

However calculation shown validity of Eq. (4.87) even when $\omega_K{\simeq}0.5-0.6$.

4.8.4. NON-EQUILIBRIUM BEHAVIOR OF RELATIVE PERMEABILITY

According to (4.87) , or (4.96) , each phase permeability is a function of two saturations, S^{II} and S^I. Due to Eq. (4.81) defining the dependence between two saturations (4.80) as $S^I{=}S^I(S^{II},t)$, it follows that Eqs. (4.87) and (4.81) describe "dynamic" phase permeabilities, while (4.96) and (4.81) describe "dynamic" phase permeabilities:

$$\widehat{K}_\alpha^{ij}(S^{II},S^I)=\widehat{K}_\alpha^{ij}(S^{II},t),\qquad \widehat{K}_\beta^{ij}(S^{II},S^I)=\widehat{K}_\beta^{ij}(S^{II},t)$$

$$\widehat{k}_\alpha^{ij}(S^{II},S^I)=\widehat{k}_\alpha^{ij}(S^{II},t),\qquad \widehat{k}_\beta^{ij}(S^{II},S^I)=\widehat{k}_\beta^{ij}(S^{II},t)$$

or:

$$\begin{cases} \widehat{k}_\gamma^{ij}(S^{II},S^I) = k_\gamma^{II}(S^{II}) + \xi_\gamma^{ij}(S^{II},S^I), \quad i,j{=}1,2,3 \\ \widehat{K}_{ik}^{II}\xi_\gamma^{kj}(S^{II},S^I) = \widehat{K}_{ij}^{I}\left[k_\gamma^I(S^I) - k_\gamma^{II}(S^{II})\right], \\ t_*(S^I)\dfrac{\partial S^I}{\partial t}P_c^{I,0} - P_c^I(S^I) = -P_c^{II}(S^{II}) \end{cases} \tag{4.104}$$

In scalar case, this system becomes more simple:

$$\begin{cases} \widehat{k}_\gamma^{ij}(S^{II},S^I) = k_\gamma^{II}(S^{II}) + \nu\left[k_\gamma^I(S^I) - k_\gamma^{II}(S^{II})\right], \\ t_*(S^I)\dfrac{\partial S^I}{\partial t}P_c^{I,0} - P_c^I(S^I) = -P_c^{II}(S^{II}) \end{cases} \tag{4.105}$$

Fig.4.17 shows the behavior of two effective relative permeabilities in time. The dashed lines correspond to the original microscale relative permeabilities, k_α^I, k_α^{II}, k_β^I and k_β^{II}. The solid lines represent the effective relative permeability. the horizontal axis corresponds to the macroscale matrix saturation S^{II}, where the flesh shows the direction of time growth.

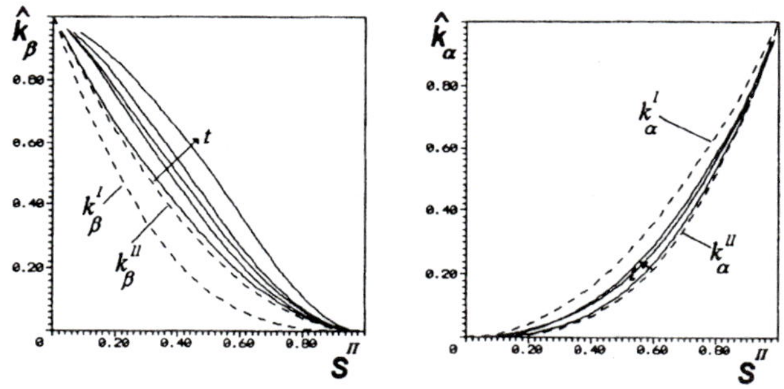

Figure 4.17. Dynamic behavior of the effective relative permeabilities. Results of calculation

These figures have been computed using (4.78a) in the following way. The exponential grow of the saturation S^{II} was accepted as the given information, which can be associated to some process of injection of the α-phase into the medium. The parameter ν was calculated basing on solution of the cell problem as will be explained in the section 4.9. The initial saturation S^I is any.

It seen, that the effective relative permeability is equal to the relative permeability of the matrix at the initial time moments, but it grows in time due to the increasing contribution of the blocks.

4.9. NUMERICAL SIMULATION OF EFFECTIVE RELATIVE PERMEABILITY

4.9.1. NUMERICAL SOLUTION OF THE CELL PROBLEMS

Splitting (4.87) , (4.96) is a very powerful tool to compute the effective phase permeabilities. Along with "extraction" of saturation described above it leads to three types of cell problems each of them is linear and is posed in homogeneous sub-domain.

Reduction of a periodic problem to the boundary-value

Problems (4.88) and (4.90) comprise periodic conditions, while standard numerical methods are developed for boundary-value problems. In general case, solution of a problem with periodic conditions is performed basing on penalty method, Fourier series and domain decomposition method. In the penalty method [19] the periodic problem is replaced by an equivalent sequence of boundary-value problems. Expansion of the solution into Fourier series, applied for instance in [59], is powerless for perforated structures. The method of domain decomposition was used in [26]. In particular cases, when some symmetry of heterogeneity is observed, the periodic problem may be replaced by a single equivalent boundary-value problem. Such a reduction have been described in [13].

In this section we will show the problems (4.88) and (4.90) may be transformed into a boundary-value in more general case, when the symmetry is required with respect to one coordinate axis only.

Let us consider a periodic problem in Y, where Y is a square in R^2, i.e., $Y = -1/2 < y_i < 1/2, \quad i = 1,2$:

$$\frac{\partial}{\partial y_i}\left(K_{ij}\left(\frac{\partial \psi_1}{\partial y_j} + \delta_{1j}\right)\right) = 0, \quad y \in Y \tag{4.106a}$$

$$\langle \psi_1 \rangle_2 = 0 \tag{4.106b}$$

$$\psi_1 \text{ is } y\text{-periodic} \tag{4.106c}$$

and a boundary-value problem in Y:

$$\frac{\partial}{\partial y_i}\left(K_{ij}\left(\frac{\partial \eta}{\partial y_j} + \delta_{1j}\right)\right) = 0, \quad y \in Y \tag{4.107a}$$

$$\eta\big|_{y_1=-1/2} = \eta\big|_{y_1=1/2} = \eta^* \tag{4.107b}$$

$$\frac{\partial \eta}{\partial y_2}\Big|_{y_2=-1/2} = \frac{\partial \eta}{\partial y_2}\Big|_{y_2=1/2} = 0 \tag{4.107c}$$

with any constant value η^*.

Problem (4.107) is correctly formulated, see [13].

Proposition:

- When the tensor K_{ij} is diagonal, i.e., K_{ij}=0, $i \neq j$, and both the functions $K_{ii}(y)$ are even with respect to y_2, then:

$$\psi_1(y) = \eta(y) + C, \quad \text{where} \quad C \equiv - \langle \eta \rangle \tag{4.108}$$

The structure of the cell may be any, only the symmetry with respect to the axis y_1 must take place, as shown in Fig. 4.18. while the symmetry

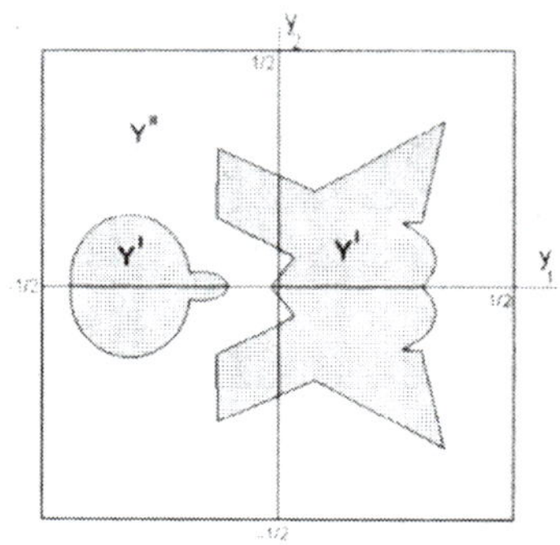

Figure 4.18. Probable structure of the heterogeneity cell

along the axis y_2 is not required.

In fact, due to symmetry of the parameters with respect to the axis y_1, the solution $\eta(y)$ is symmetric too, i.e., it is even with respect to y_2. Therefore periodicity condition (4.106c) is satisfied along the axis y_2. Due to condition (4.107b) , the function $\eta(y)$ is periodic along the axis y_1 too.

The function $\eta(y)$ satisfies Eq. (4.107a) and is periodic. Then the function $\psi_1 = \eta - \langle \eta \rangle$ satisfies Eq. (4.107a) , is y-periodic and has zero mean value in Y, i.e. satisfies the problem (4.106) . The proposition is proved.

Certainly, the constant value C in (4.108) is related to the boundary value η^* in (4.107) . Our proposition does not give this connection, but it is not necessary, because the effective permeability is defined via derivatives of the functions $\psi_k(y)$, as follows from (4.87) and (4.19) .

Thus to compute effective permeability in (4.19) , or in (4.81) , it is sufficient to solve the boundary-value problem (4.107) with any η^* instead of periodic problem (4.106) , when the condition of symmetry along one of axis is satisfied.

The function $\Psi_1=\eta(y)+y_1$ has meaning of a dimensionless pressure within a cell and satisfies the following problem:

$$\begin{aligned}
&\frac{\partial}{\partial y_i}\left(K_{ij}\frac{\partial \Psi_1}{\partial y_j}\right)=0, \quad y\in Y\\
&\Psi_1\Big|_{y_1=-1/2}=-\frac{1}{2}+\eta^*\\
&\Psi_1\Big|_{y_1=1/2}=\frac{1}{2}+\eta^*\\
&\frac{\partial \Psi_1}{\partial y_2}\Big|_{y_2=\pm 1/2}=0
\end{aligned} \tag{4.109}$$

We see the pressure drop along the axis y_1 is equal to 1, i.e., $\Delta\Psi_1=(-1/2+\eta^*)-(-1/2+\eta^*)=1$ and do not depend on η^*. In fact, the structure of flow depends only on the pressure drop and does not depends on the shift η^* of the whole pressure field.

Note that the second function, $\psi_2(y)$ can not be computed in the same way, when only the symmetry along with respect to the axis y_1 holds.

The similar theorem can be proved for the problems (4.88) and (4.90) posed in Y^{II} only. But in this case, for the function ψ_1^{II} the form of the internal boundary Γ must be also symmetric with respect to the axis y_1.

Solution of Cell Problems (4.88) - (4.90)

All the cell problems have been computed using standard solvers based on finite-difference or finite-element methods.

It is necessary to note that cell problems (4.88) - (4.90) defining the macroscale relative permeabilites are the same as problems (1.103a) - (1.103c) deduced in chapter 1 for the macroscale absolute permeability.

A symmetric cell structure was examined with a rectangular block as shown in Fig. 4.19.

Microscale permeability has been assumed to be a scalar function, i.e. $K_{ij}^{II}=K^{II}\delta_{ij}$, $K_{ij}^{I}=K^{I}\delta_{ij}$.

Solution of the problem (4.88) is shown in the left-hand part of Fig. 1.22 where the function $\Psi_1^{II}(y)=\psi_1^{II}(y)+y_1$ (b) is presented in a quarter of the cell Y.

Function Ψ_1^{II} is a dimensionless local pressure field within the subdomain Y^{II} of a cell, and, therefore, has more physical sense, than the function ψ_1^{II}.

Solutions $\Psi_1^{I}(y)=\psi_1^{I}(y)+y_1$ of cell problem (4.89) is shown in the right-hand part of Fig. 1.22.

Solution of problem (4.90) is displayed in Fig. 1.23.

Behavior of each cell function on all parts of the cell Y may be reconstructed basing on the symmetry properties. In particular, all the functions

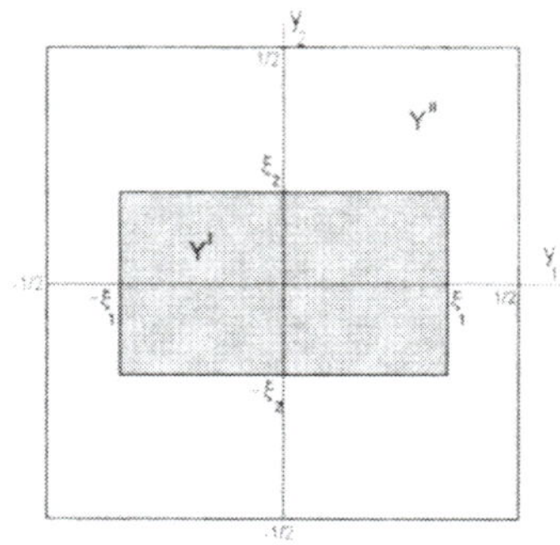

Figure 4.19. Structure of a cell used for numerical solution of the cell problems

$\psi_1^{II}(y)$, $\psi_1^{I}(y)$ and $\psi_1^{II,I}(y)$ are even with respect to the variables y_2 and y_3, but they are odd with respect to the variable y_1.

4.9.2. CALCULATION OF THE EFFECTIVE RELATIVE PERMEABILITIES. METHOD OF SPLITTING THE SATURATION AND HETEROGENEITY

In this section we show some examples of the effective relative permeabilities computed basing on equations (4.87) , or its sequences (4.96) , (4.100) , (4.101) , or (4.103) .

We present the results referring to the large values of time, after the macroscale capillary equilibrium state has reached. Transient behavior in time is shown in section 4.8.4.

The solutions of the cell problems obtained above have been used to perform integration according to (4.91) . The scalar case (4.101) is investigated. The original microscale relative permeabilities are shown in Fig. 4.20:

In these figures the solid lines correspond to the method of splitting the saturation and heterogeneity, while the dashed lines are obtained by solving the full, non-splitted, cell problem deduced in classical theory, [58], [56], and presented in the form of (4.6) and (4.7) . Several cases have been examined with various volume fraction of blocks θ and various heterogeneity degree ω_K.

Three figures 4.21 show the effective relative permeabilities for a rather small volume fraction of blocks, θ=0.3.

The next three figures 4.22 correspond to θ=0.5.

The next three figures 4.23 correspond to θ=0.7.

The last three figures 4.24 show behavior of the effective relative permeability when the volume fraction of blocks is very high, θ=0.9.

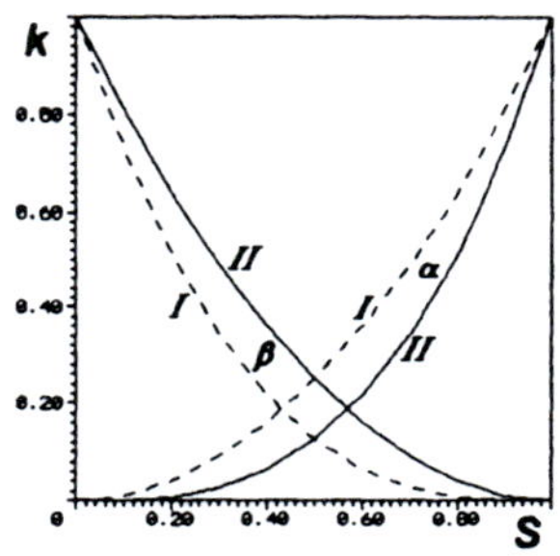

Figure 4.20. Microscale relative permeabilities used in calculations

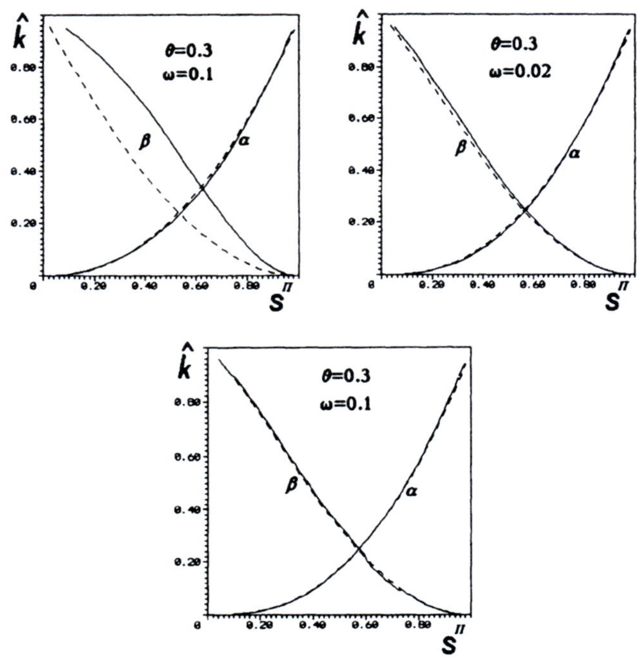

Figure 4.21. Effective relative permeabilities for a highly heterogeneous medium; the volume fraction of blocks is $\theta=0.3$

It is necessary to note that the effective relative permeability curve is not obligatory located between two original microscale curves correspondent to the matrix and to the blocks. However it is always above the microscale relative permeability for the highly permeable matrix due to the contribution of blocks. When the microscale relative permeability of the matrix is higher than that of the blocks, the effective relative permeability curve is

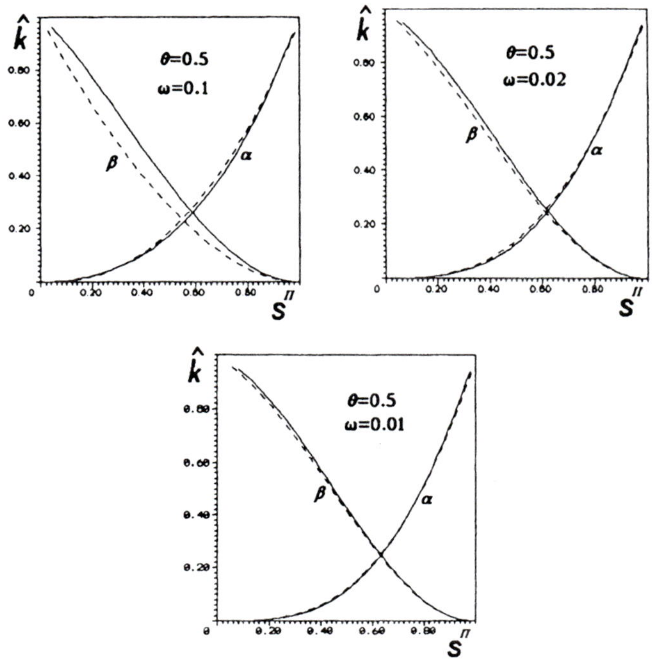

Figure 4.22. Effective relative permeabilities for a highly heterogeneous medium; the volume fraction of blocks is θ=0.5

located above both the microscale curves.

The method of splitting the saturation anf heterogeneity gives satisfactory results in various ranges of the parameter ω_K depending on the reciprocal location of the microscale relative permeability curves.

In case when the microscale relative permeability of the matrix is under that of the blocks, the method of splitting gives the good results for any ω_K<0.6 and any θ. In our example, this situation has been chosen for the α-phase.

In case when the microscale relative permeability of the matrix is above that of the blocks, the method of splitting gives the good results for ω_K<0.08 and any θ. In our example, this situation has been chosen for the β-phase.

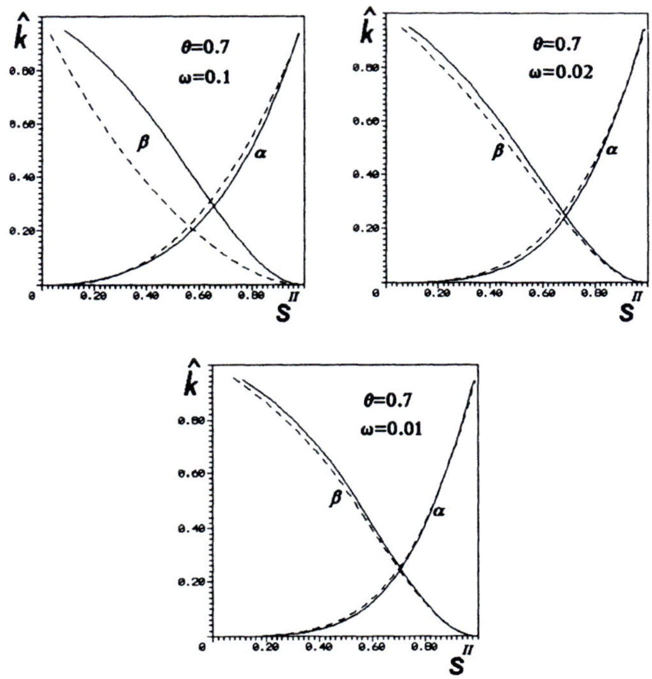

Figure 4.23. Effective relative permeabilities for a highly heterogeneous medium; the volume fraction of blocks is θ=0.7

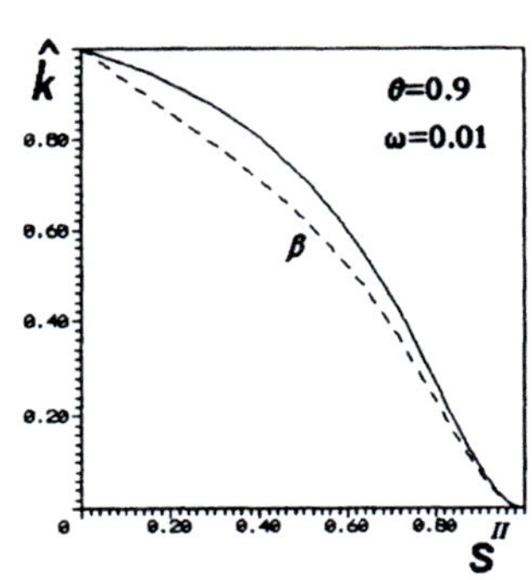

Figure 4.24. Effective relative permeabilities for a highly heterogeneous medium; the volume fraction of blocks is θ=0.9

4.10. MACROSCALE MODELS FOR PARTICULAR STRUCTURES OF TWO-PHASE FLOW

4.10.1. MACROSCALE MODEL FOR D-S SYSTEM

The model (4.77a) - (4.78) covers both the types of capillary dominating flow, i.e., D-S and D-T flow. For each of these cases the model can be simplified.

The model for D-S flow , i.e., when $\omega_K < \varepsilon$, takes the form:

$$m^{II}(1-\theta)\frac{\partial S^{II}}{\partial t} - \frac{\partial}{\partial X_i}\left(\frac{\widehat{K}_\alpha^{ik}(S^{II})}{\mu_\alpha}\frac{\partial \widehat{P}_\alpha^{II}}{\partial X_k}\right) = -m^I\theta\frac{\partial S^I}{\partial t} \tag{4.110a}$$

$$-m^{II}(1-\theta)\frac{\partial S^{II}}{\partial t} - \frac{\partial}{\partial X_i}\left(\frac{\widehat{K}_\beta^{ik}(S^{II})}{\mu_\beta}\frac{\partial\left(\widehat{P}_\alpha^{II}+\widehat{P}_c^{II}(S^{II})\right)}{\partial X_k}\right) = =m^I\theta\frac{\partial S^I}{\partial t} \tag{4.110b}$$

$$\widehat{P}_c^I(S^I,t) = \widehat{P}_c^{II}(S^{II}) \tag{4.110c}$$

$$\widehat{P}_c^I(S^I,t) \equiv P_c^I(S^I) + t_*(S^I)\frac{\partial S^I}{\partial t}P_c^{I,0}, \tag{4.110d}$$

$$\widehat{P}_c^{II}(S^{II}) \equiv P_c^{II}(S^{II}) \tag{4.110e}$$

The relaxation time $t^*(S^I)$ is defined in the similar way as (4.82) , or (4.84) .

Effective Relative Permeabilities for D-S Flow

Effective phase permeabilities are defined as:

$$\widehat{K}_\alpha^{ik}(S^{II}) \equiv \widehat{K}_{ik}^{II} k_\alpha^{II}(S^{II}); \quad i,k=1,2,3 \tag{4.111}$$

such that the effective relative permeabilities are scalar and are equal to the microscale relative permeabilities for the highly permeable medium II:

$$\widehat{k}_\alpha(S^{II}) = k_\alpha^{II}(S^{II}), \qquad \widehat{k}_\beta(S^{II}) = k_\beta^{II}(S^{II}) \tag{4.112}$$

The effective global permeability tensor is determined as:

$$\widehat{K}_{ik} = \widehat{K}_{ik}^{II} \tag{4.113}$$

Tensor $\widehat{K}_{ik}^{II}$ is defined according to (4.87b) and via cell problem (4.88)

It follows from this that low permeable blocks operate only as sources and do not any contribution ot macroscale transport coefficients. This fact corresponds well to the numerical results obtained by Bourgeat, Hidani [21] and presented in Fig. 4.25.

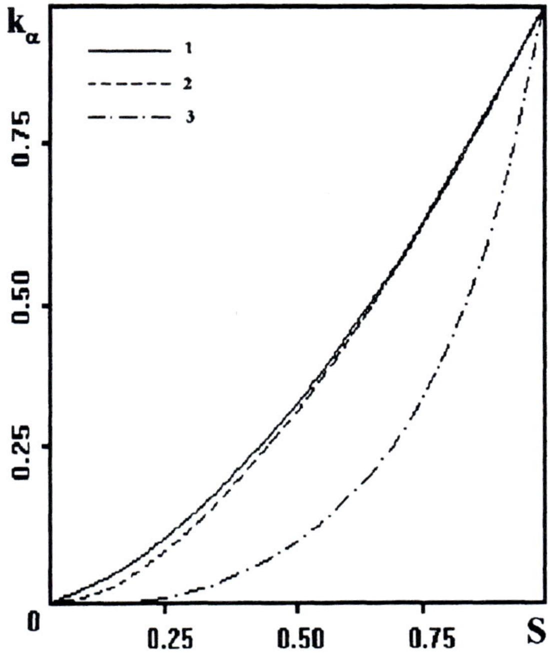

Figure 4.25. Calculation of the effective relative permeabilities for a highly heterogeneous medium
—— exact results (Bourgeat, Hidani, [21]);
- - - - relation (4.112);
– · – microscale relative permeability for dense blocks

Solid lines correspond to the effective relative phase permeabilities, obtained by numerical solution of the nonlinear cell problem, derived by Saez et al. [58], Bourgeat and Hidani [21], and Quitard wit Whitaker [56]. Dashed lines are the initial, microscale, relative phase permeability curves, given in the highly permeable domain.

Effective relative permeabilities for D-S system are always scalar. Moreover they are steady-state, although the macroscale capillary pressure in blocks is non-equilibrium.

4.10.2. MACROSCALE MODEL FOR D-T SYSTEM

When $\varepsilon<\omega_K<1$, the macroscale model of two-phase flow is another particular case of (4.77a) - (4.78) and gets the following form:

$$m^{II}(1-\theta)\frac{\partial S^{II}}{\partial t} - \frac{\partial}{\partial X_i}\left(\frac{\widehat{K}_\alpha^{ik}(S^{II})}{\mu_\alpha}\frac{\partial \widehat{P}_\alpha^{II}}{\partial X_k}\right) = -m^I\theta\frac{\partial S^I}{\partial t} \tag{4.114a}$$

$$-m^{II}(1-\theta)\frac{\partial S^{II}}{\partial t} - \frac{\partial}{\partial X_i}\left(\frac{\widehat{K}_\beta^{ik}(S^{II})}{\mu_\beta}\frac{\partial\left(\widehat{P}_\alpha^{II}+\widehat{P}_c^{II}(S^{II})\right)}{\partial X_k}\right) = $$

$$=m^I\theta\frac{\partial S^I}{\partial t} \tag{4.114b}$$

$$\widehat{P}_c^I(S^I) = \widehat{P}_c^{II}(S^{II}) \tag{4.114c}$$

$$\widehat{P}_c^I(S^I)\equiv P_c^I(S^I) \tag{4.114d}$$

$$\widehat{P}_c^{II}(S^{II})\equiv P_c^{II}(S^{II}) \tag{4.114e}$$

Basic difference from (4.77a) - (4.78) consists in that Eq. (4.114c) does not comprise explicitly the time, thus it is an equation of capillary equilibrium on the macroscale.

Effective phase permeabilities are defined in the same way as (4.87) , but they are also steady-state. Formally they depend on two saturations, as seen from (4.87) . Really they depend on one saturation only, because both the saturations S^I and S^{II} are related by Eq. (4.114c) .

4.11. NUMERICAL SIMULATION OF TWO-PHASE FLOW IN HIGHLY HETEROGENEOUS MEDIUM

4.11.1. SETTING OF 2D PROBLEM OF OIL DISPLACEMENT BY WATER

Domain

Examine the displacement of β-phase by α-phase in 2D highly heterogeneous medium. Let the heterogeneity be formed by square blocks, periodically located in the form of parallel bands (non-chess lattice). Due to the symmetry, examine the flow in a rectangle $L\times l/2$, being a half of one band, such as l is a heterogeneity period. The domain is shown on Fig. 4.26 via dimensionless variables.

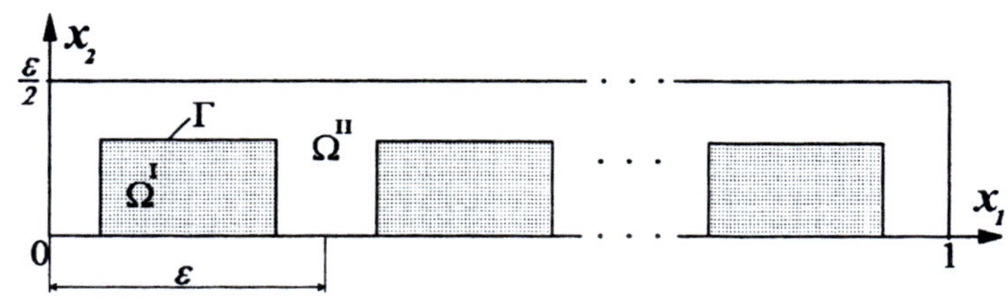

Figure 4.26. Domain of the flow

Problem of Displacement

Equations (4.16) - (4.18) have been used with the following boundary-value and initial conditions:

$$s(x,0)=s^*,\ x_1=0; \quad s(x,0)=s_*,\ x_1>0 \tag{4.115a}$$

$$\left.\frac{\partial s}{\partial n}\right|_{x_2=0,x_2=\varepsilon/2}=0 \tag{4.115b}$$

$$\left.\frac{\partial p}{\partial n}\right|_{x_2=0,x_2=\varepsilon/2}=0 \tag{4.115c}$$

$$\left.\frac{\partial s}{\partial n}\right|_{x_1=1}=0 \tag{4.115d}$$

$$s|_{x_1=0}=s^* \tag{4.115e}$$

$$p|_{x_1=0}=1 \tag{4.115f}$$

$$p|_{x_1=1}=0 \tag{4.115g}$$

Zero magnitude of the normal derivatives at the top side of the domain, ($x_2 = \varepsilon/2$), results from the periodicity condition, while the same conditions at the bottom side ($x_2 = 0$) mean the symmetry inside the unit cell. Condition (4.115d) corresponds to zero flow caused by capillary forces at the end of the medium. Relation (4.115e) determines the constant injection of α-phase in porous medium. Conditions (4.115f) and (4.115g) give the unit dimensionless pressure drop along the macroscopic length.

Method of Solution of the Microscale Problem

Usually another form of the equations is examined, introduced first in [32] (see also [42]) to avoid appearance of the functions undergoing high discontinuity like saturation. Saturation becomes discontinuous at the interface Γ due to continuity condition for capillary pressures (4.18b). Within this approach a "global pressure" $p_g = (p_\beta + p_\alpha)/2$ and a half of the capillary pressure $R = (p_\beta - p_\alpha)/2 = p_c/2$ are used instead of s, p. Although the initial condition requires that the function R be discontinuous at the initial state, this discontinuity is instantaneously disappearing in parabolic equations. After the problem is re-written in terms of new variables, it becomes elliptic respectively to the global pressure and parabolic respectively to the function R.

The problem was solved using the finit-element method.

4.11.2. RESULTS OF SIMULATION

Initial data

For all the simulations the following data have been accepted:

$$m^I = m^{II} = 0.1, \quad K^{II} = 10^{-15}\text{m}^2, \quad Pe^{II} = 1,$$

$$P_\alpha|_{x=0} = 10^6\text{Pa}, \quad P_\alpha|_{x=1} = 0, \quad \sigma = ...\text{Pa}\cdot\text{m}, \quad \theta = 0$$

$$L = 10\text{m}, \quad \varepsilon = 0.1$$

The volume fraction of a block in the cell is equal to 0.49. One block has a form of a square.

The relative phase permeabilities and the function $J(s)$ in (4.23) have been accepted as uniform in all the domain Ω and equal to:

$$k_\alpha^I = k_\alpha^{II} = \left(\frac{s - s_*}{s^* - s_*}\right)^{2.5}, \quad k_\beta^I = k_\beta^{II} = \left(\frac{s^* - s}{s^* - s_*}\right)^{2.5}, \quad J^I(s) = J^{II}(s) = \frac{s^* - s}{s^* - s_*}$$

$$s^* = 0.8, \quad s_* = 0.2$$

The pressure drop along one cell is equal to $\Delta P = 10^5$ Pa, the characteristic capillary pressure in the matrix is equal to $P_c^{II} = 10^5$ Pa, such that really $Pe^{II} = 1$.

According to relation (4.40), the Peclet number in blocks may be calculated as $Pe^I = \sqrt{\omega_K}$.

Variants of Simulations
The following variants have been chosen to simulate, corresponding to the a priori diagram of flows presented above on Fig. 4.11.

Variant 1: $\omega_K = 0.3 > \varepsilon$. Then $Pe^I = 0.57 < 1$, and we are probably in the domain of D-T systems.

Variant 2: $\omega_K = 0.03 < \varepsilon$. Then $Pe^I = 0.17 < 1$, and we are probably in D-S domain.

Variant 3: $\omega_K = 0.1 = \varepsilon$. Then $Pe^I = 0.32 < 1$, and we probably on the interface between D-T and D-S systems.

Variant 4: $\omega_K = 0.01 = \varepsilon^2$. Then $Pe^I = 0.1$. We are always in D-S domain, but with very low permeable blocks.

Flow Structure for D-T system
Typical saturation and pressure distributions are shown on Fig. 4.27 and 4.28.

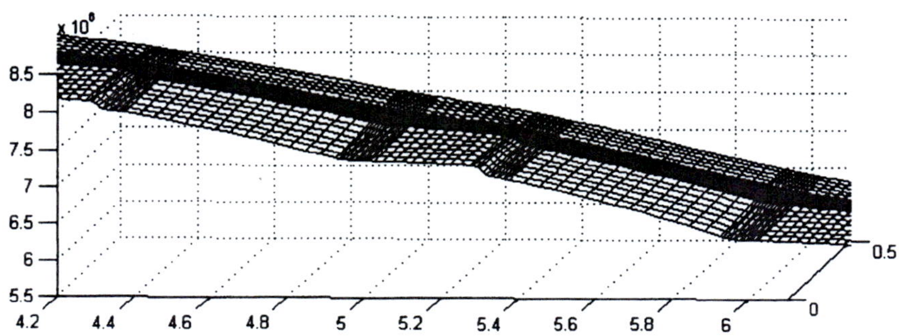

Figure 4.27. Pressure distribution for Dispersion-Translation type of flow

The pressure field is close to uniform.

Flow Structure for D-S Systems
Fig. 4.29 and 4.30 show typical saturation and pressure distributions in case, when the source type of flow predominates in blocks.

When capillary pressure in the block and in the matrix is strongly distinct, the same magnitude of capillary pressure corresponds to two strongly different values of saturation. Hence, on the interface Γ a powerful jump of saturation arises. Because the flow geometry corresponds to the source type, the saturation jump is formed around all the block. Then the block

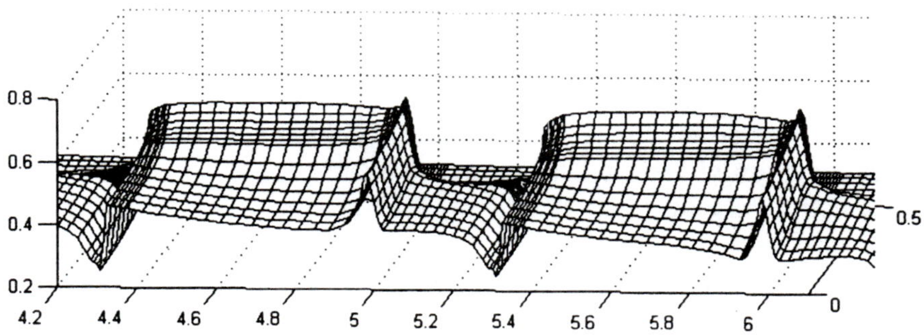

Figure 4.28. Saturation distribution for Dispersion-Translation type of flow

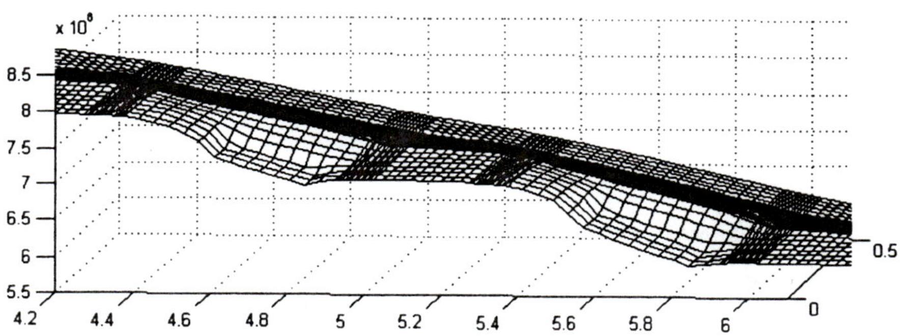

Figure 4.29. Pressure distribution for Dispersion-Source type of flow

initially containing the oil becomes enclosed by invading film of water, as shown on the Fig. 4.30, such that the flow inside the block gets a form of a

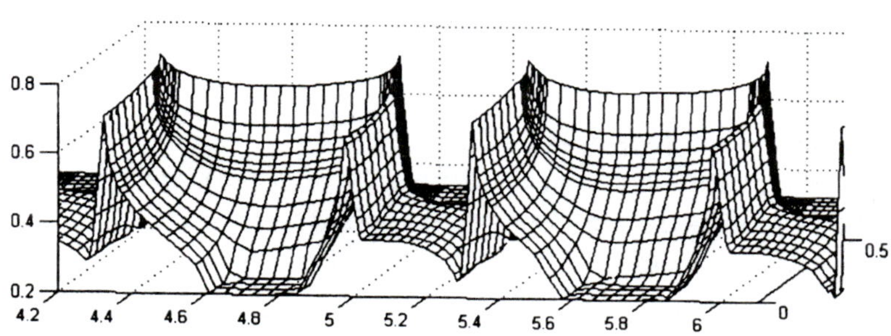

Figure 4.30. Saturation distribution for Dispersion-Source type of flow

counter-current imbibition. The equilibrium state will be reached, when the capillary pressure inside block will be equal to the pressure drop in the cell. The time of reaching the equilibrium is more larger than the time of flow around a block. Thus, the macroscale relaxation phenomena of saturation redistribution arise.

The difference between block and matrix pressure is always s all, but becomes more clear, than in case of D-T flow (4.28).

As to the saturation field, it is absolutely different from that appropriated to D-T flow (4.27). Saturation profile inside the block is uniformly symmetrical along all the axis in such a manner that any averaged gradients of saturation do not exist. It is a situation typical for Source system.

4.11.3. NUMERICAL SOLUTION OF THE MACROSCALE PROBLEM

Formulation of the macroscale problem

The same process was studied within the framework of the macroscale model (4.77a) - (4.78) .

Averaged problem (4.115a) becomes one-dimensional. Boundary-value and initial conditions are the following:

$$S^{II}\Big|_{t=0} = \begin{cases} s^*, & x=0 \\ s_*, & x>0 \end{cases} \tag{4.116a}$$

$$S^{I}\Big|_{t=0} = s_* \tag{4.116b}$$

$$S^{II}\Big|_{x=0} = s^* \tag{4.116c}$$

$$\widehat{P}_\alpha\Big|_{x=0} = 1 \tag{4.116d}$$

$$\widehat{P}_\alpha\Big|_{x=1} = 0 \tag{4.116e}$$

$$\frac{\partial S^{II}}{\partial x}\Big|_{x=1} = 0 \tag{4.116f}$$

As to the last condition, it is necessary to note that various forms of boundary-value condition at the outlet section of the medium are allowable, depending on the physical sense of the considered process. The form (4.116f) was chosen for sake of simplicity only. In any case, results obtained were analyzed in the central sections of the medium in order to avoid influence of boundary effects.

Results of Numerical Solution

Numerical solution of this 1D problem has been obtained using finite difference method.

Three modes of saturation distribution have been detected, depending on the capillary number Pe^I or Pe^{II}. They are shown in Fig. 4.31.

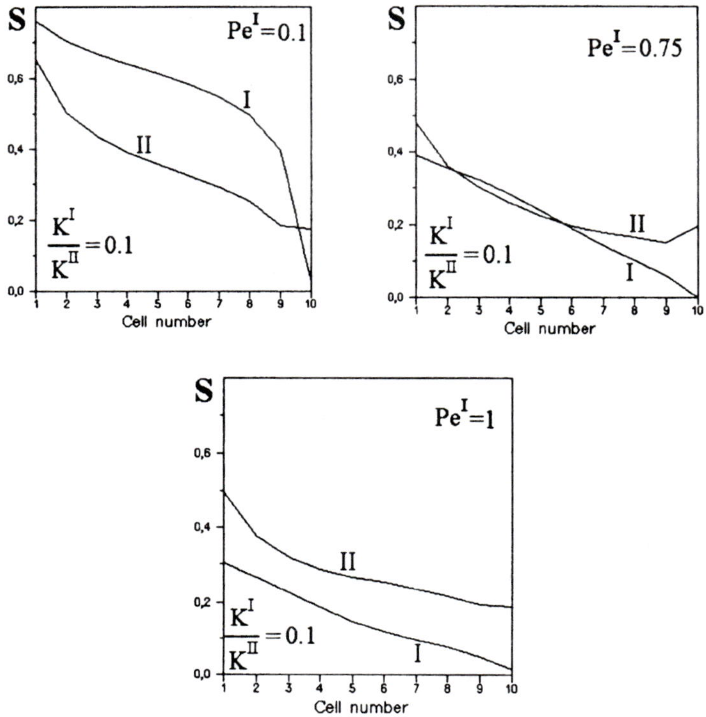

Figure 4.31. Numerical solution of the macroscale problem. Three types of saturation distribution in low permeable blocks I and highly permeable matrix II

When capillary number Pe^I is small enough, saturation in blocks is always grater than saturation in matrix. Increasing of the capillary number leads first to such a picture when both the saturations are equivalent about, and at last the matrix saturation becomes higher than the block saturation.

Comparison of the Microscale and Macroscale Solutions

Solution of the microscale problem was compared to the solution of the macroscale problem (4.77a) - (4.78) , (4.116) . Both the solutions are shown in Fig. 4.32.

The oscillating curve is the exact solution of 2D microscale problem taken along the axis x_1 at x_2=0. Solid lines correspond to the macroscale block and matrix saturations, S^I and S^{II}. The picture is fixed at the moment of water breakdown through the outlet boundary of the domain.

The goodness coincidence of the macroscale and the microscale solutions proves validity of deduced model (4.77a) - (4.78) .

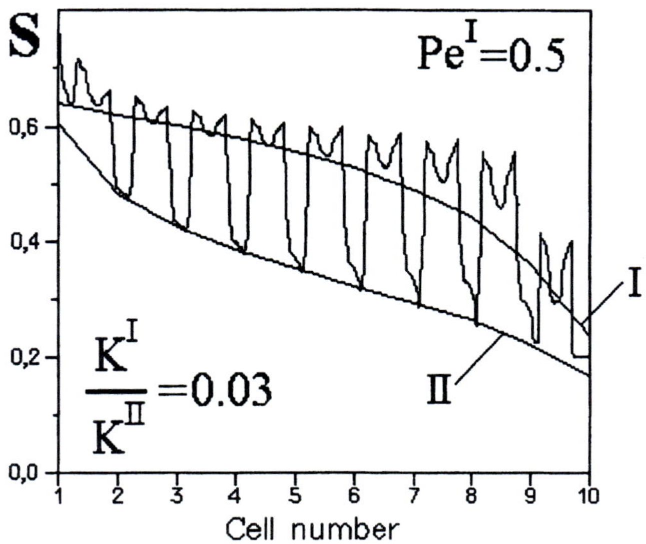

Figure 4.32. Comparison of microscale and macroscale numerical solutions. Oscillating curve is the microscale solution

CONCLUSIONS

Two-phase flow with capillary forces in double porosity medium was studied within the framework of homogenization methods. Various classifications of flow modes were proposed. One of them is based on various geometry of elementary flows in a cell, that allows to distinguish the Source-type and the Translation-type flow. Another classification is based on various modes of capillary equilibrium between the blocks and the matrix. Nonlinear and quasilinear equilibrium modes correspond to various parts of the capillary pressure curves, where nonlinear character of this curve is dominating or not.

A general macroscale model (4.77a) is obtained for the flow when the capillary forces can not be neglected, but any geometrical type of elementary flow can happen, as well as both the nonlinear or quasilinear equilibrium modes can be reached during the system evolution. An advantage of this model consists in that, it is completely free from the fast variables, on the one hand. On the other hand, the cell problem is completely free from the macroscale variable. Usually the complete splitting of the macroscale and the microscale variable may be reached in some rather simple cases, as, for instance, the case of moderate heterogeneity, which leads to a known equilibrium model of Quintard-Whitaker or Saez-Otero-Rusinek (4.5) . In our case, the model obtained describes more complicated objects.

First of all, the problem of description of nonlocal nonlinear two-phase phenomena with memory is partially solved, due to introducing of weakly nonlocal systems and an a priori selection of the weakly nonlocal modes. The idea is used, that a long-term memory operator can be converted into a local operator, when the memory is "not very long". Within the framework of such an approach, the kinetic-like operators arise instead of the convolutions, but the effect of memory is kept.

As the kinetic relationships are obtained, this model displays a non-equilibrium behavior, which is described by a nonlinear differential equation establishing some relation between the block and matrix saturations explicitly dependent on the time. The relaxation time is a non-monotonous function of the saturation, which structure is determined by a cell problem. Thus, the model includes dynamic capillary pressures.

The effective phase permeabilities (e.p.p) and the relative effective permeabilities (r.p.p.) were shown to be dependent on both the saturations. These saturations being related by a non-equilibrium equation, this property determines non-equilibrium behavior of e.p.p. and r.p.p.

When the system displays anisotropic property, e.p.p. becomes tensor functions, even when the microscale relative permeabilities are scalar.

The structure of e.p.p. is splitted in such a way, that only two cell functions should be solved to calculate the e.p.p., moreover, these cell problems are independent on saturation. So the problem to split the cell problem and the dependence on saturation is solved. This result ensures development of a very fast numerical method to up-scale the e.p.p.

Numerical 2D experiments have confirmed existence of various modes of flow detected theoretically, mathematical correctness of deduced equations, and have shown a good agreement of the macroscale behavior predicted by the model (4.77a) with solution of the equivalent microscale problem.

Unresolved Problems

Among many problems, which remains unresolved in the theory of two-phase flow through highly heterogeneous media, two problems should be solved first of all.

The first one concerns the flow with small capillary forces. Actually it is almost unreal to hope to deduce a model with a compete splitting of the fast and slow variable. But this expected "defect" of the model will be overlapped by other important effects, and, first of all, by the phenomenon of nonlinear two-phase dispersion dependent on saturation, that will give a tool to describe dynamic r.p.p. in case of neglected capillary forces.

The second problem is how to transmit all the results obtained onto disordered media. In disordered media, the scale of heterogeneity is rather indefinite notion. Moreover, due to absence of periodicity, the local problem is formulated in the infinite domain with conditions of exponential decay. The characteristic length of decay is not known a priori. It is natural to wait that this length is of order of the correlation length of the geometrical parameters of porous space. However, some numerical estimations published for instance in [8], show that this length is approximately equal to 30 correlations length, i.e. differs in order. Close magnitudes were obtained by prof. of the Saint-Etienne University A.Bourgeat and prof. of the LEPT ENSAM de Bordeaux, D.Bernard (private unpublished information).

This problem is now in developing [54].

References

1. Ahmadi, A. (1992) *Utilisation de Propriétés Equivalentes Dans les Modèles de Réservoir : Cas des Ecoulements Diphasiques Incompressibles*, Thesis, Univ. Bordeaux-I.
2. Alhanai, W., Bertin, H. and Quintard M. (1992) Two-phase flow in nodular systems: laboratory experiments, *Révue de l'Institut Français du Pétrole*, **47**, no. 1, pp. 29–44.
3. Amaziane, B. and Bourgeat, A. (1988) Effective behavior of two-phase flow in heterogeneous reservoir, numerical simulation in oil recovery, In : *IMA, Mathematics and its Applications*, M.F.Wheeler, Springer Verlag, **11**, pp. 1–22.
4. Amaziane, B., Bourgeat A. and Koebbe, J. (1991) Numerical simulation and homogenization of two-phase flow in heterogeneous porous media, *Transport in Porous Media*, no. 6, pp. 519–547.
5. Amirat, Y., Hamdache, K. and Ziani, A. (1988) Homogénéisation d'équations hyperboliques du premier ordre: Application aux milieux poreux, *Rapport Scientifique I.N.R.I.A.*, no. 803.
6. Amirat, Y., Hamdache, K. and Ziani, A. (1989) Homogénéisation d'équations hyperboliques du premier ordre et application aux écoulements miscibles en milieu poreux, *Annales Inst. H. Poincaré. Analyse non linéaire*, **6**, no. 5, pp. 397–417.
7. Amirat, Y., Hamdache, K. and Ziani, A. (1991) Kinetic formulation for a transport equation with memory, *Commun. Partial Differential Equations*, **16**, no. 8-9, pp. 1287–1311.
8. Anguy, Y. (1993) *Application de la prise de moyenne volumique a l'étude de la relation entre le tenseur de perméabilité et la microgémetrie des milieux poreux naturels*, Thesis, Univ. Bordeaux-I.
9. Arbogast, T., Douglas, J. and Hornung, U. (1990) Derivation of the double porosity model of single phase flow via homogenization theory, *SIAM J. Appl. Math.*, **21**, pp. 823–836.
10. Aronofsky, J.S., Masse, L. and Natanson, S.G. (1958) A model for the mechanism of oil recovery from the porous matrix due to water invasion in fractured reservoirs, *Trans. AIME*, **213**, pp. 17–19.
11. Auriault, J.-L. (1983) Effective macroscopic description for heat conduction in periodic composites. *Int. J. Heat and Mass Transfer*, **26**, no. 6, pp. 861–869.
12. Badea, A. and Bourgeat, A. (1995) Homogenization of two phase flow through randomly heterogeneous porous media, In. : *Mathematical Modelling of Flow through Porous media*, Proc. Int. Conf., World Scientific Pub., Singapore, pp. 15–31.
13. Bakhvalov, N.S. and Panasenko, G.P. (1989) *Homogenization of processes in periodic media*. Ed. Nauka, Moscow. (Version in English : Bakhvalov N. and Panasenko, G. *Homogenization : Averaging Processes in Periodic Media*. Kluwer Academic Publishers, Dordrecht)
14. Bertin, A., Alhanai, W. and Quintard, M. (1992) Ecoulement polyphasique dans un milieu poreux hétérogène de type nodulaire: drainage initial, *C. R. Acad. Sci. Paris*, Ser. II, **314**, pp. 431–437.
15. Barenblatt, G.I. (1971) Flow of two immiscible liquids through the homogeneous porous medium, *Izvestiya Academii Nauk SSSR, Mekhanika Zhidkosti i Gaza*, no 5, pp. 144–151.
16. Barenblatt, G.I. and Vinichenko, A.P. (1980) Non-equilibrium flow of immiscible fluids in porous media, *Uspekhi Mathematicheskikh Nauk*, no 3, pp. 35–50.
17. Barenblatt, G.I., Entov, V.M. and Ryzhik, V.M. (1972) *Theory of Non-Stationary Flow of Liquids and Gases Through Porous Media*. Nedra, Moscow.
18. Barker, J.W. and Dupouy, P. (1996) An analysis of dynamic pseudo relative permeability methods, *Proc. of 5th European Conf. of Mathematics of Oil Recovery*, 3-6 Sept., Leoben, Austria.
19. Bougemaa, A. (1998) Private communication.

20. Bourgeat, A. (1996) Two-phase flow, *in: Homogenization and Porous Media*, Ed. Hornung, U., Springer, pp. 95–128.
21. Bourgeat, A., Hidani, A. (1995) Effective model of two-phase flow in a porous medium made of different rock types, *Applicable Analysis*, **56**, pp. 381–399.
22. Bourgeat, A., Luchkaus, S. and Mikelić, A. (1997) Convergence of the homogenization process for a double porosity model of immiscible two-phase flow, *SIAM J. Appl. Math.*, to appear.
23. Bourgeat, A. and Mikelić, A. (1994) Homogenization of the two-phase immiscible flow in one dimensional porous medium, *Asymptotic Analysis*, **9**, pp. 359–380.
24. Bourgeat, A. and Panfilov, M. (1998) Effective two-phase flow through highly heterogeneous porous media: Capillary nonequilibrium effects, *Computational Geosciences*, **2**, no. 3, pp. 191–215.
25. Bramble, J. H. and Shatz, A.H. (1970) Rayleight-Ritz-Galerkin method for Dirichlet's problem using subspaces without boundary conditions, *Comm. Pure Appl. Mathem*, **23**, pp. 653–675.
26. Charpentier, I., and Maday, Y. (1995) Deux méthodes de décomposition de domaine pour la résolution d'équations aux dérivée partielles avec conditions de périodicité: application à la théorie de l'homogénéisation, *C. R. Acad. Sci. Paris*, Ser.I, **321**, pp. 359–366.
27. Civan, F. (1993) Waterflooding of naturally fractured reservoirs: an efficient simulation approach, *Paper SPE 25449*, Proc. of the SPE Production Operation Sympos., 21-23 March, Oklahoma City, USA, pp. 395–407.
28. Civan, F. (1994) A theoretically derived transfer function for oil recovery from fractured reservoirs by waterflooding, *Paper SPE/DOE 27745*, Proc. of SPE/DOE 9th Sympos. on Improved Oil Recovery, 17-20 April, Tulsa, Oklahoma, USA, pp. 87–98.
29. Dale, M., Ekrann, S., Mykkeltveit, J. and Virnovsky, G. (1994) Effective relative permeability and capillary pressure for 1D heterogeneous media, *Proc. of 4th European Conf. of Mathematics of Oil Recovery*, 7-10 June, Røros, Norway.
30. deSwaan, A. (1978) Theory of waterflooding in fractured reservoirs, *SPE J.*, pp. 87–98.
31. Douglas, J. and Arbogast, T. (1990) Dual porosity models for flow in naturally fractured reservoirs, *in: Dynamics of Fluids in Hierarchical Porous Media*, Ed. Cushman, J., Academic Press, pp. 177–206.
32. Douglas, J., Peacemen, D.W. and Rachford, Jr. (1959) A method for calculating multidimensional immiscible displacement, *Trans. of AIME*, **216**, pp. 297–308.
33. Durlofsky, L.J. (1998) Coarse scale models of two phase flow in heterogeneous reservoirs: volume averaged equations and their relationship to existing upscaling techniques, *Computational Geosciences*, **2**, no. 2, pp. 73–92.
34. Hansen, A., Roux, S., Aharony, A., Feder, J., Jøssang, T. and Hardy, H.H. (1997) Real-space renormalization estimates for two-phase flow in porous media, *Transport in Porous Media*, **29**, pp. 247–279.
35. Hassanizadeh, S.M., and Gray, W.G. (1993) Thermodynamic basis of capillary pressure in porous media, *Water Resour. Res.*, **29**, pp. 3389–3405.
36. Hassanizadeh, S.M., and Gray, W.G. (1993) Toward an improved description of the physics of two-phase flow, *Adv. Water Resour.*, **16**, pp. 53–67.
37. Hassanizadeh, S.M., and Gray, W.G. (1990) Mechanics and thermodynamics of multiphase flow in porous media including interphase boundaries, *Adv. Water Resources*, **13**, pp. 169–186.
38. Hassanizadeh S.M. (1997) Dynamic effects in the capillary pressure-saturation relationship, Proc. 4th Int. Conf. Civ. Eng., Sharif Univ. Technol., Tehran, Iran, 4-6 May, 1997, V.4: Water Resources and Environmental Engineering, pp. 141–149.
39. Kalaydjian, F. (1992) Dynamic capillary pressure curve for water/oil displacement in porous media, theory vs. experiment, *SPE Conf.*, Washington, DC, October 4-7, Paper 24813, pp. 491–506.

40. Kazemi, H., Gilman, J.R. and El-Shharkaway, A.M. (1992) Analytical and numerical solution of oil recovery from fractured reservoirs with empirical transfer functions, *SPE Reservoir Engineering*, no. 7(2), pp. 219–227.
41. King, P., Muggeridge, A.H. and Price, W.G. (1993) Renormalization calculation of immiscible flow, *Transport in Porous Media*, **12**, pp. 237–260.
42. Konovalov, A.N. (1988) *Problems of Multiphase Incompressible Flow in Porous Media*, Novosibirsk, Nauka, Siberian Department (in Russian).
43. Kumar, A., Farmer, L., Jerauld, G.R. and Li, D. (1997) Efficient upscaling from cores to simulation models, *Paper SPE 38744*, Proc. 1997 SPE Annual Technical Conf. and Exhibition, 5-8 Oct., San Antonio, Texas, USA, pp. 257–271.
44. Kutliarov, V.S. (1967) Convective Diffusion in Fractured Porous Reservoirs, *J. Prikladnoi Mathematiki and Tekhnicheskoi Phyziki*, no. 1, pp. 84–88, (in Russian).
45. Kyte, J.R. and Berry, D.W. (1975) New pseudo-functions to control numerical dispersion, *SPE J.*, August, p. 269–275.
46. Lemouzy, P., Kruel Romeu, R. and Morelon, I. (1993) A new scaling-up method to compute relative permeability and capillary pressure for simulation of heterogeneous reservoirs, *Paper SPE 26660*, Proc. 68th SPE Annual Technical Conf. and Exhibition, 3-6 October, Houston, Texas, USA.
47. Lenormand, R. (1995) Transport equations for fluid displacements in heterogeneous porous media: the MHD model, *in: Math. Modelling of Flow Through Porous Media*, Proc. Int. Conf., Eds. Bourgeat, A., Carasso, C., Luckhaus, S. and Mikelić, A., World Scientific Publishing, Singapore, pp. 248–265.
48. Nikolaevskii, V.N., Bondarev, E.A., Mirkin M.I. et al, (1968) *Motion of Hydrocarbon Mixtures in Porous Media*. Nedra, Moscow (in Russian).
49. Panfilov, M. (1990) Mean mode of porous flow in highly inhomogeneous media, *Soviet Physics Doklady*, **35**, pp. 225–227.
50. Panfilov, M. (1995) Averaged Models of Convection-Diffusion Transfer Through Highly Heterogeneous Porous Media, *in: Math. Modelling of Flow Through Porous Media*, Proc. Int. Conf., Eds. Bourgeat, A., Carasso, C., Luckhaus, S. and Mikelić, A., World Scientific Publishing, Singapore, pp. 276–300.
51. Panfilov, M. (1996) Homogenized model with capillary nonequilibrium for two-phase flow through highly heterogeneous porous media, *C. R. Acad. Sci. Paris*, Ser.II, no. 3.
52. Panfilov, M., and Bourgeat, A. (1996) Capillary Relaxation Model for Two-Phase Flow Through Dual-Porosity Media Proc, *4th Int. Sympos. On Evaluation of Reservoir Wettability and its Effects on Oil Recovery*, Sept. 11-13, 1996, Montpellier, France.
53. Panfilov, M. (1998) Upscaling Two-Phase Flow In Double Porosity Media: Nonuniform Homogenization, *in: "Recent Advances in Problems of Flow and Transport in Porous Media"* (Proc. Int. Conf.), Eds. J.M.Crolet and M.E.El-Hatri, Kluwer Acad. Publ., Dordrecht, pp. 195–215.
54. Panfilov, M. (1998) Upscaling two-phase flow through double porosity media, *in: Proc. of 6-th European Conf. on the Mathematics of Oil Recovery*, Peebles, Scotland, 8-11 Sept., paper C-28.
55. Panfilov, M. and Panfilova, I. (1996) *Averaged models of flows with heterogeneous internal structure*. Ed. Nauka, Moscow (in Russian).
56. Quintard, M. and Whitaker, S. (1988) Two Phase Flow in Heterogeneous Porous Media : The Method of Large-Scale Averaging, *Transport in Porous Media*, no. 3, pp. 357–413.
57. Smith, E.H. (1991) The influence of small-scale heterogeneity on average relative permeability, *in: Reservoir Characterization II*, Eds. Lake L.W., Caroll H.B. and Wesson T.C., Academic Press.
58. Saez, A. E., Otero, C. J. and Rusinek, I. (1989) The Effective Homogeneous Behaviour of Heterogeneous Porous Media, *Transport in Porous Media*, no. 4, pp. 212–238.

59. Suquet, P. (1990) Une métode simplifiée pour des propriétés élastiques de matériaux hétérogènes à structure périodique, *C. R. Acad. Sci. Paris*, Ser.II, **311**, pp. 769–774.
60. Stone, H.L. (1991) *Paper SPE 21207*.

TWO-PHASE FLOW IN PSEUDO-CAVITY MEDIA

5.1. INTRODUCING NOTES

In this chapter, another class of media will be studied, which is "inverse" respectively to the objects examined in all the previous chapters. These media are constructed in such a manner that the separated inclusions (blocks) are more permeable than the connected matrix, i.e.,

$$\omega_K >> 1 \tag{5.1}$$

Such objects are very similar to so called "cavity" media encountered in geology. A cavity medium is a porous rock, which contains large concavities with a characteristic scale-size much larger than the size of a pore. The flow inside a concavity should be described by Navier-Stokes equations. Due to that circumstance, the local permeability inside a concavity is not determined.

A medium, where the porous inclusions of high permeability replace the concavities, may be called as "pseudo-cavity". Although the flow inside each block is governed by equations of another type, i.e., Darcy's law, these media keep some basic features of purely cavity media. First at all, hydraulic conductivity of inclusions is much larger than conductivity of the porous matrix. Purely cavity medium may be considered as a limit of pseudo-cavity media, when the parameter ω_K tends to infinity.

Although the pseudo-cavity media are geometrically inverse respectively to the double porosity media studied in previous chapters, the hydrodynamic behavior of such systems is considerably different and will be especially examined. Connectivity of one component of the medium and disconnectedness of another one plays a considerable role. A simple change by places of media I and II leads to new, nonsymmetric, qualitative phenomena.

More known attempt to study highly heterogeneous pseudo-cavity media is associated to a series of papers of H.Bertin and M.Quintard with co-authors [1], [2], [3], where this problem was investigated in experimental way. Theoretical research was undertaken in [4] for the linear equations of elasticity theory in quasi-stationary case. However, a most important feature of highly heterogeneous media is non-stationary behavior, as shown in all the previous chapters.

Basic contradiction of pseudo-cavity media consists in the following. On the one hand, the flow through inclusions appears to be more easy due to their high permeability. On the other hand, the wetting fluid invading the medium can not penetrate from the low permeable matrix into the inclusions due to a sharp extension of pores at the interface, which reverses the direction of capillary force in such a way that the capillary force locks penetration, as shown in Fig. 5.1.

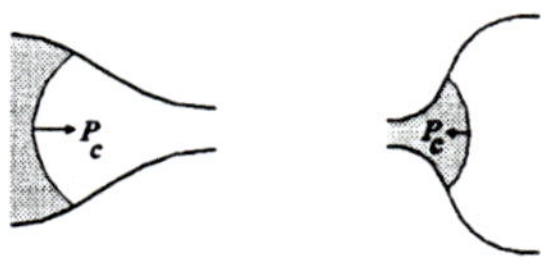

Figure 5.1. Wetting fluid (dark) is easy to passe from the left to the right through the pore constriction (left-hand picture), but can not penetrate through a pore extension (right-hand picture)

In this mode of flow, the inclusions remaining impenetrable with respect to the invading fluid, the flow inside the inclusions is one-phase.

However, if we assume that the wetting fluid has succeeded in penetrating into the inclusions (in some way), absolutely another type of flow is expected to happen. Instead of being an obstacle, the inclusion becomes a main conductor of the medium.

All these ideas are true, when the capillary forces are predominating. Otherwise, without capillary forces, the system has not any special.

In this chapter essentially non-stationary nonlinear equations of two-phase flow will be examined in case of dominating capillary forces. Various modes of flow are expected to be detected.

5.2. DEFINITION OF THE PROBLEM

5.2.1. PHYSICAL SETTING

Let us consider the flow of two incompressible phases (the $\alpha-$phase and the $\beta-$phase) in a medium $\Omega \subset I\!R^3$ made of a lowly permeable connected "matrix", Ω^I, and of a system of periodically located highly permeable "inclusions" or "blocks", Ω^{II}, such that $\Omega = \Omega^I \cup \Omega^{II}$, as shown in Fig. 5.2. Let Γ_x be the interface separating two media, Ω^I and Ω^{II}.

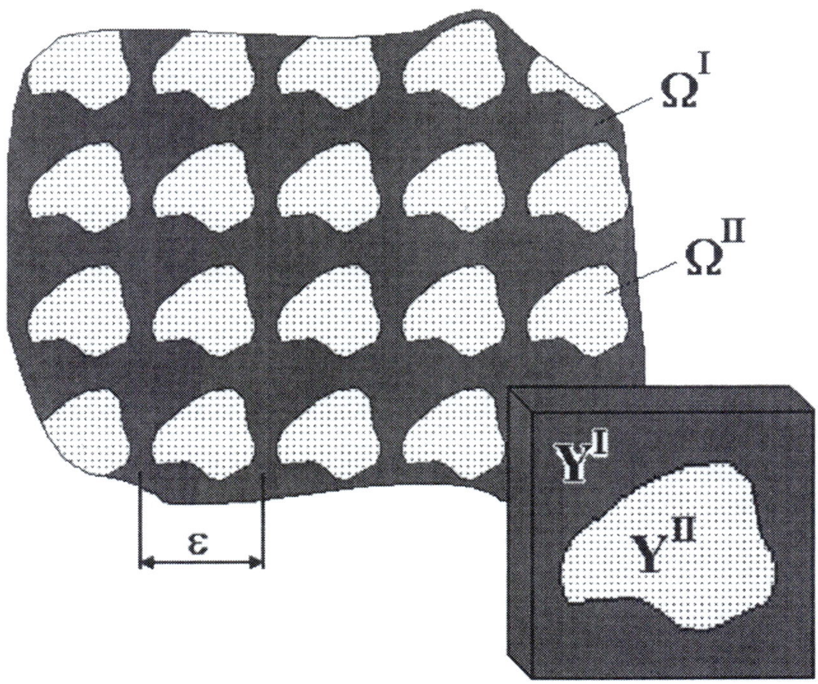

Figure 5.2. Structure of the medium with lowly permeable connected matrix I and highly permeable inclusions II

In is necessary to underline that in all this chapter the upper indexes I and II are kept for lowly and highly permeable domains correspondingly, but their geometrical location is inverse respectively to the media examined in all previous chapters.

Let P be the pressure and s be the saturation of the α–phase, X be the space vector coordinate and t be the time.

Each of such sub-domains Ω^i, $i = I, II$, is characterized by a rock permeability tensor K^i and a porosity m^i, by the relative phase permeability functions $k^i_\alpha(s)$, $k^i_\beta(s)$, and the capillary pressure function $P^i_c(s)$. The capillary pressure is the difference between the pressures in two fluids: $P_c = P_\beta - P_\alpha$. The viscosity μ of each fluid phase is supposed to be constant. The volume fractions of both the media are of the same order.

We assume first the heterogeneity is fast oscillating, such that the heterogeneity scale l is more small than the macroscopic scale L:

$$\varepsilon \equiv l/L << 1$$

Conditions of high heterogeneity with respect to the permeability and to the capillary pressure mean the following:

1) the permeability of the inclusions is must higher that the permeability of the matrix, i.e., $\omega_K = K^I/K^{II} >> 1$;

2) the orders of capillary forces in the inclusions II is must lower than in the matrix I, i.e., $\omega_c = P_c^{II,0}/P_c^{I,0} << 1$.

At last, we assume the medium is periodic, such that the heterogeneity scale ε is equal to the heterogeneity period.

5.2.2. MATHEMATICAL SETTING

Basic Characteristics of Flow

For the functions $P^i_c(s)$, $k^i_\alpha(s)$ and $k^i_\beta(s)$, where $i = I, II$, the properties (4.11) - (4.13) are assumed.

The following characteristic magnitudes are introduced:

- The minimum and the maximum pressures P_A and P_B, as in section 4.3.2.

- Averaged capillary pressures:

$$P_c^{i,0} \equiv \frac{1}{s^{i*} - s^i_*} \int\limits_{s^i_*}^{s^{i*}} P^i_c(s)ds, \quad i = I, II \tag{5.2}$$

- Mean permeabilities and porosities for each sub-domain:

$$\bar{m}^i \equiv \frac{1}{|\Omega^i|} \int\limits_{\Omega^i} m^i(x)dx, \quad \bar{K}^i \equiv \frac{1}{|\Omega^i|} \int\limits_{\Omega^i} K^i(x)dx$$

- Time of complete filling the matrix by one of the fluids (β-fluid, for instance) due to the capillary forces being equal to the capillary force in the inclusions:

$$t_* \equiv \frac{\bar{m}^I \mu_\beta L^2}{\bar{K}^I P_c^{II,0}} \tag{5.3}$$

Such a unobvious introduction of the characteristic time is caused by the fact that in this paper we are studying the mode of flow at high saturation of the α-phase in the matrix I, when the matrix capillary force P_c^I becomes of the same order as P_c^{II}.

Flow Equations

Equations describing two-phase flow in Ω with respect to the α-phase saturation s and to the α-phase dimensionless pressure p may be written in the following form, for any $\tau>0$:

In Ω^I:

$$\omega_c b^I\left(\frac{x}{\varepsilon}\right)\frac{\partial s}{\partial \tau} = Pe^I \mathrm{div}_x\left(a^I\left(\frac{x}{\varepsilon}\right)\lambda_\alpha^I(s^I)\mathrm{grad}_x p\right), \tag{5.4a}$$

$$\mathrm{div}_x\left(a^I\left(\frac{x}{\varepsilon}\right)\left[Pe^I\lambda^I(s^I)\mathrm{grad}_x p + \lambda_\beta^I(s^I)\mathrm{grad}_x p_c\right]\right) = 0 \tag{5.4b}$$

In Ω^{II}:

$$\frac{\omega_K\omega_c}{\omega_m} b^{II}\left(\frac{x}{\varepsilon}\right)\frac{\partial s}{\partial \tau} = Pe^I \mathrm{div}_x\left(a^{II}\left(\frac{x}{\varepsilon}\right)\lambda_\alpha^{II}(s^{II})\mathrm{grad}_x p\right), \tag{5.4c}$$

$$\mathrm{div}_x\left(Pe^I a^{II}\left(\frac{x}{\varepsilon}\right)\left[\lambda^{II}(s^{II})\mathrm{grad}_x p + \omega_c\lambda_\beta^{II}(s^{II})\mathrm{grad}_x p_c\right]\right) = 0 \tag{5.4d}$$

On Γ_x:

$$[p]_{\Gamma_x} = 0\,, \tag{5.4e}$$

$$\omega_c p_c^{II}|_{\Gamma_x} = p_c^I|_{\Gamma_x}\,, \tag{5.4f}$$

$$\omega_K a^I \lambda_\alpha^I \frac{\partial p^I}{\partial x_i} n_i|_{\Gamma_x} = a^{II}\lambda_\alpha^{II}\frac{\partial p^{II}}{\partial x_i} n_i|_{\Gamma_x}, \tag{5.4g}$$

$$\omega_K a^I\left(Pe^I\lambda^I\frac{\partial p^I}{\partial x_i} + \lambda_\beta^I\frac{\partial p_c^I}{\partial x_i}\right)n_i|_{\Gamma_x} = a^{II}\left(Pe^I\lambda^{II}\frac{\partial p^{II}}{\partial x_i} + \omega_c\lambda_\beta^{II}\frac{\partial p_c^{II}}{\partial x_i}\right)n_i|_{\Gamma_x}, \tag{5.4h}$$

The initial conditions :

$$s(x,y,0) = s^0(x). \tag{5.5a}$$

The boundary conditions with respect to the pressure require only that

$$p|_{\partial\Omega} = O(1) \tag{5.5b}$$

At the boundary of the domain the saturation should be given too:

$$s|_{\partial\Omega} = s_{\partial\Omega}(x,t) \tag{5.5c}$$

According to the a priori estimations explained in section 4.1.1, the macroscale model is expected to be independent on boundary condition (5.5c) .

Conditions on Γ_x and (5.5b) are explained in sections 4.1.1 and 4.3.2.

The following notations have been introduced for the dimensionless variables with $i=I,II$:

$$p^i \equiv \frac{P^i - P_B}{P_A - P_B},\ p_c^i(s) \equiv \frac{P_c^i(s)}{P_c^{i,0}},\ \tau \equiv \frac{t}{t_*},\ x \equiv \frac{X}{L},\ a^i\left(\frac{x}{\varepsilon}\right) \equiv \frac{K^i}{\bar{K}^i},\ b^i\left(\frac{x}{\varepsilon}\right) \equiv \frac{m^i}{\bar{m}^i},$$

$$\lambda_j^i(s) \equiv \frac{k_j^i(s)}{\zeta_j},\ \zeta_j \equiv \frac{\mu_\beta}{\mu_j},\ j=\alpha,\beta;\ \lambda^i(s) \equiv \lambda_\alpha^i(s) + \lambda_\beta^i(s)$$

$$\omega_K \equiv \frac{\bar{K}^I}{\bar{K}^{II}},\quad \omega_m \equiv \frac{\bar{m}^I}{\bar{m}^{II}},\quad \omega_c \equiv \frac{P_c^{II,0}}{P_c^{I,0}},\quad Pe^I \equiv \frac{P_A - P_B}{P_c^{I,0}}$$

Note that the parameter ω_c is the same as in chapter 4.

Parameters of the Problem

The parameters $a(x/\varepsilon)$, $b(x/\varepsilon)$, $\lambda(s;x/\varepsilon)$, $\lambda_\alpha(s;x/\varepsilon)$, and $\lambda_\beta(s;x/\varepsilon)$ are ε-periodic with respect to the second argument.

As in section 4.3.2, along with the functions $p_c^I(x,\tau)$ and $p_c^{II}(x,\tau)$ we use another notations for two capillary pressures as functions of saturation: $\tilde{p}_c^I(s^I)$ and $\tilde{p}_c^{II}(s^{II})$. It is clear that

$$p_c^i(x,\tau) = \tilde{p}_c^i(s(x,\tau)), \qquad i=I,II \tag{5.6}$$

Two functions $\tilde{p}_c^I(s)$ and $\tilde{p}_c^{II}(s)$ are known a priori.

As in section 4.3.2, we assume that the parameter ω_c is rather independent on ω_K or ω_m, but relation (4.24) gives some limitations for the ratio between ω_c and ω_K.

We assume furthermore:

$$0<\varepsilon\ll 1, \qquad 0<\omega_K \ll \varepsilon \ll \omega_c \ll 1, \qquad \omega_m \sim 1 \tag{5.7a}$$

For the sake of simplicity let us assume:

$$Pe^I \sim \omega_c \tag{5.7b}$$

Unit Cell and Fast Variables
As usual, to any point x in the domain Ω we associate the image $y=x/\varepsilon$ in the cell $Y=\{-1/2<y_i<1/2,\ \ i=1,2,3\}$, which is imaged in Fig. 5.2.

The surface $\Gamma=\partial Y^{II}$, which is the image of the surface Γ_x, divides the cell into two open sets Y^I (the dense connected matrix) and Y^{II} (the highly permeable inclusion). Γ is assumed to be not intersecting the external boundary of Y: $\Gamma\cap\partial Y=\emptyset$.

5.2.3. ABOUT EQUILIBRIUM MODES OF FLOW

In section 4.5.3 it was explained that in a medium with low permeable inclusions and highly permeable matrix only two equilibrium modes of flow can be reached within the process of monotonous saturation growth (displacement of the β-phase by the α-phase). Both frozen modes are non-appropriated to those media.

In pseudo-cavity media, the partially frozen mode shown in Fig. 4.12 is a normal phenomenon.

In this case the low permeable matrix I is the basic transport channel. Any perturbation can reach the highly permeable blocks II through the matrix only. Thus, the block saturation is always considerably dependent on the matrix saturation.

Any saturation growth follows first the curve I in Fig. 4.12. Any invasion through the matrix can happen when $s^I>s^I_*$, where s^I_* is a percolation threshold for the α-phase. Then three modes can be reached according to Fig. 4.12: partially frozen mode, nonlinear mode and quasi-linear mode.

When the matrix saturation s^I grows in time, the system passes consequently from the partially frozen mode, through nonlinear mode to the quasi-linear one.

5.3. DEDUCTION OF THE MACROSCALE MODEL

The quasi-linear mode of flow constitutes the main aim of this chapter. According to Fig. 4.12, the block saturation $s^I I$ undergoes variations within large scale, while the matrix saturation s^I is practically fixed near the final point s^{I*}. According to remarks of section 5.1.3, in the problem of monotonous saturation growth, the quasi-linear mode of flow corresponds to large values of time. This fact explains the chose of the characteristic time (5.3) .

The following must be accepted basing on (4.42):

$$p_c^I \sim \omega_c \tag{5.8}$$

5.3.1. TWO-SCALE FORMULATION

With the usual change of variable $x \to (x, y = x/\varepsilon)$ we are now considering the pressure and the saturation depending on both the large and the small scale, x and y: $p = p(x, y, t)$, $s = s(x, y, t)$, assuming y to be a new independent variable.

With the change of variable we have the new operators :

$$\partial_k^\varepsilon \equiv \partial_k + \varepsilon \partial_{\mathrm{x}k} \quad \text{with} \quad \partial_k \equiv \frac{\partial}{\partial y_k}, \quad \partial_{\mathrm{x}k} \equiv \frac{\partial}{\partial x_k} \text{ and } \partial_\tau \equiv \frac{\partial}{\partial \tau}$$

Then problem (5.4) - (5.5) gets the following equivalent two-scale formulation.

When $x \in \Omega$, $y \in Y^I$:

$$\begin{aligned} &Pe^I \left\{ \partial_i \left(a^I(y) \lambda_\alpha^I \partial_i p \right) + \varepsilon \left[\partial_i \left(a^I \lambda_\alpha^I \partial_{\mathrm{x}i} p \right) + \partial_{\mathrm{x}i} \left(a^I \lambda_\alpha^I \partial_i p \right) \right] + \right. \\ &\qquad \left. \varepsilon^2 \partial_{\mathrm{x}i} \left(a^I \lambda_\alpha^i \partial_{\mathrm{x}i} p \right) \right\} = \varepsilon^2 \omega_c b^I(y) \partial_\tau s, \end{aligned} \tag{5.9a}$$

$$\begin{aligned} &Pe^I \left\{ \partial_i \left(a^I \lambda^I \partial_i p \right) + \varepsilon \left[\partial_i \left(a^I \lambda^I \partial_{\mathrm{x}i} p \right) + \partial_{\mathrm{x}i} \left(a^I \lambda^I \partial_i p \right) \right] + \right. \\ &\qquad \left. \varepsilon^2 \partial_{\mathrm{x}i} \left(a^I \lambda^I \partial_{\mathrm{x}i} p \right) \right\} + \\ &\left\{ \partial_i \left(a^I \lambda_\beta^I \partial_i p_c \right) + \varepsilon \left[\partial_i \left(a^I \lambda_\beta^I \partial_{\mathrm{x}i} p_c \right) + \partial_{\mathrm{x}i} \left(a^I \lambda_\beta^I \partial_i p_c \right) \right] + \right. \\ &\qquad \left. \varepsilon^2 \partial_{\mathrm{x}i} \left(a^I \lambda_\beta^I \partial_{\mathrm{x}i} p_c \right) \right\} = 0 \end{aligned} \tag{5.9b}$$

When $x \in \Omega$, $y \in Y^{II}$:

$$Pe^I \left\{ \partial_i \left(a^{II}(y) \lambda_\alpha^{II} \partial_i p \right) + \varepsilon \left[\partial_i \left(a^{II} \lambda_\alpha^{II} \partial_{\mathrm{x}i} p \right) + \partial_{\mathrm{x}i} \left(a^{II} \lambda_\alpha^{II} \partial_i p \right) \right] + \right.$$

$$\varepsilon^2\partial_{\mathrm{x}i}\left(a^{II}\lambda_\alpha^{II}\partial_{\mathrm{x}i}p\right)\Big\}=\frac{\varepsilon^2\omega_K\omega_c}{\omega_m}b^{II}(y)\partial_\tau s, \tag{5.10a}$$

$$\begin{aligned}Pe^I&\Big\{\partial_i\left(a^{II}\lambda^{II}\partial_i p\right)+\varepsilon\left[\partial_i\left(a^{II}\lambda^{II}\partial_{\mathrm{x}i}p\right)+\partial_{\mathrm{x}i}\left(a^{II}\lambda^{II}\partial_i p\right)\right]+\\&\varepsilon^2\partial_{\mathrm{x}i}\left(a^{II}\lambda^{II}\partial_{\mathrm{x}i}p\right)\Big\}+\\\omega_c&\Big\{\partial_i\left(a^{II}\lambda_\beta^{II}\partial_i p_c\right)+\varepsilon\left[\partial_i\left(a^{II}\lambda_\beta^{II}\partial_{\mathrm{x}i}p_c\right)+\partial_{\mathrm{x}i}\left(a^{II}\lambda_\beta^{II}\partial_i p_c\right)\right]+\\&\varepsilon^2\partial_{\mathrm{x}i}\left(a^{II}\lambda_\beta^{II}\partial_{\mathrm{x}i}p_c\right)\Big\}=0\end{aligned} \tag{5.10b}$$

When $x\in\Omega$, $y\in\Gamma$:

$$[p]_\Gamma=0\ , \tag{5.11a}$$

$$\omega_c p_c^{II}|_\Gamma=p_c^I|_\Gamma\ , \tag{5.11b}$$

$$\omega_K a^I\lambda_\alpha^I\left(\partial_i+\varepsilon\partial_{\mathrm{x}i}\right)p^I n_i|_\Gamma=a^{II}\lambda_\alpha^{II}\left(\partial_i+\varepsilon\partial_{\mathrm{x}i}\right)p^{II}n_i|_\Gamma \tag{5.11c}$$

$$\begin{aligned}&\omega_K a^I\left\{Pe^I\lambda^I\left(\partial_i+\varepsilon\partial_{\mathrm{x}i}\right)p^I+\lambda_\beta^I\left(\partial_i+\varepsilon\partial_{\mathrm{x}i}\right)p_c^I\right\}n_i|_\Gamma=\\&a^{II}\left\{Pe^I\lambda^{II}\left(\partial_i+\varepsilon\partial_{\mathrm{x}i}\right)p^{II}+\omega_c\lambda_\beta^{II}\left(\partial_i+\varepsilon\partial_{\mathrm{x}i}\right)p_c^{II}\right\}n_i|_\Gamma\end{aligned} \tag{5.11d}$$

The necessary conditions of existence for a Y-periodic solution of (5.9)-(5.11) are :

$$\begin{aligned}\frac{\varepsilon\omega_K\omega_c}{\omega_m Pe^I}b^{II}\partial_\tau\left(\omega_m\langle s^I\rangle_1+\langle s^{II}\rangle_2\right)=&\partial_{\mathrm{x}i}\left\langle a^{II}\lambda_\alpha^{II}\left(\partial_i+\varepsilon\partial_{\mathrm{x}i}\right)p\right\rangle_2+\\&\omega_K\partial_{\mathrm{x}i}\left\langle a^I\lambda_\alpha^I\left(\partial_i+\varepsilon\partial_{\mathrm{x}i}\right)p\right\rangle_1,\end{aligned} \tag{5.12a}$$

$$\begin{aligned}Pe^I&\left\{\partial_{\mathrm{x}i}\left\langle a^{II}\lambda^{II}\left(\partial_i+\varepsilon\partial_{\mathrm{x}i}\right)p\right\rangle_2+\omega_K\partial_{\mathrm{x}i}\left\langle a^I\lambda^I\left(\partial_i+\varepsilon\partial_{\mathrm{x}i}\right)p\right\rangle_1\right\}+\\&\omega_c\partial_{\mathrm{x}i}\left\langle a^{II}\lambda_\beta^{II}\left(\partial_i+\varepsilon\partial_{\mathrm{x}i}\right)p_c\right\rangle_2+\\&\omega_K\partial_{\mathrm{x}i}\left\langle a^I\lambda_\beta^I\left(\partial_i+\varepsilon\partial_{\mathrm{x}i}\right)p_c\right\rangle_1=0\end{aligned} \tag{5.12b}$$

As usual, we apply the following notations:

$$\langle\cdot\rangle\equiv\int_Y(\cdot)dy \quad\text{and}\quad \langle\cdot\rangle_i\equiv\int_{Y^i}(\cdot)dy\ ,\quad i=I,II.$$

Conditions (5.12a) and (5.12b) were obtained by integrating equations (5.9)-(5.10) over the whole cell Y and by using the Green formula, with the conditions at the interface (5.11) and with $\int_Y(\partial f/\partial y_i)dy=0$, $\forall i=1,2,3$, for any $Y-$ periodic function $f(y)$.

5.3.2. ASYMPTOTIC EXPANSION FOR SOLUTION

General Structure of Expansion

It follows from (4.42) that:

$$p_c^I \sim \omega_c \qquad \text{and} \qquad p_c^{II} \sim 1 \tag{5.13a}$$

The following rescaling for the pressure results from the structure of Eqs. (5.9b) and (5.10b):

$$p \sim \omega_c / Pe^I \tag{5.13b}$$

As far as relation (5.7b) takes place, boundary condition (5.5b) is satisfied.

Then the following form of expansions is valid:

$$\begin{aligned} p^i(x,y,\tau) = \frac{\omega_c}{Pe^I}\Big\{ & p_0(x,\tau) + \omega_c p_{010}(x,\tau) + \omega_K p_{001}(x,\tau) + \ldots + \\ & \varepsilon\left[p_{100}^i(x,y,\tau) + \omega_c p_{110}^i(x,y,\tau) + \omega_K p_{101}^i(x,y,\tau) + \ldots\right] + \\ & \varepsilon^2 \ldots\Big\}, \qquad i = I, II \end{aligned} \tag{5.14a}$$

$$\begin{aligned} p_c^I(x,y,\tau) = \omega_c\Big\{ & p_{c0}(x,\tau) + \omega_c p_{c010}(x,\tau) + \omega_K p_{c001}(x,\tau) + \ldots + \\ & \varepsilon\left[p_{c100}^I(x,y,\tau) + \omega_c p_{c110}^I(x,y,\tau) + \omega_K p_{c101}^I(x,y,\tau) + \ldots\right] + \\ & \varepsilon^2 \ldots\Big\} \end{aligned} \tag{5.14b}$$

$$\begin{aligned} p_c^{II}(x,y,\tau) = & p_{c0}(x,\tau) + \omega_c p_{c010}(x,\tau) + \omega_K p_{c001}(x,\tau) + \ldots + \\ & \varepsilon\left[p_{c100}^{II}(x,y,\tau) + \omega_c p_{c110}^{II}(x,y,\tau) + \omega_K p_{c101}^{II}(x,y,\tau) + \ldots\right] + \\ & \varepsilon^2 \ldots\Big\} \end{aligned} \tag{5.14c}$$

$$s^I(x,y,\tau) = s^{I*} + \omega_c s_{010}^I(x,\tau) + \omega_K s_{001}^I(x,\tau) + \varepsilon s_{100}^I(x,y,\tau) + \ldots \tag{5.14d}$$

$$\begin{aligned} s^{II}(x,y,\tau) = & s_0^{II}(x,\tau) + \omega_c s_{010}^{II}(x,\tau) + \omega_K s_{001}^{II}(x,\tau) + \\ & \varepsilon s_{100}^{II}(x,y,\tau) + \ldots \end{aligned} \tag{5.14e}$$

Independence of the terms with zero first lower index on the fast variable y is already a result of substitution of these series in the initial equations, which may be easily verified.

Expansions for capillary pressure and saturation are not independent, because these functions are related by Eq. (5.6). To determine the relation between consequent terms of series (5.14b)-(5.14c) and (5.14d)-(5.14e), we use Taylor's expansion for the functions $\tilde{p}_c^i(s^i)$, $i{=}I, II$:

$$\tilde{p}_c^I(s^I) = \tilde{p}_c^I(s^{I*}){+}\omega_c\tilde{p}_c^{I\prime}(s^{I*})s_{010}^I{+}\omega_K\tilde{p}_c^{I\prime}(s^{I*})s_{001}^I{+}\varepsilon... \quad (5.15a)$$

$$\tilde{p}_c^{II}(s^{II}) = \tilde{p}_c^{II}(s_0^{II}){+}\omega_c\tilde{p}_c^{II\prime}(s_0^{II})s_{010}^{II}{+}\omega_K\tilde{p}_c^{II\prime}(s_0^{II})s_{001}^{II}{+}\varepsilon... \quad (5.15b)$$

Then, according to (5.6), we get relating equations for the consequent terms of expansion (5.14b) - (5.14c) for capillary pressure and of (5.14d) - (5.14e) for saturation:

$$\omega_c p_{c0}{=}\tilde{p}_c^I(s^{I*}),\ p_{c0}{=}\tilde{p}_c^{II}(s_0^{II}),\ \omega_c p_{c010}{=}\tilde{p}_c^{I\prime}(s^{I*})s_{010}^I,$$
$$p_{c010}{=}\tilde{p}_c^{II\prime}(s_0^{II})s_{010}^{II},\ \omega_c p_{c001}{=}\tilde{p}_c^{I\prime}(s^{I*})s_{001}^I,\ p_{c001}{=}\tilde{p}_c^{II\prime}(s_0^{II})s_{001}^{II} \quad (5.16)$$

Limit at $\varepsilon\to 0$, $\omega_K\to 0$ and Zero Term of Expansions

For a moderately heterogeneous medium, i.e., when $\omega_K{\sim}1$, the weak limit at $\varepsilon\to 0$ is equal to the simple averaged value. However in case of a highly heterogeneous medium, the limit behavior is more complicated. For instance, in the medium examined in all previous chapters, i.e., when the inclusions are lowly permeable, the simple averaged value may be not bounded. In this case, the limit corresponds to a nonuniformly averaged value.

In case we are studying, the limit can be determined basing on physical ideas. In fact, when $\omega_K\to 0$ the blocks become infinitely permeable. Then the pressure field inside each block remains constant and, consequently, bounded. Due to boundary condition (5.5b) the pressure field is expected to be bounded within the whole cell Y. Decreasing of the heterogeneity scale ε to zero keeps the pressure field bounded.

Because dimensional capillary pressures in media I and II are of the same order in quasi-linear mode, the capillary pressure filed should be bounded inside each cell.

As far as the averaged pressure and capillary pressure over Y always exists, we can assume it is equal to the weak limit of corresponding functions at $\varepsilon\to 0$.

Thus, in this case, notion of non-uniform averaging becomes useless, and traditional homogenization technique can be applied, although the medium is highly heterogeneous.

Asymptotic expansions for any function $f(x,y,\tau)$ may be defined more strictly in such a way, that the part of this function which does not depend on the fast variable y would be equal to the value averaged over the whole heterogeneity cell Y. Therefore, if:

$$f{=}u(x,y,\tau){+}w(x,\tau) \quad (5.17a)$$

then $\langle f\rangle=w$, or $\langle u\rangle=0$, or:

$$\langle u^{II}\rangle_2 = -\langle u^I\rangle_1 \tag{5.17b}$$

Then denoting each term of expansions (5.14) as:

$$p^i_{jkl}=u^i_{jkl}(x,y,\tau)+w_{jkl}(x,\tau),\quad p^i_{cjkl}=u^i_{cjkl}(x,y,\tau)+w_{cjkl}(x,\tau),$$

$$s^i_{jkl}=u^i_{sjkl}(x,y,\tau)+w_{sjkl}(x,\tau),\quad i=I,II;\quad j,k,l\geq 0$$

we get the following condition on the structure of the expansions:

$$\frac{\omega_c}{Pe^I}\left\{\left\langle\varepsilon u^I_{100}+\varepsilon\omega_c u^I_{110}+...\right\rangle_1+\left\langle\varepsilon u^{II}_{100}+\varepsilon\omega_c u^{II}_{110}+...\right\rangle_2\right\}=0 \tag{5.18a}$$

$$\omega_c\left\langle\varepsilon u^I_{c100}+\varepsilon\omega_c u^I_{c110}+...\right\rangle_1+\left\langle\varepsilon u^{II}_{c100}+\varepsilon\omega_c u^{II}_{c110}+...\right\rangle_2=0 \tag{5.18b}$$

5.3.3. EXPANSION OF THE PROBLEM

Substituting series (5.14) in problem (5.9a)-(5.12b) we obtain the following results for each term of expansions (5.14).

In Y^I

$$\partial_i\left(a^I\left(\partial_i p^I_{1jk}+\partial_{xi}p_{0jk}\right)\right)=0,\quad p^I_{1jk}|_\Gamma=p^{II}_{1jk}|_\Gamma;$$
$$j=0,1,2;\quad k=0,1;\quad j+k\leq 2 \tag{5.19a}$$

$$\partial_i\left(a^I\left(\partial_i p^I_{c1jk}+\partial_{xi}p_{c0jk}+\right)\right)=0,\quad p^I_{c1jk}|_\Gamma=p^{II}_{c1jk}|_\Gamma;$$
$$j,k=0,1;\quad j+k\leq 1 \tag{5.19b}$$

where $p_{000}\equiv p_0$.

In Y^{II}

$$\partial_i\left(a^{II}\left(\partial_i p^{II}_{1jk}+\partial_{xi}p_{0jk}\right)\right)=0,\quad j=0,1,2;\; k=0,1;\; j+k\leq 2; \tag{5.20a}$$

$$a^{II}\left(\partial_i p^{II}_{100}+\partial_{xi}p_0\right)n_i|_\Gamma=0,\quad \langle u^{II}_{100}\rangle_2=-\langle u^I_{100}\rangle_1; \tag{5.20b}$$

$$a^{II}\left(\partial_i p^{II}_{110}+\partial_{xi}p_{010}\right)n_i|_\Gamma=0,\quad \langle u^{II}_{110}\rangle_2=-\langle u^I_{110}\rangle_1; \tag{5.20c}$$

$$\left\{\lambda^{II}_{\alpha 0}a^{II}\left(\partial_i p^{II}_{101}+\partial_{xi}p_{001}\right)+\lambda^{II\prime}_{\alpha 0}s^{II}_{001}a^{II}\left(\partial_i p^{II}_{100}+\partial_{xi}p_0\right)\right\}n_i|_\Gamma=$$
$$\lambda^{I*}_\alpha a^I\left(\partial_i p^I_{100}+\partial_{xi}p_0\right)n_i|_\Gamma,\quad \langle u^{II}_{101}\rangle_2=-\langle u^I_{101}\rangle_1; \tag{5.20d}$$

$$\left\{\lambda_{\alpha0}^{II}a^{II}\left(\partial_i p_{111}^{II}+\partial_{xi}p_{011}\right)+\lambda_{\alpha0}^{II\prime}s_{010}^{II}a^{II}\left(\partial_i p_{101}^{II}+\partial_{xi}p_{001}\right)\right\}n_i|_\Gamma=$$
$$\left\{\lambda_\alpha^{I*}a^I\left(\partial_i p_{110}^I+\partial_{xi}p_{010}\right)+\right.$$
$$\left.\lambda_\alpha^{I\prime*}s_{010}^I a^I\left(\partial_i p_{100}^I+\partial_{xi}p_0\right)\right\}n_i|_\Gamma,\quad \langle u_{111}^{II}\rangle_2=-\langle u_{111}^I\rangle_1; \quad (5.20e)$$

$$\lambda_{\alpha0}^{II}a^{II}\left(\partial_i p_{120}^{II}+\partial_{xi}p_{020}\right)n_i|_\Gamma=0,\quad \langle u_{120}^{II}\rangle_2=-\langle u_{120}^I\rangle_1; \quad (5.20f)$$

and

$$\partial_i\left(a^{II}\left(\partial_i p_{c1jk}^{II}+\partial_{xi}p_{c0jk}\right)\right)=0,\quad j=0,1,2;\ k=0,1;\ j+k\le 2; \quad (5.21a)$$
$$a^{II}\left(\partial_i p_{c100}^{II}+\partial_{xi}p_{c0}\right)n_i|_\Gamma=0,\quad \langle u_{c100}^{II}\rangle_2=0; \quad (5.21b)$$
$$a^{II}\left(\partial_i p_{c110}^{II}+\partial_{xi}p_{c010}\right)n_i|_\Gamma=0,\quad \langle u_{c110}^{II}\rangle_2=-\langle u_{c100}^I\rangle_1; \quad (5.21c)$$
$$a^{II}\left(\lambda_{\beta0}^{II}\left(\partial_i p_{c101}^{II}+\partial_{xi}p_{c001}\right)+\lambda_0^{II}\left(\partial_i p_{101}^{II}+\partial_{xi}p_{001}\right)\right\}n_i|_\Gamma=$$
$$a^I\lambda_\alpha^{I*}\left(\partial_i p_{100}^I+\partial_{xi}p_0\right)n_i|_\Gamma,\quad \langle u_{101}^{II}\rangle_2=0; \quad (5.21d)$$

$$a^{II}\left\{\lambda_{\beta0}^{II}\left(\partial_i p_{c111}^{II}+\partial_{xi}p_{c011}\right)+\lambda_{\beta0}^{II\prime}s_{010}^{II}\left(\partial_i p_{c101}^{II}+\partial_{xi}p_{c001}\right)+\right.$$
$$\left.\lambda_0^{II}\left(\partial_i p_{111}^{II}+\partial_{xi}p_{011}\right)+\lambda_0^{II\prime}s_{010}^{II}\left(\partial_i p_{101}^{II}+\partial_{xi}p_{001}\right)\right\}n_i|_\Gamma=$$
$$a^I\left\{\lambda_\beta^{I\prime*}\left(\partial_i p_{c100}^I+\partial_{xi}p_{c0}\right)+\lambda_\alpha^{I*}\left(\partial_i p_{110}^I+\partial_{xi}p_{010}\right)+\right.$$
$$\left.\lambda^{I\prime*}s_{010}^I\left(\partial_i p_{100}^I+\partial_{xi}p_0\right)\right\}n_i|_\Gamma,\quad \langle u_{c111}^{II}\rangle_2=-\langle u_{c101}^I\rangle_1; \quad (5.21e)$$

$$a^{II}\left\{\lambda_{\beta0}^{II}\left(\partial_i p_{c120}^{II}+\partial_{xi}p_{c020}\right)+\lambda_0^{II}\left(\partial_i p_{120}^{II}+\partial_{xi}p_{020}\right)n_i|_\Gamma=0,\right.$$
$$\langle u_{c120}^{II}\rangle_2=-\langle u_{c110}^I\rangle_1; \quad (5.21f)$$

where conditions (5.11) at the interface Γ and structural conditions (5.18) are already introduced as boundary and integral conditions of corresponding problems.

5.3.4. STRUCTURE OF THE TERMS IN PRESSURE AND SATURATION SERIES

Representations for the Terms of Expansions

Let us introduce the following notations:

$$C_k\equiv\langle\psi_k^I\rangle_1,\quad D_k\equiv\langle\chi_k^I\rangle_1,\quad k=1,2,3;\quad \Phi(x,\tau)\equiv\frac{\lambda_\alpha^{I\prime*}s_{010}^I}{\lambda_{\alpha0}^{II}}-\frac{\lambda_{\alpha0}^{II\prime}\lambda_\alpha^{I*}s_{010}^{II}}{(\lambda_{\alpha0}^{II})^2};$$

$$\Psi(x,\tau)\equiv\frac{\lambda_{\alpha}^{I\prime*}\lambda_{\beta 0}^{II}-\lambda_{\beta}^{I\prime*}\lambda_{\alpha 0}^{II}}{\lambda_{\alpha 0}^{II}}s_{010}^{I}-\frac{\lambda_{\alpha}^{I*}\lambda_{\beta 0}^{II}\lambda_{\alpha 0}^{II\prime}}{\left(\lambda_{\alpha 0}^{II}\right)^{2}}s_{010}^{II}$$

It follows from the relationships of the previous section, with $i{=}I,II$:

$$p_{100}^{i}=\left(\psi_{k}^{i}(y)-C_{k}\right)\partial_{\mathrm{x}k}p_{0}+w_{100}(x,\tau), \tag{5.22a}$$

$$p_{110}^{i}=\left(\psi_{k}^{i}-C_{k}\right)\partial_{\mathrm{x}k}p_{0}+w_{110}(x,\tau), \tag{5.22b}$$

$$p_{101}^{i}=\left(\psi_{k}^{i}-C_{k}\right)\partial_{\mathrm{x}k}p_{001}+\frac{\lambda_{\alpha}^{I*}}{\lambda_{\alpha 0}^{II}}\chi_{k}^{i}(y)\partial_{\mathrm{x}k}p_{0}+w_{101}(x,\tau), \tag{5.22c}$$

$$p_{111}^{i}=\left(\psi_{k}^{i}-C_{k}\right)\partial_{\mathrm{x}k}p_{011}+\frac{\lambda_{\alpha}^{I*}}{\lambda_{\alpha 0}^{II}}\chi_{k}^{i}(y)\partial_{\mathrm{x}k}p_{010}+\Phi\chi_{k}^{i}\partial_{\mathrm{x}k}p_{0}+w_{111}, \tag{5.22d}$$

$$p_{120}^{i}=\left(\psi_{k}^{i}-C_{k}\right)\partial_{\mathrm{x}k}p_{020}+w_{120}(x,\tau), \tag{5.22e}$$

For the capillary pressure, where always $i{=}I,II$:

$$p_{c100}^{i}=\left(\psi_{k}^{i}(y)-C_{k}\right)\partial_{\mathrm{x}k}p_{c0}+w_{c100}(x,\tau), \tag{5.23a}$$

$$p_{c110}^{i}=\left(\psi_{k}^{i}-C_{k}\right)\partial_{\mathrm{x}k}p_{c010}-\frac{1}{1-\theta}C_{k}\partial_{\mathrm{x}k}p_{c0}+w_{c110}(x,\tau), \tag{5.23b}$$

$$p_{c101}^{i}=\left(\psi_{k}^{I}-C_{k}\right)\partial_{\mathrm{x}k}p_{c001}-\frac{\lambda_{\alpha}^{I*}}{\lambda_{\alpha 0}^{II}}\chi_{k}^{I}(y)\partial_{\mathrm{x}k}p_{0}+w_{c101}(x,\tau), \tag{5.23c}$$

$$\begin{aligned} p_{c111}^{II}=&\left(\psi_{k}^{II}-C_{k}\right)\partial_{\mathrm{x}k}p_{c011}-\frac{1}{1-\theta}C_{k}\partial_{\mathrm{x}k}p_{c001}+\frac{\lambda_{\beta}^{I\prime*}s_{010}^{I}}{\lambda_{\beta 0}^{II}}\chi_{k}^{II}\partial_{\mathrm{x}k}p_{c0}-\\ &\left[\frac{\Psi\chi_{k}^{II}}{\lambda_{\beta 0}^{II}}-\frac{\lambda_{\alpha}^{I*}D_{k}}{(1-\theta)\lambda_{\alpha 0}^{II}}\right]\partial_{\mathrm{x}k}p_{0}-\\ &\frac{\lambda_{\alpha}^{I*}}{\lambda_{\alpha 0}^{II}}\chi_{k}^{II}\partial_{\mathrm{x}k}p_{010}+w_{c111}(x,\tau), \end{aligned} \tag{5.23d}$$

$$\begin{aligned} p_{c120}^{II}=&\left(\psi_{k}^{II}-C_{k}\right)\partial_{\mathrm{x}k}p_{c020}-\frac{C_{k}}{1-\theta}\partial_{\mathrm{x}k}p_{c010}+\\ &\frac{\theta C_{k}}{(1-\theta)^{2}}\partial_{\mathrm{x}k}p_{c0}+w_{c120}(x,\tau) \end{aligned} \tag{5.23e}$$

Cell Problems

The functions $\psi_{k}^{I}(y)$ and $\chi_{k}^{II}(y)$, $k{=}1,2,3$ are defined as the solutions of the following cell problems:

$$\begin{cases} \partial_{i}\left(a^{I}(y)\left(\partial_{i}\psi_{k}^{I}+\delta_{ik}\right)\right)=0, \quad y\in Y^{I} \\ \psi_{k}^{I}\big|_{\Gamma}=-\,y_{k}\big|_{\Gamma}+\frac{1}{1-\theta}\langle y_{k}\rangle_{2}, \quad \psi_{k}^{I}(y)\text{ is y-periodic} \end{cases} \tag{5.24a}$$

$$\begin{cases} \partial_i \left(a^{II}(y)\partial_i \chi_k^{II}\right)=0, \quad y\in Y^{II} \\ a^{II}\partial_i\chi_k^{II} n_i|_\Gamma = a^I\partial_i\left(\psi_k^I+\delta_{ik}\right)n_i|_\Gamma \\ \langle\chi_k^{II}\rangle_2=0 \end{cases} \tag{5.24b}$$

The functions $\psi_k^{II}(y)$ and $\chi_k^I(y)$ are solutions of the following problems:

$$\begin{cases} \partial_i \left(a^{II}(y)\left(\partial_i\psi_k^{II}+\delta_{ik}\right)\right)=0, \quad y\in Y^{II} \\ a^{II}\left(\partial_i\psi_k^{II}+\delta_{ik}\right)n_i|_\Gamma=0, \\ \langle\psi_k^{II}\rangle_2=0 \end{cases} \tag{5.24c}$$

$$\begin{cases} \partial_i \left(a^{I}(y)\partial_i \chi_k^{I}\right)=0, \quad y\in Y^{I} \\ \chi_k^I|_\Gamma=\chi_k^{II}|_\Gamma, \quad \chi_k^I(y) \text{ is y-periodic} \end{cases} \tag{5.24d}$$

where θ is the volume fraction of blocks.

Problem (5.24c) for each k has an explicit solution:

$$\psi_k^{II}=-y_k+\frac{1}{1-\theta}\langle y_k\rangle_2 \tag{5.25}$$

Moreover, due to the boundary-value condition on Γ in (5.24c) the following property takes place:

$$\int\limits_{Y^{II}}\partial_i\left(a^{II}\left(\partial_i\psi_k^{II}+\delta_{ik}\right)\right)dy=\int\limits_{\Gamma}a^{II}\left(\partial_i\psi_k^{II}+\delta_{ik}\right)n_i dy\equiv 0 \tag{5.26}$$

A simple substitution of all these relations in Eqs. (5.19) - (5.21) proves they validity.

5.3.5. DEDUCTION OF MACROSCALE EQUATIONS

It follows from integral relation (5.12a) :

$$\frac{1}{\omega_m}\partial_\tau\left(\langle\omega_m S^{I*}\rangle_1+\langle S_0^{II}\rangle_2+\omega_c\left[\langle\omega_m S_{010}^I\rangle_1+\langle S_{010}^{II}\rangle_2\right]\right)=$$
$$\partial_{\mathrm{x}i}\left\langle\lambda_{\alpha0}^{II}\partial_{\mathrm{x}k}p_{001}a^{II}\zeta_{ik}^{II}(y)\right\rangle_2+\partial_{\mathrm{x}i}\left\langle\lambda_\alpha^{I*}\partial_{\mathrm{x}k}p_0 a^{II}\partial_i\chi_k^{II}\right\rangle_2+$$
$$\partial_{\mathrm{x}i}\left\langle\lambda_\alpha^{I*}\partial_{\mathrm{x}k}p_0 a^I\zeta_{ik}^I\right\rangle_1+$$

$$\omega_c\Big\{\partial_{\mathrm{x}i}\Big\langle\lambda^{II}_{\alpha 0}\partial_{\mathrm{x}k}p_{011}a^{II}\zeta^{II}_{ik}\Big\rangle_2+\partial_{\mathrm{x}i}\Big\langle\lambda^{I*}_{\alpha}\partial_{\mathrm{x}k}p_{010}a^{II}\partial_i\chi^{II}_k\Big\rangle_2+$$
$$\partial_{\mathrm{x}i}\Big\langle\lambda^{II}_{\alpha 0}\Phi\partial_{\mathrm{x}k}p_0a^{II}\partial_i\chi^{II}_k\Big\rangle_2+\partial_{\mathrm{x}i}\Big\langle\lambda^{II'}_{\alpha 0}S^{II}_{010}\partial_{\mathrm{x}k}p_{001}a^{II}\zeta^{II}_{ik}\Big\rangle_2+$$
$$\partial_{\mathrm{x}i}\Big\langle\frac{\lambda^{I*}_{\alpha}\lambda^{II'}_{\alpha 0}S^{II}_{010}}{\lambda^{II}_{\alpha 0}}\partial_{\mathrm{x}k}p_0a^{II}\partial_i\chi^{II}_k\Big\rangle_2+\partial_{\mathrm{x}i}\Big\langle\lambda^{I*}_{\alpha}\partial_{\mathrm{x}k}p_{010}a^{I}\zeta^{I}_{ik}\Big\rangle_1+$$
$$\partial_{\mathrm{x}i}\Big\langle\lambda^{I'*}_{\alpha}S^{I}_{010}\partial_{\mathrm{x}k}p_0a^{I}\zeta^{I}_{ik}\Big\rangle_1\Big\}+\ldots \tag{5.27a}$$

where $\zeta^{II}_{ik}(y)\equiv\partial_i\psi^{II}_k(y)+\delta_{ik}$, $\zeta^{I}_{ik}(y)\equiv \partial_i\psi^{I}_k(y)+\delta_{ik}$.

From (5.12b) :

$$0=\partial_{\mathrm{x}i}\Big\langle\lambda^{II}_{0}\partial_{\mathrm{x}k}p_{001}a^{II}\zeta^{II}_{ik}\Big\rangle_2+\partial_{\mathrm{x}i}\Big\langle\frac{\lambda^{I*}_{\alpha}\lambda^{II}_{0}}{\lambda^{II}_{\alpha 0}}\partial_{\mathrm{x}k}p_0a^{II}\partial_i\chi^{II}_k\Big\rangle_2+$$
$$\partial_{\mathrm{x}i}\Big\langle\lambda^{II}_{\beta 0}\partial_{\mathrm{x}k}p_{c001}a^{II}\zeta^{II}_{ik}\Big\rangle_2-\partial_{\mathrm{x}i}\Big\langle\frac{\lambda^{I*}_{\alpha}\lambda^{II}_{\beta 0}}{\lambda^{II}_{\alpha 0}}\partial_{\mathrm{x}k}p_0a^{II}\partial_i\chi^{II}_k\Big\rangle_2+$$
$$\partial_{\mathrm{x}i}\Big\langle\lambda^{I*}_{\alpha}\partial_{\mathrm{x}k}p_0a^{I}\zeta^{I}_{ik}\Big\rangle_1+$$
$$\omega_c\Big\{\partial_{\mathrm{x}i}\Big\langle\lambda^{II}_{0}\partial_{\mathrm{x}k}p_{011}a^{II}\zeta^{II}_{ik}\Big\rangle_2+\partial_{\mathrm{x}i}\Big\langle\frac{\lambda^{I*}_{\alpha}\lambda^{II}_{0}}{\lambda^{II}_{\alpha 0}}\partial_{\mathrm{x}k}p_{010}a^{II}\partial_i\chi^{II}_k\Big\rangle_2+$$
$$\partial_{\mathrm{x}i}\Big\langle\lambda^{II}_{0}\Phi\partial_{\mathrm{x}k}p_0a^{II}\partial_i\chi^{II}_k\Big\rangle_2+\partial_{\mathrm{x}i}\Big\langle\lambda^{II'}_{0}S^{II}_{010}\partial_{\mathrm{x}k}p_{001}a^{II}\zeta^{II}_{ik}\Big\rangle_2+$$
$$\partial_{\mathrm{x}i}\Big\langle\frac{\lambda^{I*}_{\alpha}\lambda^{II\prime}_{0}S^{II}_{010}}{\lambda^{II}_{\alpha 0}}\partial_{\mathrm{x}k}p_0a^{II}\partial_i\chi^{II}_k\Big\rangle_2+\partial_{\mathrm{x}i}\Big\langle\lambda^{I*}_{\alpha}\partial_{\mathrm{x}k}p_{010}a^{I}\zeta^{I}_{ik}\Big\rangle_1+$$
$$\partial_{\mathrm{x}i}\Big\langle\lambda^{I\prime*}S^{I}_{010}\partial_{\mathrm{x}k}p_0a^{I}\zeta^{I}_{ik}\Big\rangle_1+\partial_{\mathrm{x}i}\Big\langle\lambda^{I\prime*}_{\beta}S^{I}_{010}\partial_{\mathrm{x}k}p_{c0}a^{I}\zeta^{I}_{ik}\Big\rangle_1+$$
$$\partial_{\mathrm{x}i}\Big\langle\lambda^{II}_{\beta 0}\partial_{\mathrm{x}k}p_{c011}a^{II}\zeta^{II}_{ik}\Big\rangle_2+\partial_{\mathrm{x}i}\Big\langle\lambda^{I\prime*}_{\beta}\partial_{\mathrm{x}k}p_{c0}a^{II}\partial_i\chi^{II}_k\Big\rangle_2-$$
$$\partial_{\mathrm{x}i}\Big\langle\Psi\partial_{\mathrm{x}k}p_0a^{II}\partial_i\chi^{II}_k\Big\rangle_2-\partial_{\mathrm{x}i}\Big\langle\frac{\lambda^{I*}_{\alpha}\lambda^{II}_{\beta 0}}{\lambda^{II}_{\alpha 0}}\partial_{\mathrm{x}k}p_{010}a^{II}\partial_i\chi^{II}_k\Big\rangle_2+$$
$$\partial_{\mathrm{x}i}\Big\langle\lambda^{II\prime}_{\beta 0}S^{II}_{010}\partial_{\mathrm{x}k}p_{c001}a^{II}\zeta^{II}_{ik}\Big\rangle_2-$$
$$\partial_{\mathrm{x}i}\Big\langle\frac{\lambda^{I*}_{\alpha}\lambda^{II\prime}_{\beta 0}S^{II}_{010}}{\lambda^{II}_{\alpha 0}}\partial_{\mathrm{x}k}p_0a^{II}\partial_i\chi^{II}_k\Big\rangle_2\Big\}+\ldots \tag{5.27b}$$

Performing simplest transformations, we obtain at last the following macroscale equations:

$$\partial_\tau\left(\omega_m\theta S^I+(1-\theta)S^{II}\right)=\partial_{\mathrm{x}i}\left(\frac{A_{ik}\widehat{k}_\alpha(S^I)}{\mu_\alpha}\partial_{\mathrm{x}k}P\right) \tag{5.28a}$$

$$\partial_{\mathrm{x}i}\left(\frac{A_{ik}\widehat{k}_{\beta}(S^{I})}{\mu_{\beta}}\partial_{\mathrm{x}k}P_{c}\right)+$$

$$Pe^{I}\partial_{\mathrm{x}i}\left(A_{ik}\left[\frac{\widehat{k}_{\alpha}(S^{I})}{\mu_{\alpha}}+\frac{\widehat{k}_{\beta}(S^{I})}{\mu_{\beta}}\right]\partial_{\mathrm{x}k}P\right)=0 \tag{5.28b}$$

$$\tilde{p}_{c}^{II}(S^{II})=\frac{1}{\omega_{c}}\tilde{p}_{c}^{I}(S^{I})\simeq\frac{1}{\omega_{c}}\frac{d\tilde{p}_{c}^{I}}{ds}\left(s^{I*}\right)\left(S^{I}-s^{I*}\right) \tag{5.28c}$$

$$P_{c}=\tilde{p}_{c}^{I}(S^{I})\simeq\frac{d\tilde{p}_{c}^{I}}{ds}(s^{I*})\left(S^{I}-s^{I*}\right) \tag{5.28d}$$

where θ is the volume fraction of blocks; S^{I} and S^{II} are the macroscale matrix and block saturations; P is the macroscale pressure in the α-phase; P_c is the macroscale capillary pressure.

The macroscale relative permeabilities $\widehat{k}_{\alpha}(S^{I})$ and $\widehat{k}_{\beta}(S^{I})$ are defined as:

$$\widehat{k}_{\alpha}(S^{I})\equiv k_{\alpha}^{I}(S^{I})\simeq k_{\alpha}^{I}(s^{I*})+\frac{dk_{\alpha}^{I}}{ds}\left(s^{I*}\right)(S^{I}-s^{I*}), \tag{5.28e}$$

$$\widehat{k}_{\beta}(S^{I})\equiv k_{\beta}^{I}(S^{I})\simeq\frac{dk_{\beta}^{I}}{ds}\left(s^{I*}\right)(S^{I}-s^{I*}), \tag{5.28f}$$

The effective global permeability tensor A_{ik} is defined in the following way:

$$A_{ik}=\left\langle a^{I}\left(\partial_{i}\psi_{k}^{I}+\delta_{ik}\right)\right\rangle_{1}+\left\langle a^{II}\partial_{i}\chi_{k}^{II}\right\rangle_{2} \tag{5.28g}$$

where the cell functions $\psi_{k}^{I}(y)$ and $\chi_{k}^{II}(y)$, $k=1,2,3$ are solutions of cell problems (5.24) .

5.4. MACROSCALE EQUATIONS FOR QUASI-LINEAR MODE OF FLOW

5.4.1. MACROSCALE EQUATIONS

Returning back to dimensional values we can present Eqs. (5.28) in the following form:

$$\frac{\partial}{\partial t}\left(\langle m\rangle_1 S^I+\langle m\rangle_2 S^{II}\right)=\frac{\partial}{\partial x_i}\left(\frac{\widehat{K}_{ik}\widehat{k}_\alpha(S^I)}{\mu_\alpha}\frac{\partial \widehat{P}_\alpha}{\partial x_k}\right) \tag{5.29a}$$

$$-\frac{\partial}{\partial t}\left(\langle m\rangle_1 S^I+\langle m\rangle_2 S^{II}\right)=\frac{\partial}{\partial x_i}\left(\frac{\widehat{K}_{ik}\widehat{k}_\beta(S^I)}{\mu_\beta}\frac{\partial\left(\widehat{P}_\alpha+\widehat{P}_c^I(S^I)\right)}{\partial x_k}\right) \tag{5.29b}$$

$$\widehat{P}_c^I(S^I)=\widehat{P}_c^{II}(S^{II}) \tag{5.29c}$$

$$\begin{cases} \widehat{P}_c^I(S^I)\equiv P_c^I(S^I) \\ \widehat{P}_c^{II}(S^{II})\equiv P_c^{II}(S^{II}) \end{cases} \tag{5.29d}$$

where $\widehat{K}_{ik}$ is the effective global permeability tensor; $\widehat{k}_\alpha(S^I)$ and $\widehat{k}_\beta(S^I)$ are the effective relative permeabilities; $\widehat{P}_c^i(S^i)$ is the effective capillary pressure for the medium i $(i{=}I, II)$; $\widehat{P}_\alpha$ is the effective pressure in the α-phase; $\langle m\rangle_i \equiv \int_{Y^i} m(y)dy$.

5.4.2. CAPILLARY EQUILIBRIUM BEHAVIOR

It cab be seen that Eq. (5.29c) establishes a macroscale equilibrium behavior of the system. In fact, any perturbation is more rapidly spreading within the highly permeable block, than through the tight matrix. Characteristic times of capillary waves propagation through a block and through the matrix within a cell are:

$$t^{*II}\sim\frac{l^2\mu_\alpha m^{II}}{K^{II}P_c^{II,0}}, \qquad t^{*I}\sim\frac{l^2\mu_\alpha m^I}{K^I P_c^{I,0}}$$

Therefore:

$$\frac{t^{*II}}{t^{*I}}\sim\frac{K^I}{K^{II}}\frac{P_c^{I,0}}{P_c^{II,0}}\sim\frac{\omega_K}{\omega_c}$$

Because some correlation like (4.23) and (4.41) are true, then

$$\frac{t^{*II}}{t^{*I}} \sim \sqrt{\omega_K} \ll 1$$

Thus transient process in the blocks happens more rapidly, than in the matrix. Because the time scale (5.3) is equivalent to the characteristic time of saturation propagation within the low permeable matrix, then any process within the blocks may be considered as spontaneous respectively to this time scale.

However it is necessary to note, that in papers [2] behavior of such systems was explained as non-equilibrium behavior.

5.4.3. EFFECTIVE ABSOLUTE PERMEABILITY

Definition of Effective Absolute Permeability

Tensor $\widehat{K}_{ik}$ is defined in another way, than for media with lowly permeable blocks, namely:

$$\widehat{K}_{ik}=\Big\langle K^I(y)\left(\frac{\partial \psi_k^I}{\partial y_i}+\delta_{ik}\right)\Big\rangle_1+\Big\langle \omega_K K^{II}(y)\frac{\partial \chi_k^{II}}{\partial y_i}\Big\rangle_2, \quad i,k=1,2,3 \qquad (5.30)$$

where the functions $\psi_k^I(y)$ and $\chi_k^{II}(y)$, $k=1,2,3$ are solutions of two cell problems:

$$\begin{cases} \dfrac{\partial}{\partial y_i}\left(K^I(y)\left(\dfrac{\partial \psi_k^I}{\partial y_i}+\delta_{ik}\right)\right)=0, \quad y\in Y^I \\ \psi_k^I\big|_\Gamma=-\,y_k\big|_\Gamma+\dfrac{1}{1-\theta}\langle y_k\rangle_2, \quad \psi_k^I(y) \text{ is y-periodic} \end{cases} \qquad (5.31a)$$

$$\begin{cases} \dfrac{\partial}{\partial y_i}\left(K^{II}(y)\dfrac{\partial \chi_k^{II}}{\partial y_i}\right)=0, \quad y\in Y^{II} \\ \omega_K K^{II}\dfrac{\partial \chi_k^{II}}{\partial y_i}n_i\big|_\Gamma=K^I\dfrac{\partial}{\partial y_i}\left(\psi_k^I+\delta_{ik}\right)n_i\big|_\Gamma \\ \langle \chi_k^{II}\rangle_2=0 \end{cases} \qquad (5.31b)$$

where θ is the volume fraction of the blocks.

Although relations (5.30) and (5.31) comprise the small parameter ω_K, all terms are of the same order, that follows explicitly from dimensionless form of these equations, (5.28g) and (5.24a) , (5.24b) .

Correctness of the Cell Problems

Problem (5.31a) is posed in a domain Y^I which contains a hole. The Dirichlet condition is posed on the boundary of the hole, and periodicity conditions are formulated on the external boundary of this domain. As the right-hand part of the differential equation satisfies the necessary condition of existence for periodic solution, correctness of such a mixed problem is obvious.

Problem (5.31b) is a classical Neumann boundary-value problem posed in a domain Y^{II} which is bounded by the boundary Γ. Solution of such a problem exists. The last integral condition ensures uniqueness of solution.

5.4.4. PROPERTIES AND SOLUTION OF CELL PROBLEMS

Reduction of the Periodic Problem (5.31a) to the Boundary-Value

It is easy to show, that problem (5.31a) with respect to the function $\psi_k^I(y)$ is equivalent to the following classical boundary-value problem, if the function $K^{II}y$ and the interface Γ are symmetric with respect to each axis y_j, $j{\neq}k$:

$$\left\{\begin{array}{l} \dfrac{\partial}{\partial y_i}\left(K^I(y)\dfrac{\partial \Psi_k^I}{\partial y_i}\right)=0, \quad y\in Y^I \\[2ex] \Psi_k^I|_\Gamma=\dfrac{1}{1-\theta}\langle y_k\rangle_2, \\[2ex] \Psi_k^I|_{y_k=1/2}=\dfrac{1}{2}, \quad \Psi_k^I|_{y_k=-1/2}=-\dfrac{1}{2}, \\[2ex] \dfrac{\partial \Psi_k^I}{\partial y_j}|_{y_j=\pm 1/2}=0, \quad j\neq k \end{array}\right. \tag{5.32}$$

where

$$\Psi_k^I(y) \equiv \psi_k^I(y) + y_k, \qquad k=1,2,3$$

This problem is a classical mixed Dirichlet-Neumann boundary-value problem, whose solution exists and is unique.

The function $\tilde{\psi}_k^I \equiv \Psi_k^I - y_k$ satisfies the same equation as ψ_k^I and the same boundary-value condition at the interface Γ. They will be really equivalent, if the function $\tilde{\psi}_k^I$ may be proved to be periodic. Periodicity of this function is easy to be established.

First, it follows from conditions at $y_k=\pm 1/2$, that:

$$\tilde{\psi}_k^I|_{y_k=-1/2} = \tilde{\psi}_k^I|_{y_k=+1/2} = 0$$

Due to the symmetry with respect to the axes y_j, $j{\neq}k$, the solution of (5.32) and the function $\tilde{\psi}_k^I$ is even respectively to the variable y_j, $j{\neq}k$.

Thus, the function $\tilde{\psi}_k^I$ is periodic. Then it is equivalent to ψ_k^I.

Physical Sense of the Cell Problem

The last problem has clear physical sense. It describes the local dimensionless pressure field Ψ_k^I inside the low permeable sub-domain of a cell, with constant pressure at the inlet and the outlet sides ($y_k{=}\pm 1/2$), impermeable other sides, and with a constant pressure along the whole interface Γ.

In fact, when the block is very permeable, the pressure drop within it is vanishing and can be neglected. Therefore, the pressure along the whole block/matrix interface is practically constant.

Then the sense of formula (5.25) becomes clear too. According to it, the function $\Psi_k^{II}{\equiv}\ \psi_k^{II}{+}y_k$ which may be treated as the local pressure inside the block, is constant.

This constant value of the block pressure is equal to $C{\equiv}\ \langle y_k\rangle_2/(1{-}\theta)$. When the shape of the block is symmetric with respect to each axis, this constant value is equal to zero.

Numerical Solution of the Cell Problems

All the cell problems have been computed using finite-element methods. The same symmetric cell structure was examined with a rectangular block ($\xi_1{\times}\xi_2$), as in chapter 4. Microscale permeability has been assumed to be a scalar function, i.e. $K_{ij}^{II}{=}K^{II}\delta_{ij}$, $K_{ij}^{I}{=}K^{I}\delta_{ij}$.

Typical behavior of two functions $\Psi_1^I(y)$ and $\psi_1^I(y){=}\ \Psi_1^I(y){-}y_1$, being solutions of problems (5.32) and (5.31a) , is shown in Figs. 5.3 and 5.4, on a quarter of the cell Y.

Two figures 5.4a and 5.4a show the same function $\psi_1(y_1, y_2)$ but under different angle of view.

The function $\chi_1^{II}(y)$ defined via the problem in the block II is displayed in Fig. 5.5

This solutions have been obtained using finite-element method.

Independence of the Effective Permeability of ω_K

Basic property of equation (5.30) is that the effective permeability of pseudo-cavity media does not depend on the heterogeneity parameter ω_K. This fact is more explicitly seen in the dimensionless form of the same formula, (5.28g) .

Thus, when the block permeability increases to infinity, the effective permeability $\widehat{K}$ of a pseudo-cavity medium tends to a constant limiting value. The realistic dependence of $\widehat{K}$ on ω_K when $\omega_K{>}1$ is shown in Fig. 5.6 bu the curve 1.

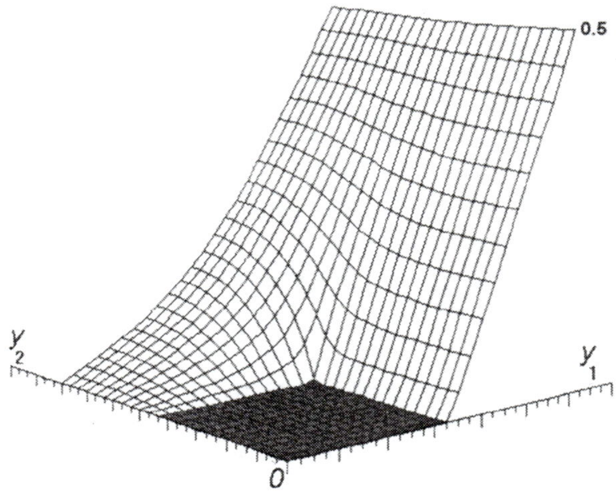

Figure 5.3. Cell function $\Psi_1^I(y)$ in 2D case. The block is a square

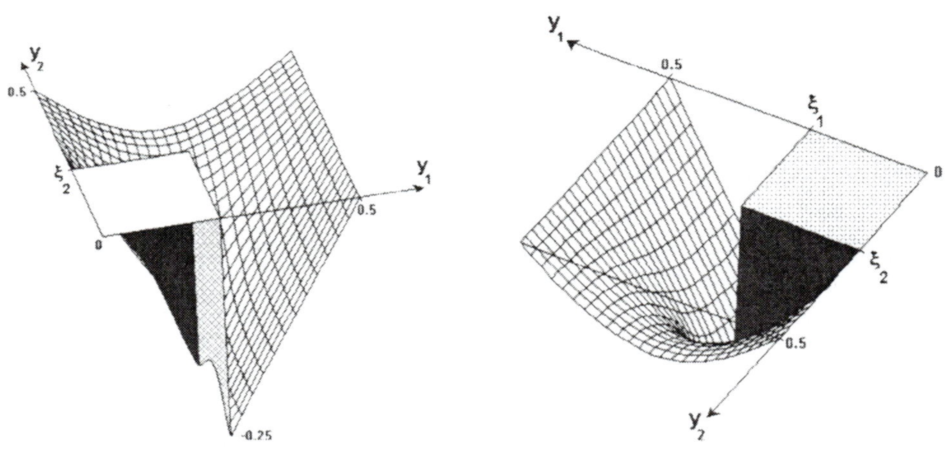

Figure 5.4. Cell function $\psi_1^I(y){=}\Psi_1^I(y){-}y_1$ in 2D case

Equation (5.30) obtained in this chapter gives the limiting line 2. The third curve is obtained using classical equation for the effective permeability of moderately heterogeneous media, which is independent on what whether $\omega_K{>}1$, or $\omega_K{<}1$:

$$\widehat{K}_{ik} = \left\langle \frac{\partial \psi_k}{\partial y_k} + \delta_{ik} \right\rangle$$

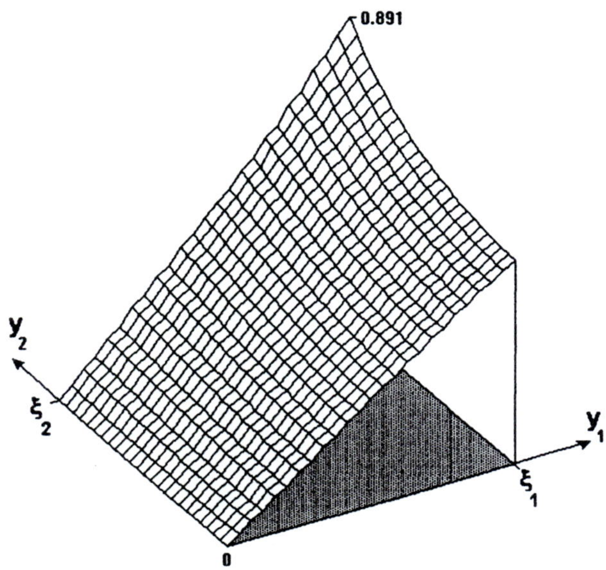

Figure 5.5. Cell function $\chi_1^{II}(y)$ inside the block Y^{II}

with the function psi_k defined as

$$\frac{\partial}{\partial y_i}\left(K(y)\left(\frac{\partial \psi_k}{\partial y_k}+\delta_{ik}\right)\right)=0, y\in Y;$$

$$\langle\psi_k\rangle=0, \qquad \psi_k \text{ is } y\text{-periodic}$$

As shown, the classical method predicts an infinite growth of the effective permeability when $\omega_K\to\infty$, that is not true.

Dependence of the Effective Permeability on Block Concentration

As usual, effective permeability depends on geometry of heterogeneity. Ignoring many geometrical details of the probable block location inside a cell, we will examine only more evident dependence of $\widehat{K}$ on the volume fraction of blocks θ. To avoid secondary geometrical effects, the medium was assumed to be isotropic on the microscale and on the macroscale, i.e., the block is a square, the microscale permeability is scalar and constant within each subdomain of a cell. Dependence $\widehat{K}(\theta)$ is presented in Fig. 5.7.

It is necessary to note, that this dependence should be invalid, when $\theta\to 1$, because the volume ratio of blocks and matrix was assumed to be of the same order within the framework of developed theory. Otherwise, a new small parameter will arise, which is the small thickness of the bands, the matrix will consist of.

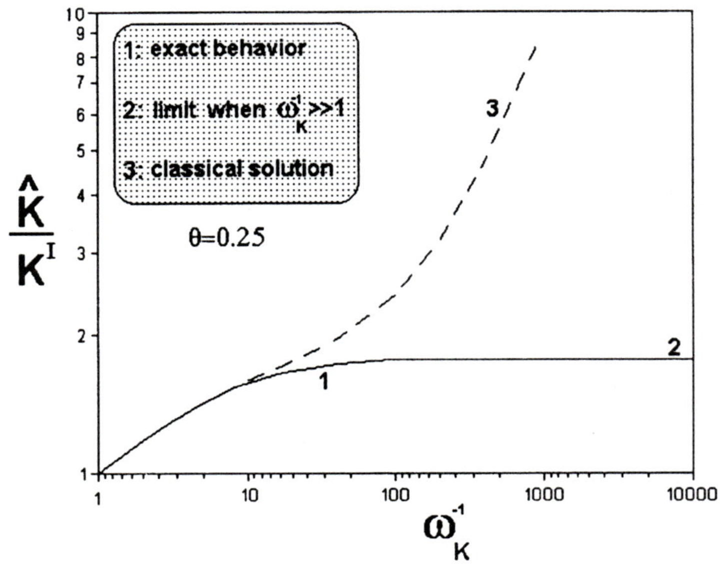

Figure 5.6. Dependence of the effective permeability on the heterogeneity degree ω_K

5.4.5. EFFECTIVE RELATIVE PERMEABILITIES

One of the most important results of this chapter is that the macroscale relative permeabilities $\widehat{k}_\alpha(S^I)$ and $\widehat{k}_\beta(S^I)$ do not depend on the microscale phase permeabilities in highly permeable blocks:

$$\widehat{k}_\alpha=\widehat{k}_\alpha(S^I)=k^I_\alpha(S^I), \qquad \widehat{k}_\beta=\widehat{k}_\beta(S^I)=k^I_\beta(S^I) \tag{5.33}$$

This result gives a powerful tool for a fast computation of effective relative permeabilities.

As in quasi-linear mode the saturation S^I is close to s^{I*}, the following relationships can be used too:

$$\widehat{k}_\alpha(S^I)\simeq k^I_\alpha(s^{I*})+\left.\frac{dk^I_\alpha}{ds}\right|_{s=s^{I*}}\left(S^I-s^{I*}\right)$$

$$\widehat{k}_\beta(S^I)\simeq\left.\frac{dk^I_\beta}{ds}\right|_{s=s^{I*}}\left(S^I-s^{I*}\right) \tag{5.34a}$$

Analogously, the effective capillary pressure is:

$$\widehat{P}^I_c(S^I)\simeq\left.\frac{dP^I_c}{ds}\right|_{s=s^{I*}}\left(S^I-s^{I*}\right) \tag{5.34b}$$

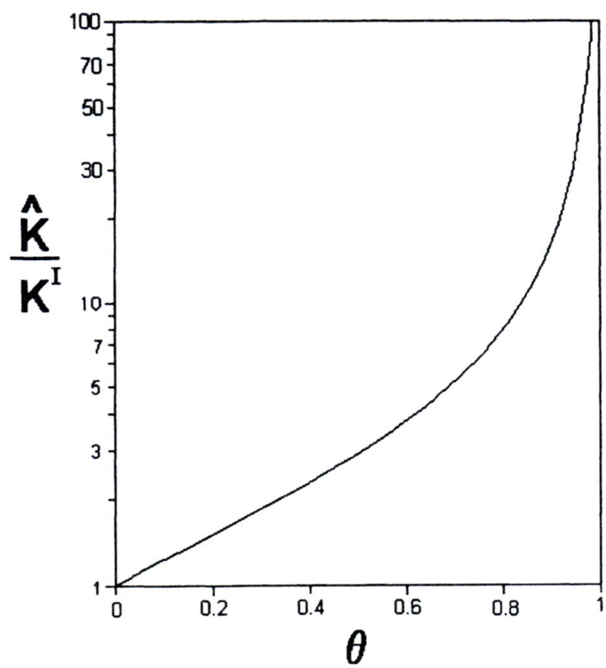

Figure 5.7. Dependence of the effective permeability of pseudo-cavity media on the volume fraction of highly permeable blocks θ

5.5. PARTIALLY FROZEN MODE. PORE-NETWORK MODELING

The frozen mode of flow seems to be rather trivial. In fact, as far as the saturation both in the blocks and in the matrix is equal to the percolation threshold for the α-phase, then the α-phase remaining immobile in the whole medium the flow becomes single-phase.

As to the partially frozen mode, the α-phase remains immobile only inside the blocks. Therefore the flow is two-phase in the tight matrix, but is one-phase in the highly permeable blocks. The character of the flow is completely clear, and the problem only consists in thorough deriving of the macroscale equations.

Instead of this, a series of network simulations will be performed to prove existence of the partially frozen mode and some features of flow.

5.5.1. PORE-NETWORK MODEL

A numerical simulator was used to study two-phase flow in the network model of porous media, thoroughly described in papers [6], [7] and in the book [5].

Within the frameworks of this model, the porous medium is considered as a connected lattice of cylindrical pores of a random radius r and a constant length l. The vertices of intersections of the pore axes form a regular lattice, as shown in Fig. 5.8.

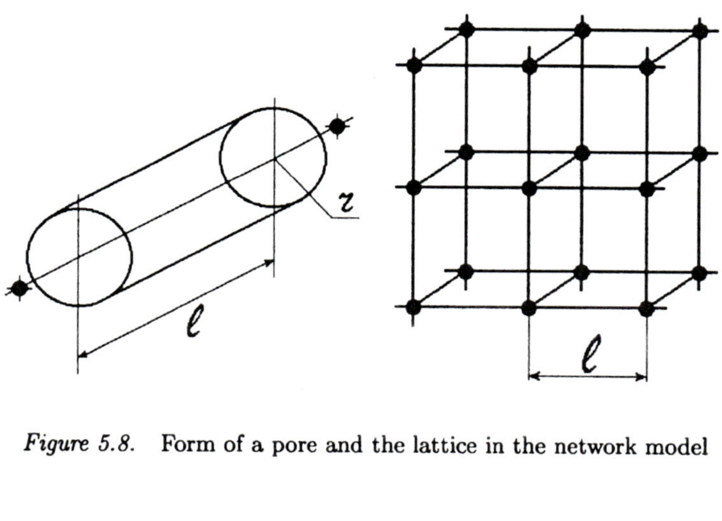

Figure 5.8. Form of a pore and the lattice in the network model

Flow inside each pore is governed by Poiseuille's law. Two immiscible fluids inside one pore are separated by a meniscus. The pressure discontinuity at a meniscus is equal to the Laplace capillary pressure. A meniscus can pass through a vertices provided that some conditions are satisfied. First at all, the displaced fluid must be connected up to the outside section of the medium. Secondly, the degree of pore restriction/constriction in the vertices must be limited in such a manner, that the capillary counter-force does not lock the meniscus advancement.

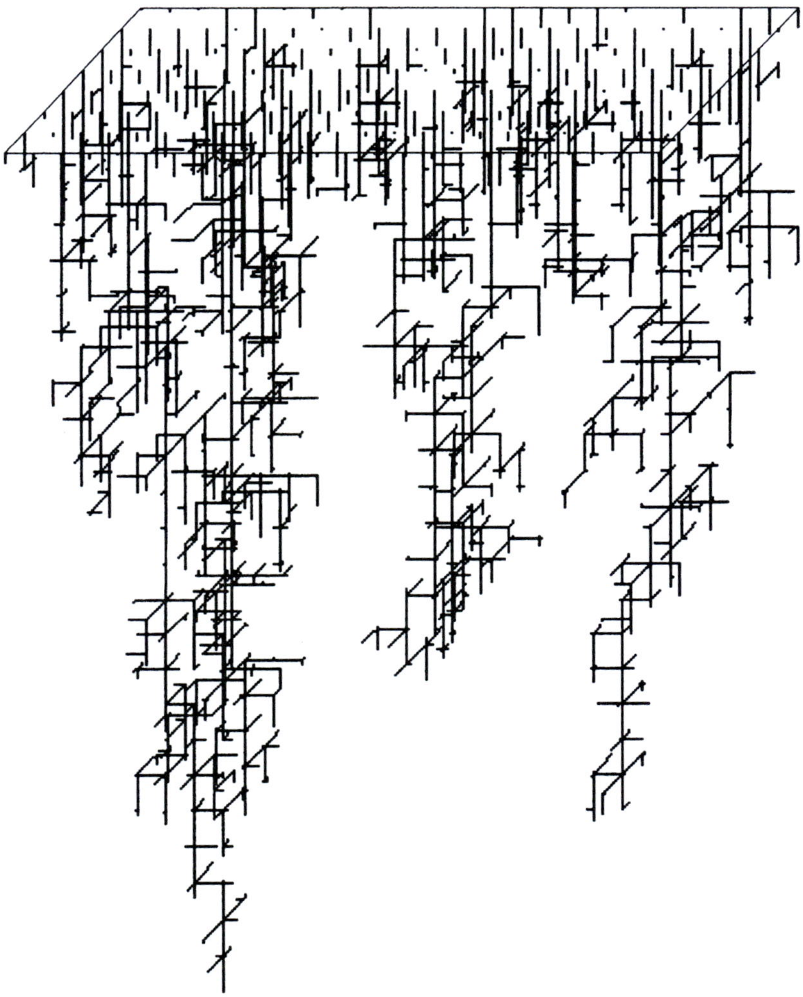

Figure 5.9. 3D pattern of invading fluid in the network model of porous medium

A percolation-difference method described in [7] and [6] was applied to solve the flow equations. According to this method, the pressure field is computed as a solution of finite-difference elliptic equations, while the saturation field is determined by means of an algorithm of meniscus system motion, which is percolation-like in its origin.

During one time step all the meniscus are displacing inside each pore in such a way, that any meniscus can not by-pass the vertices. Therefore, the spatial step of motion is always smaller than a pore length. For the meniscus, which has reached a vertices, conditions of passing through this vertices into other pore branches are checked. When the meniscus can pass through, new meniscus are born afterward the vertices. Analytical relationships are used to calculate the time needed to a meniscus to reach the adjacent vertices. This relations follow from Poiseuille's equation.

An example of 3D pattern of invading fluid is shown in Fig. 5.9, where the flow happens from the top section towards the bottom.

5.5.2. RESULTS OF SIMULATIONS

Flow when Blocks Are Impenetrable

A porous medium with two rectangular highly permeable inclusions has been examined, where the mean pore radius was ten times more large that in the connected matrix, such that the permeability ratio ω_K is of order 0,01. In Fig. 5.10 the averaged flow is directed from the top to the bottom, the capillary number Pe^I is of order 0,1. The displacing fluid is more dark. Two highly permeable blocks are blacked out.

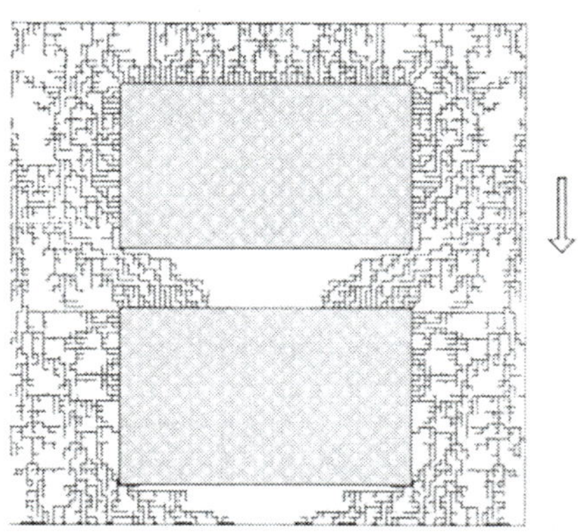

Figure 5.10. Network simulations of partially frozen mode of flow; ω_K=0.01; Pe^I=0.1. Highly permeable inclusions (two gray rectangles) remain impermeable for the invading fluid

We see that the displacing liquid (α-phase) can not invade the highly permeable blocks.

Partial Penetration of Wetting Fluid into the Blocks

When the heterogeneity degree is reduced, such that $\omega_K \sim 0.05$, the invading fluid partially penetrates into the blocks, as shown in Fig. 5.11, where four consequent time stage of invasion are presented.

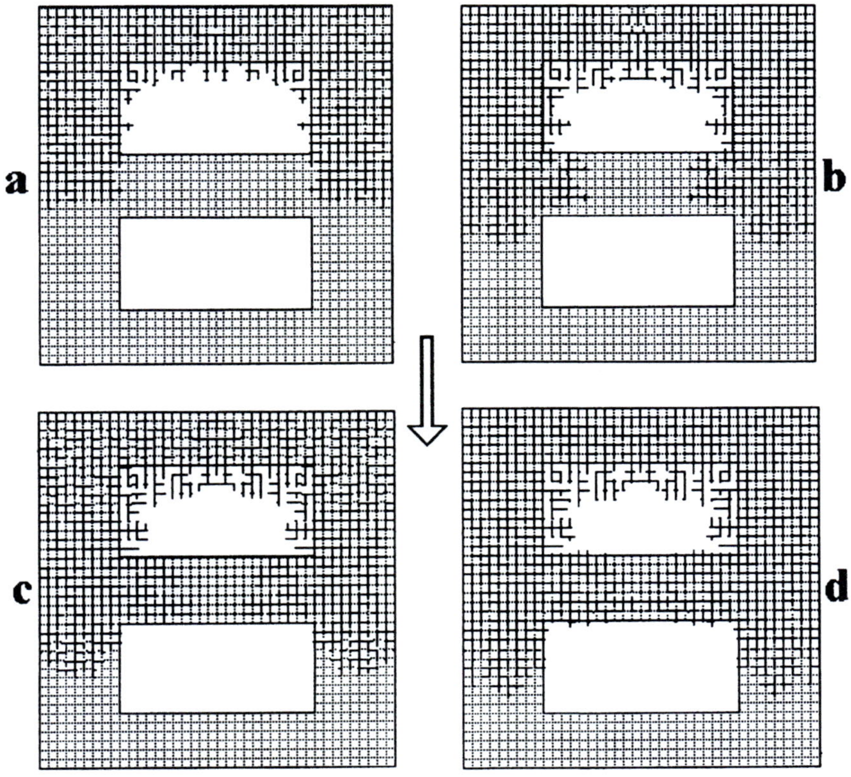

Figure 5.11. Invasion in a medium with highly permeable blocks, when ω_K=0.05. Four consequent time stages are presented. The arrow indicates the direction of invasion

About Conditions of Capillary Equilibrium Below Percolation Threshold

This simple phenomenon, having nothing of special on the pore scale of investigation, has some problem to be explained on the macroscale. In fact, let us examine the mean saturation in a block, S^{II}. Saturation growth in the time is observed, although the saturation S^{II} remains always smaller than the percolation threshold s_*. In every moment, the system is governed by capillary equilibrium law, but this equilibrium is reached under saturation smaller than the percolation threshold. Such modes of capillary equilibrium are not classical and can not be described in framework of the classical approaches.

When the macroscale saturation S^{II} is small, such that $S^{II}<s_*$, new macroscale condition of capillary equilibrium at the interface of the media should be developed. This condition should relate two saturations S^{II} and S^{I} at divers sides of the interface, which can be deduced from pore-scale modeling.

5.6. COMPARISON WITH EXPERIMENTAL RESULTS

5.6.1. EXPERIMENTAL RESULTS OF H.BERTIN AND M.QUINTARD

Experiment Description

To check the validity of the macroscale model obtained, it was compared with known experimental results published in [2] and [3].

A porous sample had the form of a long parallelepiped 0.5×0.5×2 m, which has been made of sandstone, where three highly permeable cylinders were periodically located orthogonally to the sample axis. All cylinders consisted in sand. The diameter of one cylinder was 0,28 m.

The parameters of the examined system were the following:

$$K^I{=}0.015{\cdot}10^{-12}\ \mathrm{m}^2;\quad K^{II}{=}7.4{\cdot}10^{-12}\ \mathrm{m}^2;\quad m^I{=}0.185;\quad m^{II}{=}0.437$$

$$\omega_K{=}0.002,\qquad \omega_m{=}0.4233,\qquad \theta{=}0.1846,\qquad \varepsilon{=}0.333$$

Displacement of oil by water at constant total flow rate was studied. The flow was governed by capillary forces with a very weak flow rate. The initial saturation distribution was such, that the blocks contained only dispersed water at threshold saturation $s_*^{II}{=}0.11$, while the initial matrix saturation was rather large ($s^{I,0}{=}0.45$). Other threshold saturations have been determined as $s^{II*}{=}0.81$, $s_*^I{=}0.40$, $s^{I*}{=}0.62$.

Along with the macroscopic properties of the process, the microscale saturation inside the blocks an inside the matrix were studied using x-ray registration system.

The microscale relative permeabilities and capillary pressures were measured separately for each medium I and II.

Behavior of Saturation Field

Variance in time of the saturation averaged over the whole cell, i.e.,

$$S(t) = \frac{\langle m\rangle_1 S^I + \langle m\rangle_2 S^{II}}{\langle m\rangle}$$

is shown in Fig. 5.12.

Three time stages can be observed. The first stage is a partially frozen mode, when the α-phase fills in the tight matrix I, but does not move inside the blocks II. Characteristic time of this stage is $\sim$ 10 hours. The block saturation S^{II} remains constant and equal to s_*^{II}.

The second stage corresponds to the nonlinear mode of flow, where the α-phase begins to invade the blocks too. Due to filling of blocks the average

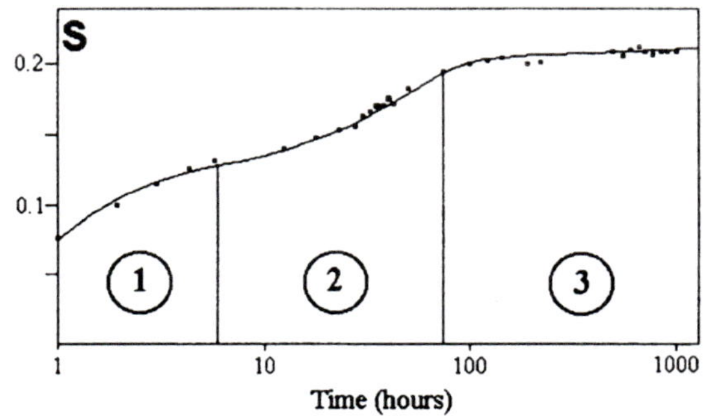

Figure 5.12. Evolution of the averaged saturation in one cell. Experimental data of H.Bertin and M.Quintard. Three time stages are emphasized (the figure is borrowed from the paper [3])

cell saturation increases. Characteristic time of the second stage os of order 100 hours.

The third stage is the quasi-linear mode of flow, when the matrix saturation varies within very thin limits near s^{I*}, while the block saturation is monotonously increasing. This is a very slow process with a characteristic time of order 1000 hours.

Note that during all the stages the system is in equilibrium state.

5.6.2. COMPARISON OF THEORETICAL AND EXPERIMENTAL DATA

The simplest theoretical results can be obtained in the following way, without complete solution of the boundary-value problem for Eqs. (5.29) .

Let the matrix saturation $S^I(t)$ as function of the time is known. Really, this function may be measured directly by x-ray technique. Then using equilibrium equation (5.29c) we can compute the block saturation $S^{II}(t)$. Because both the functions are monotonous, a dependence

$$S^{II} = S^{II}(S^I)$$

Fig. 5.13 shows the results of comparison for the developed theory with experiment, on the plot "matrix saturation vs. block saturation".

Three areas corresponds to three equilibrium modes of flow similar to those mentioned in Fig. 5.12.

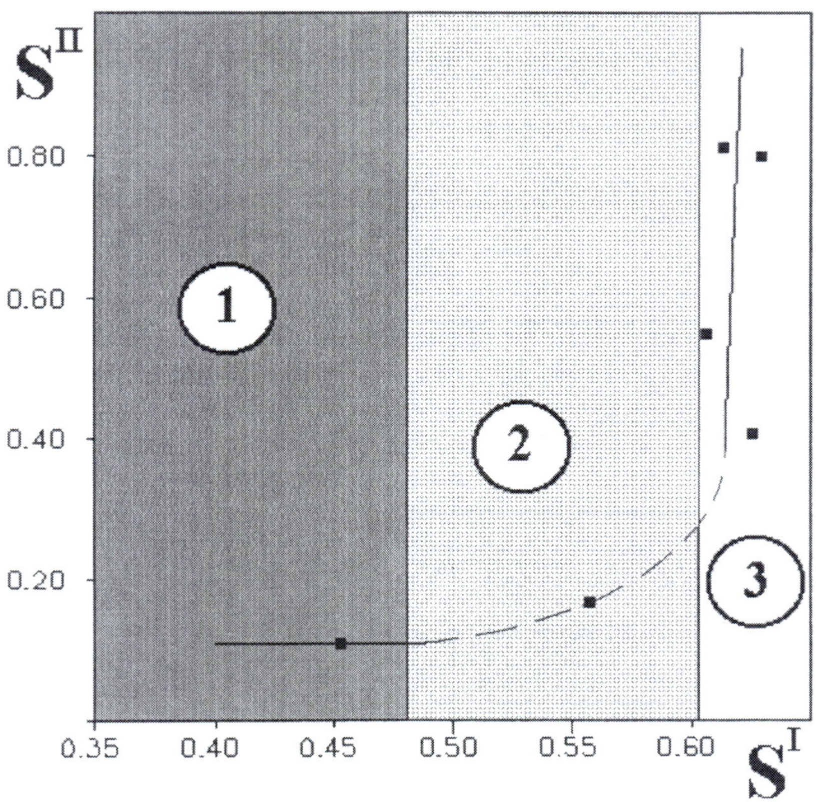

Figure 5.13. Matrix saturation versus block saturation as the result of laboratory experiments (points) and theoretical prediction (lines)

CONCLUSIONS

Two modes of two-phase flow in media with disconnected highly permeable inclusions and low permeable connected matrix were examined. Both the modes of flow are governed by capillary forces.

One of them, i.e., partially frozen mode, corresponds to those situations, when the initial saturation of wetting fluid in the inclusions is no higher than the percolation threshold. Inclusions remain impenetrable with respect to the wetting fluid, despite of their permeability is very high. This mode of flow was studied numerically using pore-network models of porous media.

The second mode corresponds to those situations, when the inclusions contain a large saturation of wetting fluid at the initial state. The flow in the whole medium is two-phase, and is governed by capillary equilibrium law. Two subclasses of such a system were shown. The first corresponds to the nonlinear mode of capillary equilibrium, when behavior of the system is governed by the nonlinear part of the capillary pressure curve. Dependence of saturation on the parameter of heterogeneity is nonlinear and singular. The second subclass is a quasi-linear mode of capillary equilibrium, which corresponds to the regular part of the capillary pressure curve. The macroscale model is deduced for the quasi-equilibrium mode, basing on homogenization method. The cell problems of new type were obtained, which were shown to be correctly formulated. NUmerical solution of the cell problems was demonstrated.

Effective absolute permeability of pseudo-cavity media is independent of the heterogeneity degree ω_K, when $\omega_K >> 1$.

Effective relative permeabilities are shown to be entirely independent of the microscale permeabilities of inclusions, and of the saturation in the inclusions. This result allows to develop a fast procedure to up-scale the relative permeabilities in such media.

Theoretical results obtained shown a good agreement compared to experimental results of H.Bertin and M.Quintard with co-authors.

Unresolved and New Problems

1. Macroscale model for nonlinear mode of flow is not deduced. Some technical problem consists in that the asymptotic expansions for saturation should contain nonlinear functions of the parameter ω_K, or ω_c, like $\omega_c^{\omega_c}$.

2. Behavior of the system within the partially frozen mode shows, that some processes of saturation growth may be observed when the saturation remains smaller than the percolation threshold. Moreover, such a process is governed by capillary equilibrium laws. Therefore the notion of capillary equilibrium may be extended below the percolation threshold saturation. The problem is to deduce new macroscale conditions of capillary equilibrium at the interface of the media, which would be valid below percolation threshold.

3. Examination of purely cavity media, i.e., homogenization of a coupled system of Stokes equations inside each concavity with Darcy's equations within the porous matrix, is the second step which is going to be developed. It is expected, that the macroscale permeability of a purely cavity medium should be equivalent to the macroscale permeability of the pseudo-cavity media with $\omega_K \to \infty$. Then, Ed. (5.30) can occur to be a correct limit for purely cavity media. However, this problem is open know.

References

1. Bertin, A., Alhanai, W. and Quintard, M. (1992) Ecoulement polyphasique dans un milieu poreux hétérogène de type nodulaire: drainage initial, *C. R. Acad. Sci. Paris*, Ser. II, **314**, pp. 431-437.
2. Bertin, A., Alhanai, W., Ahmadi, A. and Quintard, M. (19) Two-phase flow in heterogeneous nodular porous media, *? ? ?*, pp. 471-487.
3. Alhanai, W., Bertin, A. and Quintard, M. (1992) Two-phase flow in nodular systems: laboratory experiments, *Revue de l'Institut Français du Pétrole*, **47**, no. 1, pp. 29-44.
4. Bakhvalov, N.S. and Panasenko, G.P. (1989) *Homogenization of processes in periodic media*. Ed. Nauka, Moscow. (Version in English : Bakhvalov N. and Panasenko, G. *Homogenization : Averaging Processes in Periodic Media*. Kluwer Academic Publishers, Dordrecht)
5. Panfilov, M. and Panfilova, I. (1996) *Averaged models of flows with heterogeneous internal structure*. Ed. Nauka, Moscow (in Russian).
6. Pnafilova, I. and Muller J. (1998) Two-scale method for the saturation transport simulation in the network model of porous media, In: *"Recent Advances in Problems of Flow and Transport in Porous Media"* (Proc. Int. Conf.), Eds. J.M.Crolet and M.E.El-Hatri, Kluwer Acad. Publ., Dordrecht, pp. 231-243.
7. Panfilova, I. (1998) Gravity invasion of non-wetting fluid in porous medium: Classification of flow modes, *Russian Izvestiya Academii Nauk, Ser. Mekhanika Zhidkosti i Gaza*, no. 5.

Theory and Applications of Transport in Porous Media

Series Editor:
Jacob Bear, *Technion – Israel Institute of Technology, Haifa, Israel*

1. H.I. Ene and D. Polisševski: *Thermal Flow in Porous Media.* 1987
ISBN 90-277-2225-0
2. J. Bear and A. Verruijt: *Modeling Groundwater Flow and Pollution.* With Computer Programs for Sample Cases. 1987 ISBN 1-55608-014-X; Pb 1-55608-015-8
3. G.I. Barenblatt, V.M. Entov and V.M. Ryzhik: *Theory of Fluid Flows Through Natural Rocks.* 1990 ISBN 0-7923-0167-6
4. J. Bear and Y. Bachmat: *Introduction to Modeling of Transport Phenomena in Porous Media.* 1990 ISBN 0-7923-0557-4; Pb (1991) 0-7923-1106-X
5. J. Bear and J-M. Buchlin (eds.): *Modelling and Applications of Transport Phenomena in Porous Media.* 1991 ISBN 0-7923-1443-3
6. Ne-Zheng Sun: *Inverse Problems in Groundwater Modeling.* 1994
ISBN 0-7923-2987-2
7. A. Verruijt: *Computational Geomechanics.* 1995 ISBN 0-7923-3407-8
8. V.N. Nikolaevskiy: *Geomechanics and Fluidodynamics.* With Applications to Reservoir Engineering. 1996 ISBN 0-7923-3793-X
9. V.I. Selyakov and V.V. Kadet: *Percolation Models for Transport in Porous Media.* With Applications to Reservoir Engineering. 1996 ISBN 0-7923-4322-0
10. J.H. Cushman: *The Physics of Fluids in Hierarchical Porous Media: Angstroms to Miles.* 1997 ISBN 0-7923-4742-0
11. J.M. Crolet and M. El Hatri (eds.): *Recent Advances in Problems of Flow and Transport in Porous Media.* 1998 ISBN 0-7923-4938-5
12. K.C. Khilar and H.S. Fogler: *Migration of Fines in Porous Media.* 1998
ISBN 0-7923-5284-X
13. S. Whitaker: *The Method of Volume Averaging.* 1999 ISBN 0-7923-5486-9
14. J. Bear, A.H.-D. Cheng, S. Sorek, D. Ouazar and I. Herrera (eds.): *Seawater Intrusion in Coastal Aquifers. Concepts, Methods and Practices.* 1999 ISBN 0-7923-5573-3
15. P.M. Adler and J.-F. Thovert: *Fractures and Fracture Networks.* 1999
ISBN 0-7923-5647-0
16. M. Panfilov: *Macroscale Models of Flow Through Highly Heterogeneous Porous Media.* 2000 ISBN 0-7923-6176-8

Kluwer Academic Publishers – Dordrecht / Boston / London

CPSIA information can be obtained at www.ICGtesting.com
229669LV00002B/101-102/P

9 780792 361763